In the sacred memory of
Babu Kishan Chand and Revti Devi
to whom I was more than a son

Hansraj Gupta

SELECTED TOPICS IN NUMBER THEORY

SELECTED TOPICS IN NUMBER THEORY

Hansraj Gupta

Honorary Professor, Panjab University, Chandigarh, India

ABACUS PRESS

First published in 1980 by

ABACUS PRESS

Abacus House,
Speldhurst Road,
Tunbridge Wells,
Kent,
TN4 0HU,
England

British Library Cataloguing in Publication Data

Gupta, H
Selected topics in number theory.
1. Numbers, Theory of
I. Title
512′.7 QA241

ISBN 0-85626-177-7

Typeset in Great Britain by
Pintail Studios Ltd
Ringwood, Hampshire
and printed and bound by
Tien Wah Press (Pte.) Ltd, Singapore

TO THE READER

To the Reader

Books live longer than their authors. This is what provided me with the urge to write or rather compile this book from my own published and unpublished work to a large extent. I am entirely responsible for the choice of topics discussed in the book. It is very difficult for me to say if my choice has been a happy one, but I am very hopeful that you will not dislike it.

Though I received my education through the medium of the English language mostly, English is not my mother-tongue. It is quite possible that in places you may find me rather obscure, but a little effort on your part will, I am sure, make my meaning clear. In several places I have not given all the steps in the proof of a theorem. This will give you a good exercise in understanding the subject if you fill the 'blanks', so to speak, giving references to the theorems and results used. At the end of the book I have given a list of books and papers which will help you in continuing your study of the subject. Of the twelve chapters in the book as many as four (Chapters 7–10) deal with partitions: therefore the papers concerned with this topic are listed separately. The presentation is on the whole expository and some of my friends who have looked through parts of the work in typescript before publication have been kind enough to say a good word about it.

Of all branches of mathematics, number theory is the one that has not only provided food for serious thought but also created enthusiasm for the subject among young and old alike. I mentioned to a small boy (a school lad) that if he would take any four consecutive numbers greater than 11, at least one of them would be divisible by a prime greater than 11. He immediately started checking my statement mentally. I know of a very good applied mathematician who started taking keen interest in number theory, or rather the properties of natural numbers, towards the end of his life. If you like to enthuse your own son or daughter about the study of mathematics, give him or her some enticing results from numbers. For example:

$$45^2 = 2025 \quad \text{and} \quad 45 = 20 + 25; \qquad 99^2 = 9801 \quad \text{and} \quad 99 = 98 + 01$$

Any amount of such material is available in Albert H. Beiler's 'Recreations in the Theory of Numbers' (Dover, 1964). It is not without sense that number theory has been called 'The Queen of Mathematics'.

For a large part of my life, I was working in a small township away from what was then the only University in the Panjab. There was no library worth the name in or around the place. I was, therefore, rediscovering for myself many known results. For example, I did not know that symmetric functions representing the sums of the products of the first n natural numbers taken r at a time, were Stirling's Numbers of the first kind. The work presented in Chapters 5 and 6 was done mostly during the years 1931–1935 and formed part of my thesis for the Ph.D. degree. I hardly made any reference to the work done by others on the topic. It is very possible that what I have claimed to be new may be just a result discovered by a previous mathematician and only rediscovered by me. A great part of the material presented in these two chapters was published in a small monograph by the University of Lucknow in 1938. It has been included here by the kind permission of the University. Chapter 9 presents a proof by Professor G. Szekeres of a conjecture which Auluck, Chowla and myself had made in 1942. The version of his proof given here was kindly examined by him and approved. The proof has been included here with his kind permission. My thanks to Professor Szekeres.

My thanks are due also to numerous workers in the field, some of whose results I have included in this book. I am particularly thankful to Professor M. V. Subbarao for his many useful suggestions during my stay at Edmonton, where a part of this book was written in 1969–1970, and to several friends and students who read through parts of the book and helped me in clarifying certain points.

I shall thank any of my readers heartily if they point out to me any mistakes or misprints that they may find in the printed work.

Allahabad
India
August 1979

H. Gupta

CONTENTS

To the Reader		5
Chapter 1.	Preliminary	9
Chapter 2.	Primitive Roots and Indices	37
Chapter 3.	Power Residues, Quadratic Reciprocity and Sum of Squares	70
Chapter 4.	Class Algebra of k-ic Residues	119
Chapter 5.	Stirling Numbers and Bernoulli Numbers	137
Chapter 6.	Congruence Properties of Symmetric Functions	155
Chapter 7.	Partition Functions of Various Types, Generating Functions, Graphs and Identities	188
Chapter 8.	Partitions: Inequalities and Congruences	232
Chapter 9.	An Asymptotic Formula for $p(n, k)$	255
Chapter 10.	Partition of j-partite Numbers	283
Chapter 11.	Some Diophantine Equations	305
Chapter 12.	Permutations of the First n Natural Numbers	332
Miscellaneous Problems		361
Some Interesting Facts of Number Theory		364
Bibliography		365
Index		391

CHAPTER 1

PRELIMINARY

1. Introduction

We assume that the reader has already taken a course in Elementary Number Theory. In this preliminary chapter, we just give a résumé of the results with which the reader is surely familiar but to which he may want to refer now and then during the course of his study of this book. We do not prove all the results.

2. Natural numbers

The numbers $1, 2, 3, \ldots, n, \ldots$, the so-called natural numbers, can be placed in three classes:

(i) Those which have just one divisor;
(ii) Those that have exactly two divisors; and
(iii) Those that have more than two divisors.

Class (i) consists of only one element namely 1. Natural numbers which fall in class (ii) are

$$2, 3, 5, 7, 11, 13, 17, 19, \ldots$$

and are called primes. We usually use p to denote a prime. The rest of the natural numbers are placed in class (iii) and are called composite.

3. The fundamental theorem of arithmetic

It is easy to prove inductively that every natural number >1 can be expressed as a product of primes. What is really important is to realize that, disregarding the order in which the primes are written, the expression for any given n as a product of primes is unique. This is the fundamental theorem of arithmetic. For the proof, we need the following

LEMMA. *If the theorem is true for a certain $m > 1$ and p is any prime divisor of m, then p must appear in the expression for m as a product of primes.*

Proof. If $m = p$, there is nothing to prove. Otherwise, let

$$m = p \,.\, m_1, \quad \text{where} \quad m_1 > 1.$$

Let

$$m_1 = p_1 p_2 p_3 \ldots p_k,$$

where the p's are not necessarily distinct. Then

$$m = p p_1 p_2 p_3 \ldots p_k$$

is *an* expression for m as a product of primes. But we have assumed that the fundamental theorem holds for m, that is, but for the order in which the p's are written, this expression is unique. Hence this is *the only* expression for m as a product of primes. Since p appears in the expression, the lemma is proved.

Incidentally we have proved here that the expression for m_1 is also unique.

Proof of the fundamental theorem by H. Davenport. If possible, assume that n is the least number which has at least two *distinct* representations as a product of primes, say

$$n = p_1 p_2 p_3 \ldots p_k, \quad k \geqslant 2; \tag{1}$$

and

$$n = p'_1 p'_2 p'_3 \ldots p'_j, \quad j \geqslant 2. \tag{2}$$

Evidently then no p can be a p', for if $p_1 = p'_1$ then n/p_1, which is certainly less than n, will have two distinct expressions as a product of primes. Without loss of generality, we can assume that

$$p_1 \leqslant p_2 \leqslant p_3 \leqslant \cdots \leqslant p_k;$$

and

$$p'_1 \leqslant p'_2 \leqslant p'_3 \leqslant \cdots \leqslant p'_j.$$

Hence, we must have

$$n \geqslant p_1^2, \quad n \geqslant p_1'^2. \tag{3}$$

Since $p_1 \neq p'_1$, the sign of equality cannot hold in both places in (3). Therefore

$$n > p_1 p'_1. \tag{4}$$

Let $n - p_1 p'_1 = N$. Since $p_1 \mid n$ and also $p'_1 \mid n$, both p_1 and p'_1 divide N. Since $N < n$, the theorem holds for N and the lemma shows that we can write

$$N = p_1 p'_1 m. \tag{5}$$

This gives

$$n = p_1 p'_1 (m + 1) = p_1 p'_1 q_1 q_2 \ldots q_h, \tag{6}$$

say, where the q's are primes, not necessarily distinct from others on the right of (6). From (1) and (6), we now have

$$p'_1 q_1 q_2 \ldots q_h = n/p_1 = p_2 p_3 \ldots p_k. \tag{7}$$

We have thus obtained two distinct expressions as product of primes for n/p_1, distinct because p_1' appears on the left of (7) but not in the expression on its right. This contradicts the minimality of n and the theorem follows.

Thus, every natural number $n > 1$ can be expressed uniquely in the *canonical* form

$$n = p_1^{c_1} p_2^{c_2} \dots p_k^{c_k}$$

where each $c \geqslant 1, k \geqslant 1$, and we take

$$p_1 < p_2 < \dots < p_k.$$

Remarks

1. Defining the empty product of primes as 1, the fundamental theorem can be stated in the form:

Disregarding the order in which the primes are written, every natural number can be expressed uniquely as a product of primes.

2. Does the fundamental theorem need a proof at all? Has any one ever come across an example where the theorem has failed? The following example will convince the reader about the necessity of a proof.

Take the set

$$\{1, 4, 7, 10, 13, 16, 19, 22, 25, 28, 31, \dots\} \tag{8}$$

of natural numbers. It is easy to see that the set is closed under multiplication. Call an element of the set a 'prime' if it has exactly two divisors in the set. Thus, the primes of the set are

$$4, 7, 10, 13, 19, 22, 25, 31, \dots.$$

We observe that 100 belongs to the set and it can be written as a product of the primes of the set in two distinct ways:

$$100 = 4.25; \quad \text{also} \quad 100 = 10.10.$$

Exercises

1. Which is the next element of the set in (8) which has at least two distinct representations as a product of the primes of the set?
2. Consider the set consisting of all the numbers of the form $4n + 1$, where $n = 0, 1, 2, 3, \dots$. Define the 'primes' of this set and find the least element of the set having at least two distinct representations as a product of the primes of the set.

4. Infinitude of primes

THEOREM 1. *The number of primes is infinite.*

Proof. Suppose the system of natural numbers has only k primes

$$p_1, p_2, p_3, \dots, p_k.$$

Divide this set of primes into two disjoint subsets in any manner. Let

$$n = P_1 + P_2$$

where P_1 is the product of the primes in the first subset and P_2 of those in the second. (If one of the subsets is empty we take 1 to be the product of the primes in that subset.) Since the number of primes is certainly more than $1, n > p_k$.

If a certain prime p belongs to the first subset, it divides P_1 but it does not divide P_2. If it belongs to the second subset, it divides P_2 but not P_1. Thus, none of the primes $p_1, p_2, \ldots, p_k$ divides n. Hence either n is itself a prime $> p_k$, or it has a prime divisor not in our list of primes. This proves the theorem.

Exercises

1. Find all the primes that can be obtained in the above manner from the set $\{2, 3, 5, 7\}$.
2. Let p_k^* denote the kth prime with $p_1^* = 2$. Find the largest k for which the above process applied to the set $\{p_1^*, p_2^*, \ldots, p_k^*\}$ yields only primes.
3. Show that $p_k^* \leqslant 2^{2^{k-1}}$.
4. For $k > 1$, show that the numbers $k! + h, h = 2, 3, \ldots, k$, are all composite. How big can the difference between two consecutive primes be?

5. The p-potency of n

The highest power of a prime p which divides a given natural number n is said to be its potency to the base p and written $\text{pot}_p\, n$. This is zero only if $p \nmid n$, otherwise it $\geqslant 1$. We extend this definition to rational numbers by taking

$$\text{pot}_p \frac{n}{m} = \text{pot}_p\, n - \text{pot}_p\, m. \tag{1}$$

The reader can now prove the following results for himself.

$$\text{pot}_p(nm) = \text{pot}_p\, n + \text{pot}_p\, m;$$

$$\text{pot}_p\, n^k = k\, \text{pot}_p\, n, \text{ for all integers } k.$$

Thus

$$\text{pot}_3\, 54 = 3, \quad \text{pot}_2\, 54 = 1, \quad \text{pot}_5\, 54 = 0,$$

$$\text{pot}_5\, 975 = 2; \text{pot}_7(686/35) = 3 - 1 = 2.$$

Exercises

1. Can the concept of potency be extended to composite bases? Give reasons for your answer.
2. Find the values of

   (i) $\text{pot}_3\, 891$; (ii) $\text{pot}_5(11!)$; (iii) $\text{pot}_2(98/24)$.

6. The potency of $n!$ to any base p

In what follows, we write $[x]$ for the integer for which

$$[x] \leqslant x < [x] + 1$$

x being any real number positive or negative. Thus, we have

$$[8/3] = 2, \quad [-8/3] = -3, \quad [28\sqrt{3}] = 48, \quad [-e] = -3.$$

The function $[x]$ is spoken of as the greatest integer function.

We can now find the formula for $\text{pot}_p(n!)$ as follows.

$$\begin{aligned}\text{pot}_p(n!) &= \text{pot}_p\, 1 + \text{pot}_p\, 2 + \text{pot}_p\, 3 + \cdots + \text{pot}_p\, n \\ &= \text{pot}_p\, p + \text{pot}_p(2p) + \text{pot}_p(3p) + \cdots + \text{pot}_p([n/p]\, p),\end{aligned}$$

Since

$$\text{pot}_p(jp) = \text{pot}_p\, p + \text{pot}_p\, j = 1 + \text{pot}_p\, j,$$

we get

$$\text{pot}_p(n!) = [n/p] + \text{pot}_p([n/p]\,!).$$

Using this reduction formula repeatedly, we obtain

$$\text{pot}_p(n!) = [n/p] + [n/p^2] + \cdots + [n/p^k]$$

where $p^k \leqslant n < p^{k+1}$, or what is the same thing,

$$\text{pot}_p(n!) = \sum_{j=1}^{\infty} [n/p^j].$$

For example

$$\text{pot}_7(1000!) = 142 + 20 + 2 = 164.$$

This means that the highest power of 7 which divides 1000! is the 164-th.

7. The p-potency of the combinatory function $\binom{p^c}{r}$

THEOREM 2. *For any given prime p and* $r < p^c$,

$$\text{pot}_p \binom{p^c}{r} = c - \text{pot}_p\, r.$$

Proof. The theorem is trivially true for $r = 1$. For $r > 1$, the following simple proof is due to D. H. Lehmer. By definition

$$\binom{p^c}{r} = \frac{p^c}{r} \cdot \frac{p^c - 1}{1} \cdot \frac{p^c - 2}{2} \cdot \cdots \cdot \frac{p^c - (r-1)}{r-1}.$$

Observe that for $0 < r < p^c$,

$$\text{pot}_p(p^c - j) = \text{pot}_p\, j, \quad j = 1, 2, \ldots, r-1;$$

while

$$\text{pot}_p(p^c/r) = c - \text{pot}_p\, r.$$

Hence the theorem.

8. The p-potency of $\binom{n}{r}$, $r<n$

Written in the scale of p, let

$$n=(a_h, a_{h-1}, \ldots, a_1, a_0)_p.$$

This simply means that

$$n=a_h p^h + a_{h-1} p^{h-1} + \cdots + a_1 p + a_0,$$

where

$$0 \leqslant a_j < p, \quad \text{for} \quad j=0,1,2,\ldots,h-1;$$

and

$$1 \leqslant a_h < p.$$

Similarly, let

$$r=(b_i, b_{i-1}, \ldots, b_1, b_0)_p.$$

Since $r<n$, we have $i \leqslant h$.

THEOREM 3. *The p-potency of* $\binom{n}{r}$, $r<n$, *is the same as the number of values of* j, $0 \leqslant j \leqslant h$, *for which*

$$(a_j, a_{j-1}, \ldots, a_0)_p - (b_j, b_{j-1}, \ldots, b_0)_p \tag{1}$$

is negative. (We take $b_j=0$ *for each* $i<j \leqslant h$.)

The proof depends on the fact that

$$[n/p^{j+1}] - [r/p^{j+1}] - [(n-r)/p^{j+1}] = 1 \text{ or } 0$$

according as (1) is or is not negative, and is left to the reader. A practical method of determining the potency of $\binom{n}{r}$ is the following. In the representations for n and r in the scale of p, replace

a_t by 1 and b_t by 0, if $a_t > b_t$;

a_t and b_t both by 0, if $a_t = b_t$;

and

a_t by 0 and b_t by 1, if $a_t < b_t$;

also change p to 10. We thus get two numbers n^* and r^* in place of n and r. The p-potency of $\binom{n}{r}$ is now the same as the total number of eights and nines in the number $(n^* - r^*)$.

Example. Let us evaluate $\text{pot}_5 \binom{188}{39}$.

We have

$$188 = (1223)_5, \quad 39 = (124)_5.$$

Hence

$$n^* = (1100)_{10}, \quad r^* = (001)_{10}.$$

Now

$$n^* - r^* = 1099,$$

and we get

$$\text{pot}_5 \binom{188}{39} = 2.$$

8.1. An alternative method

As before, let

$$n = (a_h, \ldots, a_2, a_1, a_0)_p.$$

Define

$$A(n, p) = \sum_{k=0}^{h} a_k.$$

Notice that $A(n, p)$ denotes the sum of the digits in the representation of n in the scale of p. Then, it will be readily seen that

$$n - A(n, p) = \sum_{k=0}^{h} a_k(p^k - 1)$$

so that

$$\frac{n - A(n, p)}{p - 1} = \sum_{k=1}^{h} a_k(p^{k-1} + \cdots + p^2 + p + 1)$$

$$= \sum_{k=1}^{h} (a_h p^{h-k} + \cdots + a_{k+1} p + a_k)$$

$$= \sum_{k=1}^{h} [n/p^k]. \quad \text{[J. L. Lagrange]} \tag{2}$$

The right side of (2) is nothing but the p-potency of $n!$.

Hence

$$\text{pot}_p \binom{n}{r} = \{A(r, p) + A(n - r, p) - A(n, p)\}/(p - 1). \tag{3}$$

Evidently (3) implies that

$$\text{pot}_p\binom{n}{r} = \text{pot}_p\binom{n}{n-r}.$$

Exercises

1. Deduce theorem 2 from (3).
2. Write 50! in the canonical form.
3. Express $\binom{29}{5}$ as a product of primes.
4. Using each of the two methods given in the text, find

$$\text{(i) pot}_5\binom{625}{20}; \quad \text{(ii) pot}_7\binom{98}{7}; \quad \text{(iii) pot}_7\binom{12}{6}.$$

5. For the p-potency of $\binom{n}{r}$ to be more than 0, is it necessary that p divide n?
6. If $n = (\ldots, a_2, a_1, a_0)_p$, $r = (\ldots, b_2, b_1, b_0)_p$ show that

$$\binom{n}{r} \equiv \binom{a_0}{b_0}\binom{a_1}{b_1}\binom{a_2}{b_2} \ldots \pmod p. \quad \text{[E. Lucas 1887]}$$

9. The functions $d(n)$, $\sigma(n)$ and $\sigma_r(n)$

Recall that $d(n)$ denotes the number of divisors (positive) of n, $\sigma(n)$ the sum of these divisors and $\sigma_r(n)$ the sum of their rth powers. Thus

$$d(n) = \sigma_0(n), \quad \sigma(n) = \sigma_1(n).$$

If the canonical expression for n is

$$n = p_1^{c_1} p_2^{c_2} \ldots p_k^{c_k},$$

then, we have

$$d(n) = (c_1 + 1)(c_2 + 1) \ldots (c_k + 1); \tag{1}$$

and

$$\sigma_r(n) = \prod_{j=1}^{k} (p_j^{r(c_j+1)} - 1)/(p_j^r - 1). \tag{2}$$

Note that (1) can be deduced from (2).

10. Euler's totient function $\phi(n)$.

The function $\phi(n)$ denotes the number of positive integers which do not exceed n and are prime to it.

Writing (a, b) for the g.c.d. of two positive integers a and b (here and throughout this book), we have by definition

$$\phi(n) = \sum_{\substack{m \leqslant n \\ (m,n)=1}} 1.$$

Using the symbol $<'$ for less than and prime to, we can write

$$\phi(n) = \sum_{m <' n} 1, \quad \text{for} \quad n > 1; \tag{1}$$

and we take $\phi(1) = 1$,

THEOREM 4. *For each $n \geqslant 1$,*

$$\sum_{d \mid n} \phi(d) = n.$$

Proof. Consider the set of rational fractions

$$F = \left\{ \frac{r}{n} : \quad r = 1, 2, 3, \ldots, n \right\}.$$

Reduce each fraction in F to its lowest terms and get F^*, another set of n fractions. It is easy to see that all fractions in F^* are distinct (as were those in F) in the sense that no two of them are equal in value.

If for any given r ($\leqslant n$),

$$\frac{r}{n} = \frac{h}{k}$$

in its lowest terms, then

$$(h, k) = 1, \quad h \leqslant k \quad \text{and} \quad k \mid n. \tag{2}$$

(In fact if $(r, n) = g$, then $r = gh, n = gk$.) Moreover, any fraction satisfying the three conditions in (2) for any given n, belongs to F^*. It will thus be seen that the number of fractions in F^* which have the denominator k is $\phi(k)$ and the theorem follows readily.

Exercise

Take $n = 30$, and verify the statements in the proof.

11. Multiplicative functions

DEFINITION. *A function $f(n)$ defined for positive integral values of n, is said to be multiplicative, if*

$$f(mn) = f(m) \,.\, f(n) \text{ provided } (m, n) = 1.$$

The function is said to be strongly multiplicative, if for all natural numbers m and n, we have

$$f(mn) = f(m) \,.\, f(n).$$

The three functions in section 9 are multiplicative but not strongly multiplicative. We proceed to prove

THEOREM 5. *Euler's function ϕ is multiplicative.*

In other words, for $h = mn, (m, n) = 1$,

$$\phi(h) = \phi(m)\,\phi(n).$$

The proof depends on the following

LEMMA. *If $(m, n) = 1$, then as t_1 runs through a complete set of divisors of m and t_2 through a complete set of divisors of n, $t = t_1 t_2$ runs through a complete set of divisors of mn.*

In fact for any divisor t of mn, the divisor t_1 of m is given by (t, m) and the divisor t_2 of n is given by (t, n). Moreover, $(t_1, t_2) = 1$.

Proof of theorem 5. The proof is by induction on h. The theorem holds for $h = 1$. Assume that it holds for

$$h = 2, 3, \ldots, mn - 1.$$

Let t be any divisor of mn and t_1 and t_2 the unique divisors of m and n respectively for which $t = t_1 t_2$. Then our assumption implies that for any such t, t_1 and t_2, we have

$$\phi(t) = \phi(t_1)\,\phi(t_2)$$

except only in one case that is when $t_1 = m$ and at the same time we have $t_2 = n$. Hence

$$\sum_{t_1 | m} \phi(t_1) \,.\, \sum_{t_2 | n} \phi(t_2) = \left\{ \sum_{t | mn} \phi(t) - \phi(mn) \right\} + \phi(m)\phi(n).$$

Theorem 4 now gives

$$m \,.\, n = \{ mn - \phi(mn) \} + \phi(m)\phi(n);$$

so that $\phi(mn) = \phi(m)\phi(n)$ provided $(m, n) = 1$.

The following example will illustrate the main step in the argument.

Example. Take $m = 6, n = 25$ and assume that the theorem holds for each $h \leqslant 149$. From theorem 4, we have

$$6 = \phi(1) + \phi(2) + \phi(3) + \phi(6);$$
$$25 = \phi(1) + \phi(5) + \phi(25).$$

Multiplying the two relations, we get

$$\begin{aligned} 6.25 = {} & \phi(1)\phi(1) + \phi(1)\phi(2) + \phi(1)\phi(3) + \phi(1)\phi(6) \\ & + \phi(5)\phi(1) + \phi(5)\phi(2) + \phi(5)\phi(3) + \phi(5)\phi(6) \end{aligned}$$

$$
\begin{aligned}
&+\phi(25)\phi(1)+\phi(25)\phi(2)+\phi(25)\phi(3)+\phi(25)\phi(6)\\
&=\phi(1)+\phi(2)+\phi(3)+\phi(6)+\phi(5)+\phi(10)+\phi(15)\\
&\quad+\phi(30)+\phi(25)+\phi(50)+\phi(75)+\phi(25)\phi(6)\\
&=\left\{\sum_{t\mid 150}\phi(t)-\phi(150)\right\}+\phi(6)\phi(25)\\
&=\{150-\phi(150)\}+\phi(6)\phi(25).
\end{aligned}
$$

This gives the desired result

$$\phi(150)=\phi(6)\phi(25).$$

Our method of proving that the function ϕ is multiplicative is very general and can be used for several functions.

11.1. Formula for $\phi(n)$

To evaluate $\phi(n)$, it will now be enough to consider the case when n is a power of a prime p.

Let $n = p^c$. The only numbers not exceeding p^c which are not prime to it are

$$1p, 2p, 3p, \ldots, p^{c-1}p.$$

These are p^{c-1} in number. Hence

$$\phi(p^c)=p^c-p^{c-1}=p^c(1-p^{-1}).$$

This readily gives

$$\phi(n)=n\,.\prod_p(1-p^{-1})$$

where p runs over the distinct prime divisors of n. Thus

$$\phi(180)=180\,.\,(1-1/2)(1-1/3)(1-1/5)=48.$$

Exercises

1. Prepare for ready reference, a table of values of $\phi(n)$ for values of n up to 250, and use it to find the solutions (if any) of the equations

   (i) $\phi(n)=14$; (ii) $\phi(n)=24$.
2. Show that for $n > 2$, $\phi(n)$ is even.
3. Show that for any given m there exists a least n for which $m \mid \phi(n)$. What is this n for $m = 15$?

12. A generalization of the Möbius function

Let the function $\mu_r(n)$ be defined for all positive integral values of n and $r \geqslant 0$, by the relation

$$\sum_{t\mid n}\mu_r(t)=1 \text{ or } 0$$

according as n is or is not an rth power of an integer >0. Then proceeding on the same lines as for the totient function ϕ, it can be shown that μ_r is a multiplicative function.

(1) For $r = 0$, from definition, we have

$$\mu_0(1) = 1;$$

$$\mu_0(1) + \mu_0(p) = 0, \text{ for each prime } p;$$

so that

$$\mu_0(p) = -1;$$

$$\mu_0(1) + \mu_0(p) + \mu_0(p^2) = 0,$$

which gives

$$\mu_0(p^2) = 0.$$

Proceeding in this manner, we get

$$\mu_0(p^c) = 0, \quad \text{for each } c > 1.$$

Using the multiplicativity of μ_r for $r = 0$, we now have

$$\mu_0(n) = \begin{cases} 0, \text{ if } n \text{ has a square divisor } >1; \\ (-1)^k, \text{ if } n \text{ is a product of } k \text{ distinct primes.} \end{cases}$$

This is the well-known Möbius function which we shall hereafter denote by $\mu(n)$.

(2) For $r = 1$, proceeding as in (1) above, we get

$$\mu_1(n) = \begin{cases} 1 \text{ for } n = 1; \\ 0 \text{ for each } n > 1. \end{cases}$$

(3) In the case of $r = 2$, from the definition we get

$$\mu_2(1) = 1, \quad \mu_2(p^c) = (-1)^c \text{ for every prime } p \text{ and } c \geqslant 1.$$

Hence for

$$n = p_1^{c_1} p_2^{c_2} \ldots p_k^{c_k}$$

where the p's are distinct primes, we have

$$\mu_2(n) = (-1)^C$$

with

$$C = c_1 + c_2 + \cdots + c_k.$$

This $\mu_2(n)$ is known as Liouville's function and is strongly multiplicative.

(4) For $r > 2$, we leave it to the reader to show that

$$\mu_r(p^c) = \begin{cases} 1, & \text{if } c = rj, \quad j \geqslant 0; \\ -1, & \text{if } c = rj + 1, \quad j \geqslant 0; \\ 0, & \text{otherwise.} \end{cases}$$

12.1. The sum function M_r

With the function μ_r is connected the sum function M_r defined by the relation

$$M_r(n) = \sum_{j=1}^{n} \mu_r(j).$$

For Liouville's function, G. Pólya had conjectured in 1919 that

$$M_2(n) \leqslant 0 \quad \text{for} \quad n \geqslant 2.$$

The conjecture, however, failed for $n = 906180359$.

Exercise
Is μ_1 strongly multiplicative? Is μ_r strongly multiplicative for any values of $r > 2$?

13. The Möbius inversion formula

Let $F(n)$ be any number-theoretic function (i.e. defined for all positive integral values of n) and let

$$G(n) = \sum_{t|n} F(t).$$

Then

$$F(n) = \sum_{t|n} \mu(t)\, G(n/t).$$

Proof. We first observe that $F(n)$ can be expressed as a linear function of G's. In fact, by definition

$$G(1) = F(1), \qquad F(1) = G(1);$$

$$G(2) = F(1) + F(2), \qquad F(2) = G(2) - G(1);$$

and so on. Now

$$G(n/t) = \sum_{t_1|(n/t)} F(t_1).$$

Therefore

$$\mu(t)G(n/t) = \sum_{t_1|(n/t)} \mu(t)F(t_1);$$

so that

$$\sum_{t|n} \mu(t)G(n/t) = \sum_{t|n} \sum_{t_1|(n/t)} \mu(t)F(t_1)$$

$$= \sum_{t_1|n} \mu(t)F(t_1)$$

$$= \sum_{t_1|n} \sum_{t|(n/t_1)} \mu(t)F(t_1)$$

$$= \sum_{t_1|n} F(t_1) \sum_{t|(n/t_1)} \mu(t).$$

But from the definition of the Möbius function, we have

$$\sum_{t|(n/t_1)} \mu(t) = 1, \quad \text{if } \frac{n}{t_1} = 1;$$
$$= 0, \quad \text{otherwise.}$$

Therefore

$$\sum_{t|n} \mu(t)G(n/t) = F(n).$$

This proves what is known as the Möbius inversion formula. The following examples illustrate its usefulness.

Example 1. We have

$$\sigma(n) = \sum_{t|n} t.$$

The inversion formula gives

$$n = \sum_{t|n} \sigma(t)\mu(n/t) = \sum_{t|n} \mu(t)\sigma(n/t).$$

Thus for $n = 6$, we have

$$\sum_{t|6} \mu(t)\sigma(6/t) = \mu(1)\sigma(6) + \mu(2)\sigma(3) + \mu(3)\sigma(2) + \mu(6)\sigma(1),$$
$$= 1.12 - 1.4 - 1.3 + 1.1 = 6.$$

Example 2. From theorem 4,

$$n = \sum_{t|n} \phi(t).$$

Hence

$$\phi(n) = \sum_{t|n} \mu(t) \,.\, \frac{n}{t} = n \,.\, \sum_{t|n} \frac{\mu(t)}{t}.$$

For $n = 12$, this gives

$$\phi(12) = 12 \{1 - \tfrac{1}{2} - \tfrac{1}{3} + \tfrac{1}{6}\} = 4.$$

This is certainly true since the only numbers $<'$ 12 are 1, 5, 7, 11.

Exercises

1. Apply the inversion formula to the divisor function

$$d(n) = \sum_{t|n} 1.$$

2. If the function F is multiplicative, show that the function G given by the relation

$$G(n) = \sum_{t|n} F(t)$$

is also multiplicative.

3. Show that

$$\sum_{t|n} \frac{\mu(t)}{t} = \prod_{p} (p-1)/p$$

where p runs over the prime divisors of n.

4. If

$$G(n) = \prod_{t|n} F(t)$$

show that

$$\log F(n) = \sum_{t|n} \mu(n/t) \log G(t).$$

14. The 'hat' function

The 'hat' function Λ is defined by the relations

$$\Lambda(n) = \begin{cases} \log p, & \text{if } n \text{ is a positive power of prime } p; \\ 0, & \text{otherwise.} \end{cases}$$

This means that

$$\Lambda(1) = 0;$$
$$\Lambda(2) = \Lambda(4) = \Lambda(8) = \cdots = \Lambda(2^c) = \log 2;$$
$$\Lambda(3) = \Lambda(9) = \Lambda(27) = \cdots = \Lambda(3^c) = \log 3;$$
$$\Lambda(p) = \Lambda(p^2) = \Lambda(p^3) = \cdots = \Lambda(p^c) = \log p;$$

for each $c \geqslant 1$.

THEOREM 6. *For $n \geqslant 1$,*

$$\sum_{t|n} \Lambda(t) = \log n.$$

Proof. Let

$$n = p_1^{c_1} p_2^{c_2} \ldots p_k^{c_k}$$

be the canonical decomposition of n. Then the only divisors of n for which $\Lambda(t)$ is not zero are

$$p_j, p_j^2, \ldots, p_j^{c_j}; \quad j = 1, 2, \ldots, k.$$

Hence

$$\sum_{t|n} \Lambda(t) = \sum_{j=1}^{k} c_j \log p_j = \log n.$$

Exercise
Show that

$$\Lambda(n) = \sum_{t|n} \mu(t) \log(n/t).$$

15. Residue systems modulo *m*

A set of integers forms a complete residue system modulo m if and only if it has exactly m members no two of which are congruent modulo m. The two most useful sets with this property are:

$$\{0, 1, 2, \ldots, m-1\};$$

and

$$\{-[m/2], \ldots, -2, -1, 0, 1, 2, \ldots, [m/2]\} \text{ when } m \text{ is odd;}$$
$$\{-(m-2)/2, \ldots, -2, -1, 0, 1, 2, \ldots, m/2\} \text{ when } m \text{ is even.}$$

On the other hand, a set of integers forms a reduced residue system modulo m, if and only if each of its $\phi(m)$ members is prime to m and no two of them are congruent modulo m. Except for $m = 1, 2$ a reduced residue system modulo m has an even number of members.

A useful property of these residue sets is the following. If each element of a complete (or reduced) residue set is multiplied by a number g prime to m, the set so obtained is still a complete (or reduced) residue set modulo m.

The reader will recall that the Euler–Fermat Theorem

$$g^{\phi(m)} \equiv 1 \pmod{m}, \quad (g, m) = 1$$

is a direct consequence of this property of reduced residue sets.

16. The linear congruence

In its most general form, the linear congruence is

$$ax + b \equiv 0 \pmod{m} \tag{1}$$

where a and b are given integers and $m > 1$. Without loss of generality, we can further assume that a is positive. If $(a, m) = g$, the congruence cannot have a solution unless $g \mid b$. Assuming that $g \mid b$, let

$$a = a_1 g, \quad b = b_1 g, \quad m = m_1 g.$$

Then, the solutions of (1) in the range $[0, m-1]$ (which is all that we require) are the same as the solutions of the congruence

$$a_1 x + b_1 \equiv 0 \pmod{m_1} \tag{2}$$

in the said range. Since $(a_1, m_1) = 1$, there exists a unique positive integer $c <' m_1$ such that

$$ca_1 \equiv 1 \pmod{m_1}$$

(c is called the associate of a_1 modulo m_1). Multiplying the two sides of (2) by c, we get

$$\begin{aligned} x &\equiv -cb_1 \pmod{m_1} \\ &\equiv h \pmod{m_1}, \text{ say} \end{aligned}$$

where h is a non-negative integer $< m_1$. All the incongruent solutions of (1) are now given by

$$h, h + m_1, h + 2m_1, \ldots, h + (g-1)m_1. \tag{3}$$

With our choice of h, all these g solutions lie in the range $[0, m-1]$. Thus, (1) has either no solution – that is when $g \nmid b$, or it has exactly g solutions incongruent modulo m, but congruent modulo m_1. It will thus be realized that the congruences (1) and (2) are not equivalent. For (2) the solutions listed in (3) are not considered to be distinct while for (1) they are distinct.

17. The Chinese remainder theorem

One is sometimes faced with the problem of finding the least natural number x which when divided by m_1 leaves the remainder r_1, when divided by m_2 the remainder r_2, and so on when divided by m_j the remainder r_j.

When the m's are mutually prime to each other in pairs, there is always a solution to the problem. We assume that they are. The problem then reduces to that of solving the simultaneous congruences:

$$x \equiv r_i \pmod{m_i}, \quad i = 1, 2, 3, \ldots, j.$$

Take the first two congruences

$$x \equiv r_1 \pmod{m_1} \tag{1}$$

$$x \equiv r_2 \pmod{m_2} \tag{2}$$

Congruence (1) is satisfied if we take

$$x = m_1 x_1 + r_1$$

where x_1 is an integer. This will satisfy (2) if for some x_1

$$m_1x_1 + (r_1 - r_2) \equiv 0 \pmod{m_2}. \tag{3}$$

Since $(m_1, m_2) = 1$, (3) has a unique solution modulo m_2. If this is

$$x_1 \equiv c_1 \pmod{m_2}, \quad \text{where} \quad 0 \leqslant c_1 < m_2;$$

then for some integer x_2, we have

$$x_1 = m_2x_2 + c_1;$$

so that

$$x = m_1m_2x_2 + m_1c_1 + r_1$$

satisfies the two congruences (1) and (2) simultaneously. We now take up the third congruence and get

$$m_1m_2x_2 + (m_1c_1 + r_1 - r_3) \equiv 0 \pmod{m_3}$$

and the process is continued. In the end we get for x an expression which is unique modulo $m_1m_2m_3 \ldots m_j$. The Chinese remainder theorem provides a solution to the problem in a more tangible form.

Let

$$M = m_1m_2 \ldots m_j;$$
$$M_i = M/m_i.$$

Evidently then $(m_i, M_i) = 1$. Note also that M_i is divisible by each of the m's other than m_i. Since $(m_i, M_i) = 1$, there exists an $N_i < m_i$, such that

$$N_iM_i \equiv 1 \pmod{m_i}.$$

We assert that

$$x_0 = r_1M_1N_1 + r_2M_2N_2 + \cdots + r_jM_jN_j \tag{4}$$

provides a solution for our set of congruences, though not necessarily the least. Since each term on the right of (4) with the exception of $r_1M_1N_1$ is divisible by m_1, we have

$$x_0 \equiv r_1M_1N_1 \pmod{m_1}$$
$$\equiv r_1 \pmod{m_1}.$$

Similarly, it can be shown that

$$x_0 \equiv r_i \pmod{m_i} \text{ for } i = 2, 3, \ldots, j;$$

and our assertion follows.

We next prove that any number x for which each of the j congruences holds, satisfies the congruence

$$x \equiv x_0 \pmod{M}.$$

Since

$$x \equiv r_i \,(\text{mod } m_i) \text{ for each } i \leqslant j,$$

$$x \equiv x_0 \,(\text{mod } m_i) \text{ for each } i \leqslant j.$$

This means that $x - x_0$ is divisible by each of the m_i's, $i \leqslant j$; and therefore by their l.c.m. which in this case is nothing but M. This proves that the given system of congruences has a unique solution modulo M.

Example. Find the least natural number which when divided by 7, 10 and 11 leaves in order the remainders 1, 6 and 2.

Here, we have to find a common solution of the congruences

$$x \equiv 1 \,(\text{mod } 7); \quad x \equiv 6 \,(\text{mod } 10); \quad x \equiv 2 \,(\text{mod } 11).$$

We can let

$$x = 7x_1 + 1.$$

Then we must have

$$7x_1 \equiv 5 \,(\text{mod } 10).$$

Multiplying both sides of this congruence by 3, the associate of 7 modulo 10, we get

$$x_1 \equiv 15 \equiv 5 \,(\text{mod } 10).$$

Let

$$x_1 = 10x_2 + 5,$$

so that

$$x = 70x_2 + 36.$$

This will satisfy the third congruence of the set, if and only if

$$70x_2 \equiv -34 \,(\text{mod } 11);$$

i.e.

$$4x_2 \equiv -1 \,(\text{mod } 11).$$

Again multiplying the sides of the congruence by 3, the associate of 4 modulo 11, we get

$$x_2 \equiv -3 \equiv 8 \,(\text{mod } 11).$$

This means that for some integer x_3, we have

$$x_2 \equiv 11x_3 + 8.$$

Hence

$$x = 770x_3 + 596 \equiv 596 \,(\text{mod } 770).$$

Therefore, the required number is 596.

Using the Chinese remainder theorem, we will have

$$M = 770, \quad M_1 = 110, \quad M_2 = 77, \quad M_3 = 70.$$

These give

$$N_1 = 3, \quad N_2 = 3, \quad N_3 = 3.$$

(It is a mere chance that the N's have turned out to be equal.) Hence

$$\begin{aligned} x_0 &= 1.110.3 + 6.77.3 + 2.70.3 = 2136 \\ &\equiv 596 \pmod{770}. \end{aligned}$$

The required number is 596 as before.

Remark

A solution to such problems may exist even when the m's are not mutually prime in pairs.

Exercises

1. Find the least number which when divided by 2, 3, 4, 5 and 6 leaves the remainder 1 in each case but is divisible by 7.
2. Find the least number which when divided by 7, 11 and 12 leaves in order the remainders 5, 9 and 10 but is divisible by 13. Find also the least number of four digits which satisfies these conditions.

18. A useful theorem

Let

$$f(x) = a_0 x^r + a_1 x^{r-1} + \cdots + a_{r-1} x + a_r$$

be a polynomial in x of degree $r \geqslant 1$, with coefficients $a_0, a_1, \ldots, a_r$ all integers (not necessarily positive), $a_0 > 0$. Then, we have

THEOREM 7. *For p prime, the congruence*

$$f(x) \equiv 0 \pmod{p}$$

cannot have more than r solutions incongruent mod p, *unless each of the a's is divisible by p.*

Proof. If the congruence has no solution, there is nothing to prove. If x_1 is a root of the congruence, then the congruence can be written

$$f(x) - f(x_1) \equiv 0 \pmod{p}.$$

Since

$$f(x) - f(x_1) = (x - x_1) f_1(x)$$

where $f_1(x)$ is a polynomial in x of degree $r - 1$ with integral coefficients. Any other solutions of the congruence

$$f(x) \equiv 0 \pmod{p}$$

are solutions of the congruence

$$f_1(x) \equiv 0 \pmod{p}.$$

This new congruence can be handled in the same manner as the original one. Every time we find a solution, we can diminish the degree of the congruence by one. Since the linear congruence with a prime modulus cannot have more than one solution except when the prime modulus divides a_0, the theorem follows.

It must be remembered that the theorem may not hold if the modulus is not a prime. For example, the congruence

$$(x-2)(x-3) \equiv 0 \pmod{12}$$

has four solutions: 2, 3, 6 and 11 modulo 12. This is because of the fact that

$$ab \equiv 0 \pmod{p}$$

implies that either a or $b \equiv 0 \pmod{p}$. But if the modulus is a composite number, this sort of argument does not hold ground.

Exercises

1. Solve the congruences:

(i) $x^3 - 1 \equiv 0 \pmod{7}$;
(ii) $x^3 - 4 \equiv 0 \pmod{13}$;
(iii) $x^4 - 3 \equiv 0 \pmod{13}$;
(iv) $(x-1)(x-2)(x-3)(x-4) \equiv 0 \pmod{10}$;
(v) $x^4 - 1 \equiv 0 \pmod{8}$.

2. Show that the congruence

$$x^4 - 1 \equiv (x-1)(x-2)(x-3)(x-4) \pmod{5}$$

holds for all values of x. Why is it so?

19. The inclusion and exclusion principle

Suppose we are asked to find the number of numbers not exceeding 30 and prime to it.

The total number of numbers $\leqslant 30$ is of course 30. From these, we exclude multiples of 2, multiples of 3 and multiples of 5.

(i) Multiples of 2 are:

2, 4, 6, 8, 10, 12, 14, 16, 18, 20, 22, 24, 26, 28, 30;

(ii) Multiples of 3 are:

3, 6, 9, 12, 15, 18, 21, 24, 27, 30;

(iii) Multiples of 5 are:

5, 10, 15, 20, 25, 30.

In doing so, we will have excluded multiples of 2.3, 2.5, and 3.5 each twice instead of doing so only once and each multiple of 2.3.5 thrice instead of once. We, therefore, include multiples of 6, 10, 15 each once. These are

(1) 6, 12, 18, 24, 30; (2) 10, 20, 30; (3) 15, 30.

In doing so, we have now included 30 thrice, instead of doing so only twice. We must, therefore, exclude this multiple of 2.3.5 once. This process of exclusion and inclusion leaves us with

$$30-(15+10+6)+(5+3+2)-1=8$$

numbers. The required answer is thus 8. As we already know, this means that $\phi(30)=8$. Even in the general case, the principle involves no other technique.

Take N objects and let $a_1, a_2, a_3, \ldots, a_k$ be k attributes. Denote the set of objects which possess the attribute a_i by A_i and the number of elements in A_i by N_i. Evidently, the members of the set $A_i \cap A_j$, $i<j$, possess both the attributes a_i and a_j. Write $A_{ij}=A_i \cap A_j$ and let N_{ij} denote the number of elements in the set A_{ij}. These notations are readily extended. Thus, we write

$$A_{hij} \quad \text{for} \quad A_h \cap A_{ij}$$

and denote the number of elements in A_{hij} by N_{hij} and there is no loss of generality in assuming that $h<i<j$. With the above notation, we now prove

THEOREM 8. *The number of objects which possess none of the k attributes is given by*

$$N-\sum N_h+\sum N_{hi}-\sum N_{hij}+\cdots$$

where the sigmas indicate that all combinations of the k attributes are considered. Thus in

$$\sum N_{hij}$$

h, i, j run through all combinations of the k attributes taken three at a time.

Proof. Take any object which possesses none of the k attributes. This object is counted just once in the first term N and it is not counted at all in any of the other terms. Now consider an object which has r ($\leqslant k$) of the k attributes. This object is counted once in the first term, $\binom{r}{1}$ times in the second, $\binom{r}{2}$ times in the third, and so on. Thus it is counted

$$1-\binom{r}{1}+\binom{r}{2}-\binom{r}{3}+\cdots+(-1)^r\binom{r}{r}$$
$$=(1-1)^r=0 \text{ times}$$

in the entire expression given in the theorem. This proves the theorem.

Example. Let us consider the following well-known problem. A person writes n letters and puts down the addresses on n envelopes. The problem is to find the number of ways in which the letters can be put in the envelopes so that none of them reaches the right person.

We define the attribute a_i as the presence of the ith letter in the correct envelope. There are in all $n!$ ways in which the n letters can be placed in the covers. It is easy to see that

$$N = n!, \quad N_i = (n-1)!, \quad N_{ij} = (n-2)!,$$

and so on. Hence, the required number of ways is

$$n! - \binom{n}{1}(n-1)! + \binom{n}{2}(n-2)! - \cdots + (-1)^n \binom{n}{n}(n-n)!$$
$$= n! \sum_{j=0}^{n} (-1)^j/j!.$$

The problem can also be solved without using the inclusion and exclusion principle in the following manner.

Let u_j denote the number of ways in which j of the n letters will be wrongly addressed. Then, we have

$$u_n + \binom{n}{1} u_{n-1} + \binom{n}{2} u_{n-2} + \cdots + \binom{n}{r} u_{n-r} + \cdots + \binom{n}{n} u_0 = n!. \tag{1}$$

Note that $\binom{n}{r} u_{n-r}$ is the number of ways in which r of the letters are delivered to the right persons and the remaining $(n-r)$ to wrong ones. Our result (1) holds for all positive integers n and we take $u_0 = 1$ because there is only one way of addressing all the letters correctly.

Replacing n by 1, 2, 3, . . . in succession, (1) can be used to get the values of $u_1, u_2, u_3, \ldots$. We thus have

$$u_1 = 0, \quad u_2 = 1, \quad u_3 = 2,$$

and so on. Now, we can write (1) in the form

$$\frac{u_n}{n!} + \frac{u_{n-1}}{(n-1)!1!} + \frac{u_{n-2}}{(n-2)!2!} + \cdots + \frac{u_0}{0!n!} = 1. \tag{2}$$

Here it should be remembered that (2) holds for each $n \geqslant 0$. Hence, for $|x| < 1$, we have

$$\sum_{r=0}^{\infty} u_r x^r/r! \,.\, \sum_{s=0}^{\infty} x^s/s! = 1 + x + x^2 + \cdots = 1/(1-x).$$

This gives

$$(1-x)\sum_{r=0}^{\infty} u_r x^r/r! = \exp(-x)$$
$$= 1 - \frac{x}{1!} + \frac{x^2}{2!} - \cdots + \frac{(-x)^r}{r!} + \cdots .$$

Hence

$$u_r - ru_{r-1} = (-1)^r, \quad \text{for each } r \geqslant 1 \text{ with } u_0 = 1.$$

We have thus obtained a nice recurrence formula for u_r. Also

$$\sum_{r=0}^{\infty} u_r x^r/r! = (1 + x + x^2 + \cdots)\exp(-x)$$

so that

$$u_r = r! \sum_{j=0}^{r} (-1)^j/j! \tag{3}$$

which is exactly what the inclusion and exclusion principle gave. From (3) it readily follows that

$$u_r = [r!e^{-1} + \tfrac{1}{2}].$$

(Here e is the base of the natural logarithms.)

20. Use of inverse functions in summation

In what follows, $f(x)$ is a function of a real variable x defined for $x \geqslant 0$ and satisfying the conditions

(i) $f(x_1) < f(x_2)$ for $0 \leqslant x_1 < x_2$;
(ii) $0 < f(1) \leqslant 1$:

and $f^{-1}(x)$ denotes the inverse of $f(x)$. Small letters other than x and y denote natural numbers. It might be observed that of the two conditions stated above, the second hardly imposes any restriction on $f(x)$, for if

$$t < f(1) \leqslant (t+1), \quad t > 0;$$

one can consider the function $f(x) - t$ in place of $f(x)$.

Our object here is to prove

THEOREM 9. *For any n and $m = [f(n)]$,*

$$S(n) = \sum_{j=1}^{n} [f(j)] + \sum_{k=1}^{m} [f^{-1}(k)] = nm + d,$$

where d denotes the number of values of $j(\leqslant n)$ for which $f(j)$ is an integer.

Proof. Draw the curve

$$y = f(x), \quad \tfrac{1}{2} \leqslant x \leqslant n.$$

Consider the rectangle ABCD formed by the lines

$$x = \tfrac{1}{2}, \quad y = f(\tfrac{1}{2}), \quad x = n, \quad y = f(n).$$

Evidently, $\sum_{j=1}^{n} [f(j)]$ is the number of lattice points inside and on the boundary of the region marked J in the figure. Similarly $\sum_{k=1}^{m} [f^{-1}(k)]$ is the number of lattice points inside or on the boundary of the region K in the figure.

Hence $S(n)$ is the number of lattice points inside and on the rectangle ABCD + the number of lattice points on the curve $y = f(x), \frac{1}{2} \leqslant x \leqslant n$. Thus

$$S(n) = nm + d.$$

Though the Eisenstein technique employed in the proof is well-known the theorem proved here does not appear to have been mentioned in the literature before.

Example 1. Let

$$f(x) = ax/b, \quad a < b, \quad (a, b) = 1.$$

Then for $m = [an/b]$, we have

$$\sum_{j=1}^{n} [aj/b] + \sum_{k=1}^{m} [bk/a] = nm + [n/b].$$

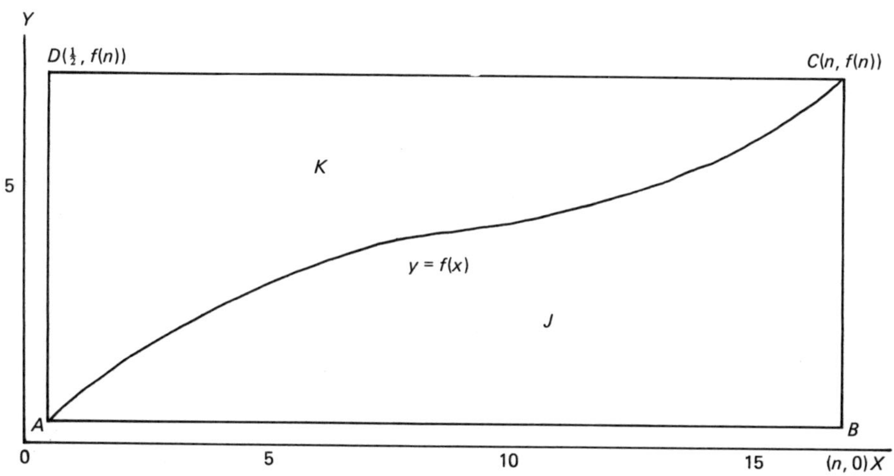

In particular if a and b are odd integers, $0 < a <' b$, and we write

$$a' = (a-1)/2, \quad b' = (b-1)/2;$$

then

$$\sum_{j=1}^{b'} [aj/b] + \sum_{k=1}^{a'} [bk/a] = a'b'.$$

Example 2. Let

$$f(x) = \sqrt{x}, \quad f^{-1}(x) = x^2.$$

Then for $m = [\sqrt{n}]$, the theorem gives

$$\sum_{j=1}^{n} [\sqrt{j}] + \sum_{k=1}^{m} k^2 = nm + m.$$

Hence

$$\sum_{j=1}^{n} [\sqrt{j}] = nm + m - \{m(m+1)(2m+1)/6\}.$$

Exercises

1. Consider the cases

   (i) $f(x) = x$; (ii) $f(x) = \sqrt[3]{x}$.

2. Evaluate

$$\sum_{j=1}^{50} [\sqrt[3]{j}].$$

Problems

1. If

$$a_1 + a_2 + \cdots + a_k = n$$

where the a's and n are natural numbers, then show that

$$n!/(a_1!\,a_2! \ldots a_k!)$$

is an integer.

2. For $r > 1$ and $n \geqslant 1$, show that

$$\frac{1!2! \ldots (r-1)!(nr)!}{n!(n+1)! \ldots (n+r-1)!}$$

is an integer.

3. Study the table

$$
\begin{array}{rrrrrrc}
4 & 7 & 10 & 13 & 16 & 19 & \dots \\
7 & 12 & 17 & 22 & 27 & 32 & \dots \\
10 & 17 & 24 & 31 & 38 & 45 & \dots \\
13 & 22 & 31 & 40 & 49 & 58 & \dots \\
16 & 27 & 38 & 49 & 60 & 71 & \dots \\
19 & 32 & 45 & 58 & 71 & 84 & \dots \\
.. & .. & .. & .. & .. & .. & \dots
\end{array}
$$

and show that if a number m does not appear in the table (which extends to infinity in both directions) then $2m + 1$ is a prime. Conversely if p is any odd prime then $(p - 1)/2$ does not appear in the table. (In other words, the table yields all odd primes.)

4. If g is the g.c.d. of

$$\binom{n}{1}, \binom{n}{2}, \binom{n}{3}, \dots, \binom{n}{n-1}$$

then g is a prime and n is a power of this prime.

5. Find the p-potency of

$$\binom{p^c - 1}{r}, \quad 0 < r < p^c - 1.$$

6. Prepare a table of values of

$$M_3(n) = \sum_{j=1}^{n} \mu_3(j)$$

for values of n up to 1000 and state if it has any peculiarities.

7. If $(a, b) = 1$, then integers (not necessarily positive) can always be found such that $ah + bk = 1$.

8. Find the largest number of four digits which when divided by 5, 9 and 13 will leave in order the remainders 3, 2 and 5.

9. Show that the l.c.m. of

$$\binom{n}{1}, \binom{n}{2}, \binom{n}{3}, \dots, \binom{n}{n}$$

is given by

$$p_1^{r_1} p_2^{r_2} p_3^{r_3} \dots p_k^{r_k}/(n + 1)$$

where p_j denotes the jth prime with $p_1 = 2$, and r_j is given by the inequality

$$p_j^{r_j} \leqslant n + 1 \leqslant p_j^{r_j+1}. \qquad \text{[Ian S. Williams]}$$

(Here p_k is the largest prime $\leqslant n + 1$.) Consider in particular the cases $n = 21$ and $n = 22$.

10. If $P(a_1, a_2, a_3, \dots, a_m)$ is a lattice point in m dimensional space and g is the g.c.d. of the a's. Then the segment of the line joining P to the origin O has exactly $g + 1$ lattice points on it. Find the coordinates of these lattice points.

11. Let h and k be relatively prime positive integers,

$$
\begin{aligned}
((x)) &= x - [x] - \tfrac{1}{2}, \quad \text{if } x \text{ is not an integer;} \\
&= 0, \text{ if } x \text{ is an integer.}
\end{aligned}
$$

If

$$S(h, k) = \sum_{j=1}^{k} \frac{j}{k}((jh/k)),$$

prove Dedekind's reciprocity formula

$$S(h, k) + S(k, h) = -\frac{1}{4} + \frac{1}{12}\left\{\frac{h}{k} + \frac{k}{h} + \frac{1}{hk}\right\}.$$

CHAPTER 2

PRIMITIVE ROOTS AND INDICES

1. The quest

If one were required to simplify an expression such as

$$\frac{\sqrt[3]{\cdot 03}}{4\cdot 4} \times \frac{7^3}{5\cdot 7}$$

or solve an equation of the type

$$5x^3 = 13$$

one would readily have recourse to a table of logarithms and get an answer to a suitable number of decimal places. The accuracy will certainly depend upon how good the tables are. One could ask, 'Is there a similar procedure available for the solution of congruences?' Our answer is in the affirmative. A congruence may of course have no solution or it may have one or more than one solution. The table of logarithms have a base usually 10 or e. In the case of congruences a choice of base will depend upon the modulus itself, as we shall see later. The index, which will correspond to the logarithm, might be a non-negative integer or a row matrix with non-negative elements.

Our object in this chapter will be to explain how such a system can be constructed, firstly when the modulus is a prime $p \geqslant 3$ or a power of such a prime, then for powers of 2 and finally for any modulus n.

2. Properties of the exponent

We have seen that for $(a, n) = 1$

$$a^{\phi(n)} \equiv 1 \pmod{n}.$$

We ask if $\phi(n)$ is the least positive integer for which this holds. In general this is not true. Take for example $n = 7, a = 2$, then

$$2^3 \equiv 1 \pmod{7}$$

while $\phi(7) = 6$.

DEFINITION. *If d be the least positive integer for which*

$$a^d \equiv 1 \pmod{n}, \quad (a, n) = 1$$

then we say that a appertains to the exponent d mod n.

Sometimes d is also spoken of as the order of modulo n. To start with, we prove some properties of d.

THEOREM 1. *If a appertains to d* mod *n and*

$$a^k \equiv 1 \pmod{n}, \quad k > 0$$

then $d \mid k$.

Proof. Evidently $k \geqslant d$. If $d \nmid k$, let

$$k = qd + r, \quad 0 < r < d;$$

then

$$a^{qd+r} \equiv 1 \pmod{n}. \tag{1}$$

Since

$$a^d \equiv 1 \pmod{n},$$

(1) implies that

$$a^r \equiv 1 \pmod{n}. \tag{2}$$

Since $0 < r < d$, (2) is impossible and the theorem follows.

COROLLARY. $d \mid \phi(n)$.

THEOREM 2. *If a appertains to d* mod *n, then*

$$1, a, a^2, a^3, \ldots, a^{d-1}$$

are incongruent mod *n*.

Proof. If possible, suppose for some $j, k, 0 \leqslant j < k < d$,

$$a^j \equiv a^k \pmod{n}.$$

Then

$$a^{k-j} \equiv 1 \pmod{n}.$$

Since $(k - j) < d$, this cannot hold and the theorem follows.

Remarks

From this theorem it is clear that if $d < \phi(n)$, then the powers of a cannot give a reduced residue system modulo n. On the other hand, if $d = \phi(n)$, that is if a appertains to $\phi(n)$ modulo n, the numbers

$$1, a, a^2, a^3, \ldots, a^{\phi(n)-1}$$

form a reduced residue system modulo n. In this case, a could serve as a base and we could define the index of any number m prime to n by the relation

$$\mathrm{Ind}_a m \equiv t(\bmod \phi(n)),$$

when $m \equiv a^t(\bmod n)$.

If such an a exists, we call it a primitive root of n. Unluckily this does not always happen and other modes of constructing a system of indices have to be found.

THEOREM 3. *If a appertains to d* mod n, *then* a^λ *appertains to* d_1 mod n, *where*

$$d_1 = d/(\lambda, d).$$

Proof. Since $(a, n) = 1$, we have

$$(a^\lambda, n) = 1.$$

If a^λ appertains to d_1 modulo n, then

$$(a^\lambda)^{d_1} \equiv 1 \ (\bmod n),$$

that is

$$a^{\lambda d_1} \equiv 1 \ (\bmod n).$$

Since a appertains to d modulo n,

$$d \mid \lambda d_1. \tag{3}$$

The least positive value of d_1 for which (3) can hold, is

$$d_1 = \frac{d}{(\lambda, d)}.$$

Moreover for this value of d_1

$$(a^\lambda)^{d_1} \equiv 1 \ (\bmod n).$$

Hence the theorem.

COROLLARY. *If a appertains to d* mod n, *then the members of the set of integers*

$$\{a^\lambda : (\lambda, d) = 1\}$$

all appertain to d.

Exercises

1. If two different integers a_1 and a_2 appertain to d_1 and d_2 modulo n and $(d_1, d_2) = 1$, then $a_1 a_2$ appertains to $d_1 d_2$ modulo n.
2. If a appertains to d with respect to an odd prime p, then $(a + p)$ also appertains to d modulo p.
3. Find the exponents to which 2 and 3 appertain modulo 11. Find also the integers which appertain to 10 modulo 11. How many incongruent primitive roots has 11?
4. If a appertains to d modulo p, where p is a prime of the form $4k + 3$, show that $(-a)$ appertains to $d/2$ or $2d$ modulo p according as d is even or odd.

3. Primitive roots of powers of odd primes

In this section, we prove two theorems giving the number of primitive roots of p^u, $u \geqslant 1$, p being an odd prime.

THEOREM 4. *Every odd prime p has $\phi(p-1)$ primitive roots.*

Proof. Let a be any integer less than and prime to p. If a appertains to d modulo p, then since

$$a^{p-1} \equiv 1 \pmod p$$

d must be a divisor of $(p-1)$.

Let $d_1 < d_2 < d_3 < \cdots < d_k$ be all the divisors of $(p-1)$, also let $\psi(d_j)$ denote the number of integers (less than and prime to p) which appertain to d_j. Then, since each of the integers

$$1, 2, 3, \ldots, p-1 \tag{1}$$

appertains to one and only one of the d_j's, we must have

$$\sum_{j=1}^{k} \psi(d_j) = p - 1. \tag{2}$$

We further know that

$$\sum_{j=1}^{k} \phi(d_j) = p - 1. \tag{3}$$

If no integer of set (1) appertains to a given d_j, then

$$\psi(d_j) < \phi(d_j). \tag{4}$$

If an integer a of the set (1) appertains to d_j, then a is a root of the congruence

$$x^{d_j} \equiv 1 \pmod p. \tag{5}$$

The integers which appertain to d_j must, therefore, be found among the roots of (5), namely

$$1, a, a^2, \ldots, a^{d_j - 1}.$$

Since a^λ appertains to d_j if and only if

$$(\lambda, d_j) = 1,$$

we must have, in this case,

$$\psi(d_j) = \phi(d_j). \tag{6}$$

Thus

$$\psi(d_j) \leqslant \phi(d_j). \tag{7}$$

But

$$\sum_{j=1}^{k} \psi(d_j) = \sum_{j=1}^{k} \phi(d_j). \tag{8}$$

Hence we must have

$$\psi(d_j) = \phi(d_j). \tag{9}$$

In words, this means that the number of integers of the set (1) appertaining to d_j is $\phi(d_j)$. In particular there are $\phi(p-1)$ integers which appertain to $p-1$. These are the primitive roots of p.

Exercise
Find the primitive roots of 17 and verify theorem 4.

THEOREM 5. *If p is an odd prime, p^u has $\phi(\phi(p^u))$ primitive roots.*

For our proof, we shall need the following

LEMMA. *Among the primitive roots of p, there exists at least one, say g, for which*

$$\text{pot}_p(g^{p-1} - 1) = 1.$$

Proof. Let a be any primitive root of p and suppose

$$\text{pot}_p(a^{p-1} - 1) > 1. \tag{10}$$

Let $g_j = a + jp$, then g_j is also a primitive root of p and we have

$$\begin{aligned} g_j^{p-1} &= (a + jp)^{p-1} \\ &\equiv a^{p-1} + (p-1)(jp)a^{p-2} \pmod{p^2}. \end{aligned} \tag{11}$$

Since by (10)

$$a^{p-1} \equiv 1 \pmod{p^2},$$

(11) gives

$$g_j^{p-1} - 1 \equiv -(jp)a^{p-2} \pmod{p^2}.$$

Hence for $(j, p) = 1$,

$$g_j^{p-1} - 1 \equiv 0 \pmod{p} \quad \text{but} \quad \not\equiv 0 \pmod{p^2}.$$

This proves the lemma.

Remark
g_j can be made to take even or odd values by a proper choice of j with $(j, p) = 1$.

Proof of theorem 5. Let g be a primitive root of p such that

$$\text{pot}_p(g^{p-1} - 1) = 1 \tag{12}$$

then we assert that g is a primitive root of p^u. In view of (12), we have for some h

$$g^{p-1} = 1 + hp,$$

where $(h, p) = 1$. Suppose g appertains to d modulo p^2. Then

$$\begin{aligned} g^d &\equiv 1 \pmod{p^2} \\ &\equiv 1 \pmod{p}. \end{aligned} \tag{13}$$

Since g is a primitive root of p and, therefore, appertains to $(p - 1)$ modulo p, $(p - 1) \mid d$. Let $d = (p - 1)\beta$, then

$$\begin{aligned} g^{(p-1)\beta} &= (1 + hp)^\beta \\ &\equiv 1 + \beta hp \pmod{p^2} \end{aligned}$$

Therefore, the least value of β for which

$$g^{(p-1)\beta} \equiv 1 \pmod{p^2}$$

is given by $\beta = p$. Thus g appertains to $p(p - 1) = \phi(p^2)$ modulo p^2. It is, therefore, a primitive root of p^2. Moreover

$$\text{pot}_p(g^{\phi(p^2)} - 1) = 2 \tag{14}$$

because

$$\begin{aligned} g^{\phi(p^2)} &= (1 + hp)^p \\ &= 1 + hp^2 + \binom{p}{2} h^2 p^2 + \cdots \\ &\equiv 1 + hp^2 \pmod{p^3}. \end{aligned}$$

The rest of the proof is based on induction. Let g be a primitive root of p^α, $\alpha = 2, 3, \ldots, u - 1$, such that

$$\text{pot}_p(g^{\phi(p^\alpha)} - 1) = \alpha; \tag{15}$$

then for some h,

$$g^{\phi(p^{u-1})} = 1 + hp^{u-1}, \quad (h, p) = 1.$$

If g appertains to d modulo p^u, we have

$$\begin{aligned} g^d &\equiv 1 \pmod{p^u} \\ &\equiv 1 \pmod{p^{u-1}}. \end{aligned}$$

Since g is a primitive root of p^{u-1}, $\phi(p^{u-1}) \mid d$. Let

$$d = c\phi(p^{u-1}),$$

then

$$\begin{aligned} g^{c\phi(p^{u-1})} &= (1 + hp^{u-1})^c, \quad u \geqslant 3 \\ &= 1 + chp^{u-1} \pmod{p^{2u-2}} \\ &\equiv 1 + chp^{u-1} \pmod{p^{u+1}}. \end{aligned}$$

The least value of c for which

$$g^{c\phi(p^{u-1})} - 1 \equiv 0 \pmod{p^u}$$

is, therefore, given by $c = p$. Hence g appertains to $p\phi(p^{u-1}) = \phi(p^u)$ modulo p^u. Moreover (15) holds for $\alpha = u$. This proves our assertion that g is a primitive root of p^u provided

$$\text{pot}_p(g^{p-1} - 1) = 1.$$

Since g^λ will appertain to $\phi(p^u)$ if and only if

$$(\lambda, \phi(p^u)) = 1, \quad \lambda \leqslant \phi(p^u),$$

the number of primitive roots of p^u is as stated in the theorem.

Exercises

1. If g is a primitive root of p^u, $p \geqslant 3$, then either g or $g + p^u$ is a primitive root of $2p^u$.
2. If n has one primitive root, then it has exactly $\phi(\phi(n))$ primitive roots.
3. Show that g is or is not a primitive root of n according as the numbers

$$1, g, g^2, \ldots, g^{\phi(n)-1}$$

do or do not form a reduced residue set modulo n.

4. Find a common primitive root of 7 and 11.
5. Find if 3 is a primitive root of 7^3.
6. Find if 15 has a primitive root.

The reader will note that theorems (4) and (5) simply state that powers of all odd primes have primitive roots. They also give the number of primitive roots in these cases. Given one primitive root, the proof of theorem (4) indicates how all the primitive roots can be obtained. No formula for primitive roots is, however, known. W. W. Norton and Company of New York, has recently published a table of indices from which the primitive roots can be directly read out in the case of primes below 2000.

The laws of indices are exactly similar to those of logarithms and can be proved in the same manner. Here we just state them for reference.

If g is a primitive root of n and $(m_1, n) = 1 = (m_2, n)$, then, modulo $\phi(n)$,

$$\text{Ind}_g(m_1 m_2) \equiv \text{Ind}_g(m_1) + \text{Ind}_g(m_2);$$

$$\text{Ind}_g(m^k) \equiv k\,\text{Ind}_g(m);$$

and if g_1 and g_2 are two distinct primitive roots of n, then for $(m, n) = 1$, again modulo $\phi(n)$, we have

$$\text{Ind}_{g_1}(m) \,.\, \text{Ind}_{g_2}(g_1) \equiv \text{Ind}_{g_2}(m).$$

The following example will indicate how primitive roots and indices can be utilized for solving certain congruences when the modulus is a power of an odd prime.

Example. Given that 3 is a primitive root of 31, prepare a table of indices and solve the congruence

$$5x^3 \equiv 18 \pmod{31}.$$

Solution. We have the following tables for base 3.

Ind→ ↓	0	1	2	3	4	5	6	7	8	9
	1	3	9	27	19	26	16	17	20	29
1	25	13	8	24	10	30	28	22	4	12
2	5	15	14	11	2	6	18	23	7	21

Num→ ↓	0	1	2	3	4	5	6	7	8	9
		30	24	1	18	20	25	28	12	2
1	14	23	19	11	22	21	6	7	26	4
2	8	29	17	27	13	10	5	3	16	9
3	15									

The given congruence can be written in the form

$$\begin{aligned} 3\,\text{Ind}_3 x &\equiv \text{Ind}_3 18 - \text{Ind}_3 5 \pmod{30} \\ &\equiv 26 - 20 \equiv 6 \pmod{30}. \end{aligned}$$

Hence

$$\begin{aligned} \text{Ind}_3 x &\equiv 2 \pmod{10} \\ &\equiv 2, 12, 22 \pmod{30}, \end{aligned}$$

so that

$$x \equiv 9, 8, 14 \pmod{31}.$$

We might observe that the primitive roots of 31 are the numbers whose indices are prime to 30. Such indices are

$$1, 7, 11, 13, 17, 19, 23, 29.$$

Hence the primitive roots of 31 are

$$3, 17, 13, 24, 22, 12, 11, 21.$$

Exercises

1. Prepare the index tables for the prime modulus 41 to the base 6. Find from your tables (i) the other primitive roots of 41; (ii) the exponent to which 3 appertains modulo 41; (iii) the value of $\text{Ind}_7\, 23$.
2. Use the tables given in this section to find the value of $\text{Ind}_{11}\, 27$ for the modulus 31.
3. Solve the following congruences:

$$\begin{array}{rrl}
\text{(i)} & 3x^2 \equiv 7 & (\text{mod } 17); \\
\text{(ii)} & x^2 \equiv 2 & (\text{mod } 11); \\
\text{(iii)} & x^4 \equiv 23 & (\text{mod } 41); \\
\text{(iv)} & 2x^3 \equiv 3 & (\text{mod } 25); \\
\text{(v)} & 11x^5 \equiv 5 & (\text{mod } 27); \\
\text{(vi)} & 5x \equiv 7 & (\text{mod } 31).
\end{array}$$

4. Find the remainder when

$$12^7 + 17^2$$

is divided by 31.

4. Non-existence of primitive roots

In the preceding section we have shown that primitive roots exist for numbers of the form p^u and $2p^u$, $p \geqslant 3$. It can be readily verified that 3 is a primitive root of 2 and of 4. We proceed to show that primitive roots do not exist in any other case. For this purpose we need the following lemmas.

LEMMA. *If*

$$a \equiv 3 \quad \text{or} \quad -3 \pmod 8,$$

then for $u \geqslant 1$,

$$\text{pot}_2(a^{2^u} - 1) = u + 2.$$

Proof. When $u = 1$, let

$$a = 8k \pm 3;$$

then

$$\begin{aligned} a^2 &= 64k^2 \pm 48k + 9 \\ &\equiv 1 \pmod 8 \quad \text{but} \quad \not\equiv 1 \pmod{16}. \end{aligned}$$

Hence

$$\text{pot}_2(a^2 - 1) = 3.$$

Thus the lemma holds for $u = 1$. Suppose it holds for $u = 1, 2, 3, \ldots, t - 1$. Then

$$\text{pot}_2(a^{2^{t-1}} - 1) = t + 1;$$

that is for an odd h,

$$a^{2^{t-1}} = 1 + h2^{t+1}.$$

Then we have

$$\begin{aligned} a^{2^t} &= (1 + h2^{t+1})^2 \\ &= 1 + h2^{t+2} + 2^{2t+2} \\ &\equiv 1 \pmod{2^{t+2}} \quad \text{but} \quad \not\equiv 1 \pmod{2^{t+3}}. \end{aligned}$$

The lemma readily follows by induction.

LEMMA. *If*

$$a \equiv 1 \quad \text{or} \quad -1 \pmod 8$$

then for $u \geqslant 1$,

$$\text{pot}_2(a^{2^u} - 1) \geqslant u + 3.$$

Proof. Let

$$a = 8k \pm 1.$$

Then

$$\begin{aligned} a^2 &= 64k^2 \pm 16k + 1 \\ &\equiv 1 \pmod{16}. \end{aligned}$$

The lemma can be proved by induction without any difficulty.

As an immediate consequence of these lemmas we have

THEOREM 6. *For* $u \geqslant 3$, 2^u *has no primitive roots.*

Proof. Odd numbers congruent to 3 or −3 modulo 8, appertain to 2^{u-2} modulo 2^u while those congruent to 1 or −1 modulo 8 appertain to 2^w modulo 2^u for some $w \leqslant u - 3$. Thus no odd number can appertain to 2^{u-1} modulo 2^u.

LEMMA. *If a appertains to* d_1 mod n_1 *and to* d_2 mod n_2 *and* $(n_1, n_2) = 1$, *then a appertains to* $\{d_1, d_2\}$| (*the l.c.m. of* d_1 *and* d_2) mod $n_1 n_2$.

Proof. Since $(a, n_1) = 1 = (a, n_2)$, we must have

$$(a, n_1 n_2) = 1.$$

Suppose a appertains to d modulo $n = n_1 n_2$. Then

$$\begin{aligned} a^d &\equiv 1 \pmod n \\ &\equiv 1 \pmod{n_j}, \quad j = 1, 2. \end{aligned}$$

Hence $d_1 \mid d$ and $d_2 \mid d$. The least value of d must, therefore, be a multiple of $\{d_1, d_2\}$. Since

$$a^{\{d_1, d_2\}} \equiv 1 \pmod{n_j}, \quad j = 1, 2;$$

and $(n_1, n_2) = 1$;

$$a^{\{d_1, d_2\}} \equiv 1 \pmod n$$

we must have

$$d = \{d_1, d_2\}$$

THEOREM 7. *The number $n = mp^u$, where $(m, p) = 1, u \geqslant 1$ and $p \geqslant 3$, has no primitive root if $m > 2$.*

Proof. Take any a prime to n. Let a appertain to d_1 modulo m, to d_2 modulo p^u and to d modulo n. Then from the lemma above

$$d = \{d_1, d_2\}.$$

Since $d_1 \mid \phi(m)$ and $d_2 \mid \phi(p^u)$, we must have

$$d \leqslant \{\phi(m), \phi(p^u)\}.$$

For $m > 2$, $\phi(m)$ and $\phi(p^u)$ are both even. Hence

$$d \leqslant \tfrac{1}{2}\phi(m)\phi(p^u) = \tfrac{1}{2}\phi(n).$$

The theorem follows directly.

As a direct consequence of theorems 6 and 7 we have

THEOREM 8. *Primitive roots exist only for the numbers $2, 4, p^u$ and $2p^u$ where $p \geqslant 3$ and $u \geqslant 1$.*

Exercises

1. Show that the least value of k, for which the congruence

$$x^k \equiv 1 \pmod{n}$$

holds for all integral values of x prime to n, is given by

$$k = |\{\phi(2^\lambda), \phi(p_1^{u_1}), \phi(p_2^{u_2}), \ldots, \phi(p_r^{u_r})|\}$$

where

$$n = 2^{u_0}p_1^{u_1}p_2^{u_2}\ldots p_r^{u_r}; \quad u_0 \geqslant 0; \quad u_j \geqslant 1, \quad j = 1, 2, \ldots, r;$$

is the canonical decomposition of n, and

$$\lambda = u_0 - 1 \quad \text{or} \quad u_0$$

according as $u_0 > 2$ or $u_0 \leqslant 2$.

2. Find the value of k in exercise 1, when n = 546, 343 and 144.
3. For a given k, what is the greatest value of n, for which the congruence

$$x^k \equiv 1 \pmod{n}$$

is satisfied by all integral values of x prime to n?

This question is best answered in terms of the *p-piquancy of n* which we define as follows.

If $\phi(p^u) \mid n$ but $\phi(p^{u+1}) \nmid n, u \geqslant 0, p \geqslant 2$; then we say that the p-piquancy of n is u or that n is u-piquant in p and write

$$\text{piq}_p(n) = u.$$

Thus

$$\mathrm{piq}_7(588) = 3$$

because

$$\phi(7^3) = 294 \mid 588 \quad \text{but} \quad \phi(7^4) \nmid 588.$$

The answer to our problem now is that the required n is given by

$$n = 2^h \prod p_i^{u_i},$$

where p_i runs over all odd primes,

$$u_i = \mathrm{piq}_{p_i}(k),$$

and

$$h = \mathrm{piq}_2(k) \quad \text{or} \quad 1 + \mathrm{piq}_2(k)$$

according as $\mathrm{piq}_2(k) = 1$ or >1. The proof of this statement is left to the reader.

Example. Let $k = 78$.

The primes for which the piquancy of 78 is positive are 2, 3, 7 and 79. Thus

$$\mathrm{piq}_2(78) = 2,$$

$$\mathrm{piq}_3(78) = 2,$$

$$\mathrm{piq}_7(78) = 1,$$

and

$$\mathrm{piq}_{79}(78) = 1.$$

Hence

$$n = 2^3 \,.\, 3^2 \,.\, 7 \,.\, 79$$
$$= 39816.$$

Exercises

1. Evaluate

$$\mathrm{piq}_{11}(100),\ \mathrm{piq}_{11}(121),\ \mathrm{piq}_{11}(1210).$$

2. Find the greatest value of n for which the congruence

$$x^k \equiv 1 \pmod{n}$$

is satisfied for every x prime to n, when $k = 12; 72; 56$.

3. If

$$\mathrm{piq}_p(n) = u$$

where $u \geqslant 1$ and $p \geqslant 2$, then show that an integer x exists such that

$$\mathrm{pot}_p(x^n - 1) = u + \left[\frac{2}{p}\right].$$

5. The indices for powers of 2

The idea of indices can be extended to moduli of the form $2^u, u \geqslant 3$, in the following manner.

Since 3 appertains to 2^{u-2} modulo 2^u when $u \geqslant 3$, the two sets of numbers

$$3^k \quad \text{and} \quad -3^k, \quad k = 1, 2, 3, \ldots, \phi(2^{u-1}),$$

taken together form a reduced residue set modulo 2^u. It will be noted that no member of the first set is congruent to any member of the second, and vice versa. In fact, each member of the first set is either congruent to 1 or to 3 modulo 8, while each member of the second set is congruent to 5 or 7 modulo 8.

For $(n, 2) = 1$, we define the index of n as follows. Let

$$n \equiv (-1)^j 3^k \pmod{2^u}$$

where $j = 0$ or 1, then

$$\text{Ind } n \equiv [j, k] (\text{mod}\, [2, \phi(2^{u-1})]);$$

where we write

$$[j, k] \equiv [r, s] (\text{mod}[a, b])$$

to mean

$$j \equiv r \pmod{a}; \quad k \equiv s \pmod{b}.$$

To explain how these new type of indices can be used, we solve the congruence

$$11x^5 \equiv 7 \pmod{32}.$$

Solution. The table of indices is in two parts.

For $j = 0$,

k	1	2	3	4	5	6	7	8
n	3	9	27	17	19	25	11	1

For $j = 1$,

k	1	2	3	4	5	6	7	8
n	29	23	5	15	13	7	21	31

We have now to consider the congruence

$$\begin{aligned} 5 \text{ Ind } x &\equiv \text{Ind } 7 - \text{Ind } 11 \pmod{[2, 8]} \\ &\equiv [1, 6] - [0, 7] \pmod{[2, 8]} \\ &\equiv [1, -1] \pmod{[2, 8]}. \end{aligned}$$

Since each element of the index must be so written that it may be divisible by 5, we rewrite the congruence as

$$5 \text{ Ind } x \equiv [5, 15](\text{mod}[2, 8]).$$

Hence

$$\text{Ind } x \equiv [1, 3](\text{mod}[2, 8]).$$

This gives

$$x \equiv 5 \ (\text{mod } 32).$$

Exercise

A reduced residue set modulo 2^u, $u \geqslant 3$ is provided by the members of the two sets

$$5^k \quad \text{and} \quad -5^k, \quad k = 1, 2, 3, \ldots, 2^{u-2}.$$

Use this fact to define index of n modulo 2^u and solve the congruences

(1) $11x^5 \equiv 7 \ (\text{mod } 32)$
(2) $5x^3 \equiv 17 \ (\text{mod } 64)$
(3) $7x^2 \equiv 31 \ (\text{mod } 32)$
(4) $9x \equiv 3^5 - 5^3 \ (\text{mod } 16)$.

6. Indices for all n

Proceeding on the same lines we can now extend the notion of primitive roots and indices to all odd moduli $n \geqslant 3$.

Let the prime decomposition of n be

$$n = p_1^{u_1} p_2^{u_2} p_3^{u_3}, \ldots, p_m^{u_m},$$

where p's are odd primes. Let g_h be a primitive root of $p_h^{u_h}$, then using the Chinese remainder theorem we can find a number g such that

$$g \equiv g_h (\text{mod } p_h^{u_h}), \quad h = 1, 2, 3, \ldots, m. \tag{1}$$

We shall call g a pseudo primitive root of n and define the index of any number t prime to n with respect to the base g as a one row matrix in the following manner. Let

$$t \equiv g^{a_h} (\text{mod } p_h^{u_h}), \quad h = 1, 2, 3, \ldots, m,$$

then

$$\text{Ind}_g t \equiv [a_1, a_2, \ldots, a_m](\text{mod}[\phi(p_1^{u_1}), \phi(p_2^{u_2}), \ldots, \phi(p_m^{u_m})])$$

or simply

$$\text{Ind}_g t \equiv [a_h] (\text{mod} [\phi(p_h^{u_h})]).$$

Here

$$[a_1, a_2, \ldots, a_m] \equiv [b_1, b_2, \ldots, b_m](\text{mod}[c_1, c_2, \ldots, c_m])$$

is to be taken to mean that

$$a_h \equiv b_h \pmod{c_h}.$$

These generalized indices can be used in the same manner as the ordinary indices defined earlier.

For even values of n, only a slight modification is necessary. Let

$$n = 2^{u_0} p_1^{u_1} p_2^{u_2} \ldots p_m^{u_m}$$

be the prime decomposition of n. The pseudo primitive root of n is now selected so as to satisfy not only the conditions (1) but also the further condition

$$g \equiv \begin{cases} 1 \pmod 2), & u_0 = 1, \\ 3 \pmod 4), & u_0 = 2, \\ 3 \pmod 8). & u_0 \geqslant 3. \end{cases} \tag{2}$$

Let $(t, n) = 1$ and

$$\begin{aligned} t &\equiv (-1)^j 3^k \pmod{2^{u_0}} \\ &\equiv g^{a_h} \pmod{p_h^{u_h}} \end{aligned}$$

then we define

$$\operatorname{Ind}_g t \equiv [j, k; a_h] (\operatorname{mod}[2, 2^d; \phi(p^{u_h})])$$

where

$$d = u_0 - 1 \quad \text{or} \quad u_0 - 2$$

according as u_0 is or is not <3.

The use of the generalized indices will be clear from the following.

Example. Solve the congruence

$$3x^7 \equiv 23 \pmod{308}$$

Solution. We notice that 3 is a primitive root of 7, and 2 is a primitive root of 11. A pseudo primitive root of 308 is, therefore, given by the congruences

$$g \equiv \begin{cases} 3 \pmod 7) \\ 2 \pmod{11}) \\ 3 \pmod 4). \end{cases}$$

One value of g is 255. The congruence takes the form

$$\begin{aligned} 7 \operatorname{Ind} x &\equiv \operatorname{Ind} 23 - \operatorname{Ind} 3 (\operatorname{mod}[2, 2; 6, 10]), \\ &\equiv [0, 1; 2, 0] - [0, 1; 1, 8] (\operatorname{mod}[2, 2; 6, 10]) \\ &\equiv [0, 0; 1, 2] (\operatorname{mod}[2, 2; 6, 10]) \\ &\equiv [0, 0; 7, 42] (\operatorname{mod}[2, 2; 6, 10]) \end{aligned}$$

so that

$$\text{Ind}\, x \equiv [0, 0; 1, 6] \pmod{[2, 2; 6, 10]}.$$

This gives

$$x \equiv 185 \pmod{308}.$$

Exercise

Using generalized indices, solve the congruences

(1) $x^3 \equiv 11 \pmod{56}$,
(2) $3x^4 \equiv 7 \pmod{75}$,
(3) $x^5 \equiv 13 \pmod{60}$,
(4) $7x^4 \equiv 16 \pmod{45}$,
(5) $3x^3 \equiv 23 \pmod{44}$.

7. Methods of finding primitive roots of primes

In an earlier section, we have stated that there is no formula which will give a primitive root of a prime p. We can, however, offer two methods which can be used for finding the primitive roots of a given prime p.

The first is a method of elimination and is based on the fact that if a appertains to d modulo p, $d < \phi(p)$, then

$$a^\lambda, \quad \lambda = 1, 2, 3, \ldots, d,$$

cannot be primitive roots of p.

We list the numbers

$$1, 2, 3, \ldots, p - 1.$$

We then take $a = 2$ and find the exponent d to which it appertains modulo p. If $d = p - 1$, then 2 is a primitive root of p and we are through. If $d < p - 1$, we eliminate from the list all the numbers which are congruent to

$$2, 2^2, 2^3, \ldots, 2^d$$

modulo p. We next try one of the numbers not so eliminated and repeat the process till $\phi(p - 1)$ numbers are left. These are, then, the primitive roots of p.

Example. Let us find the primitive roots of 41.

We list the numbers

$$1, 2, 3, \ldots, 40.$$

Since 2 appertains to 20 modulo 41, we eliminate from our list the numbers

$$2, 4, 8, 16, 32, 23, 5, 10, 20, 40, 39, 37, 33, 25, 9, 18, 36, 31, 21, 1.$$

We next consider 3. We find that 3 appertains to 8 modulo 41. This enables us to eliminate the numbers

$$3, 9, 27, 40, 38, 32, 14, 1.$$

Of these 1, 9, 32 and 40 have already been eliminated. We are, thus, left with the $\phi(40)$ numbers

$$6, 7, 11, 12, 13, 15, 17, 19, 22, 24, 26, 28, 29, 30, 34, 35;$$

which are the primitive roots of 41.

In the second method, we break up the set of primes less than p into two subsets, one consisting of those $<\sqrt{p}$ and the other of those $>\sqrt{p}$. We first set up simple congruences modulo p, connecting the primes of the second set (and possibly some of the first set also) with those of the first. We then translate these congruences in terms of indices to an unspecified base. A study of the latter congruences leads to some important relations among the indices themselves. We explain the method with the help of the following example.

Example. Take $p = 41$, then the members of the first set are 2, 3 and 5 and those of the second 7, 11, 13, 17, 19, 23, 29, 31 and 37.

The congruences are presented below in a tabular form. We write x for Ind 2 and y for Ind 3. It will be remembered that irrespective of the base

$$\begin{aligned}\text{Ind}(-1) &\equiv (p-1)/2 \pmod{p-1} \\ &\equiv 20 \pmod{40}.\end{aligned}$$

mod 41	mod 40	
$2^3.5 \equiv -1$	Ind 5 $\equiv 20 - 3x$	(1)
$2.7 \equiv -3^3$	Ind 7 $\equiv 20 - x + 3y$	(2)
$11.3 \equiv -2^3$	Ind 11 $\equiv 20 + 3x - y$	(3)
$13.3 \equiv -2$	Ind 13 $\equiv 20 + x - y$	(4)
$17.2 \equiv -7$	Ind 17 $\equiv -2x + 3y$	(5)
$19.2 \equiv -3$	Ind 19 $\equiv 20 - x + y$	(6)
$23.2 \equiv 5$	Ind 23 $\equiv 20 - 4x$	(7)
$29 \equiv -2^2.3$	Ind 29 $\equiv 20 + 2x + y$	(8)
$31 \equiv -2.5$	Ind 31 $\equiv -2x$	(9)
$37 \equiv -2^2$	Ind 37 $\equiv 20 + 2x$	(10)

Also

$5 \equiv -2^2.3^2$	Ind 5 $\equiv 20 + 2x + 2y$	(11)
$11 \equiv -2.3.5$	Ind 11 $\equiv -2x + y$	(12)

From (1) and (11) we have

$$5x + 2y \equiv 0 \pmod{40}. \tag{13}$$

Also (3) and (12) give

$$5x - 2y \equiv 20 \pmod{40}. \tag{14}$$

From (13) and (14), we readily obtain

$$x \equiv 2 \pmod 4$$

$$y \equiv 5 \pmod{10}.$$

The corresponding sets of values are now shown by (13) to be

$$\left.\begin{aligned} x &\equiv 6, 14, 22, 30, 38 \\ y &\equiv 5, 25 \end{aligned}\right\} \pmod{40} \tag{15}$$

and

$$\left.\begin{aligned} x &\equiv 2, 10, 18, 26, 34 \\ y &\equiv 15, 35 \end{aligned}\right\} \pmod{40} \tag{16}$$

Since

$$(\text{Ind } 2, 40) = 2$$

and

$$(\text{Ind } 3, 40) = 5$$

2 and 3 cannot be primitive roots of 41.

Similar reasoning shows that 23, 31 and 37 too are not primitive roots of 41.

In the case of the primes 7, 11, 13, 17, 19 and 29 values of x and y can be selected from the sets (15) or (16) so that the index will be congruent to either 1 or −1 modulo 40. Hence 7, 11, 13, 17, 19 and 29 are all prime primitive roots of 41.

Thus, taking $x = 26$, $y = 15$, we have

$$\left.\begin{aligned} \text{Ind } 7 &\equiv -1 \\ \text{Ind } 11 &\equiv 3 \\ \text{Ind } 13 &\equiv 31 \\ \text{Ind } 17 &\equiv 33 \\ \text{Ind } 19 &\equiv 9 \\ \text{Ind } 29 &\equiv 7 \end{aligned}\right\} \pmod{40}. \tag{17}$$

It will be noted that all these indices are prime to 40.

The indices of composite numbers can be readily computed from (17). Thus

$$\text{Ind } 6 \equiv \left\{\begin{aligned} &\text{Ind } 2 + \text{Ind } 3 \\ &26 + 15 \\ &1 \end{aligned}\right\} \pmod{40}.$$

This shows that 6 also is a primitive root of 41. Other composite primitive roots of 41 can be similarly obtained.

The reader will do well to find the values of x and y for which

$$\text{Ind } 11 \equiv 1 \quad \text{or} \quad -1 \pmod{40}.$$

Remark

The second method is discussed in detail in the Introduction to the Table of Indices and Primitive Roots published by the Royal Society of London.

Exercises

1. Use the method of elimination to find the primitive roots of 37 and 73.
2. Using the method of indices find the least prime primitive roots of all primes below 100.

8. A property of the primitive roots

In this section we prove an interesting property of the primitive roots of an odd prime p. But before we can do this, we must prove

THEOREM 9. *If $f(j)$ is an arithmetical function, defined for all positive integral values of j, and*

$$S'(n) = \sum_{j <' n} f(j),$$

then

$$S'(n) = \sum_{d \mid n} \mu(d)\{f(d) + f(2d) + f(3d) + \cdots + f(n)\}. \tag{1}$$

Proof. We notice that $f(j)$ occurs on the right side of (1) every time that j is a multiple of d where d itself is a divisor of n. Hence

$$\sum_{d \mid n} \mu(d)\{f(d) + f(2d) + f(3d) + \cdots + f(n)\}.$$

$$= \sum_{d \mid n} \left\{\mu(d) \sum_{k=1}^{[n/d]} f(kd)\right\}$$

$$= \sum_{j=1}^{n} \left\{f(j) \sum_{\substack{d \mid j \\ d \mid n}} \mu(d)\right\}$$

$$= \sum_{j=1}^{n} \left\{f(j) \sum_{\substack{d \mid (j,n) \\ j \leqslant n}} \mu(d)\right\}.$$

Now

$$\sum_{d \mid m} \mu(d) = 1 \quad \text{or} \quad 0$$

according as $m = 1$ or $m > 1$. Hence

$$\sum_{j=1}^{n}\left\{f(j)\sum_{d\mid(j,n)}\mu(d)\right\} = \sum_{\substack{j\leqslant n\\(j,n)=1}} f(j)$$

and this is the theorem.

The property of the primitive roots to which we have alluded can be stated in the form

THEOREM 10. *If $S'(p)$ denotes the sum of all the incongruent primitive roots of an odd prime p, then*

$$S'(p) \equiv \mu(p-1)(\operatorname{mod} p),$$

where μ is the Möbius function as usual.

Proof. We have

$$S'(p) \equiv \sum_{j<' p-1} g^j (\operatorname{mod} p),$$

where g is a primitive root of p. Theorem 9 now gives

$$S'(p) \equiv \sum_{d\mid(p-1)}\left\{\mu(d)\sum_{k=1}^{[(p-1)/d]} g^{kd}\right\}(\operatorname{mod} p)$$

$$\equiv \sum_{d\mid(p-1)}\left\{\mu(d)\frac{g^{p-1}-1}{g^d-1}g^d\right\}(\operatorname{mod} p).$$

Since g is a primitive root of p,

$$g^{p-1} - 1 \equiv 0 (\operatorname{mod} p),$$

while for $d < (p-1)$,

$$g^d - 1 \not\equiv 0 (\operatorname{mod} p).$$

Hence

$$S'(p) \equiv \mu(p-1)g^{p-1} (\operatorname{mod} p)$$

$$\equiv \mu(p-1)(\operatorname{mod} p).$$

Theorem 10 which we have just proved is due to Gauss. The following is a generalization of the theorem by Raymond Moller [*American Math. Monthly*, **59** (1952), 226–230].

THEOREM 11. *The sum of the rth powers of the numbers which appertain to d mod p*

$$\equiv \frac{\phi(d)}{\phi(d_1)}\mu(d_1)(\operatorname{mod} p),$$

where

$$d_1 = \frac{d}{(r, d)}.$$

Proof. Let t be any number which appertains to d modulo p. Then the set of all numbers which appertain to d modulo p is given by

$$T = \{t^\lambda : \lambda <' d\}.$$

From theorem 3, we know that t^r appertains modulo p to d_1, where

$$d_1 = \frac{d}{(r, d)}.$$

Hence the members of set

$$R = \{t^{\lambda r} : \lambda <' d\},$$

when divided by p, leave as remainders the numbers given by the set

$$J = \{t^{rj} : j <' d_1\} \tag{2}$$

each repeated $\dfrac{\phi(d)}{\phi(d_1)}$ times.

The set J is the same modulo p as the set of numbers

$$K = \{a^k : k <' d_1\}, \quad \text{where } a = t^r \pmod p.$$

The sum of the elements of K is given by

$$\sum_{h \mid d_1} \mu(h)\{a^h + a^{2h} + \cdots + a^{d_1}\} = \sum_{h \mid d_1} \mu(h) \cdot \frac{a^{d_1} - 1}{a^h - 1} \cdot a^h.$$

Now

$$a^h \not\equiv 1 \pmod p$$

for any $h < d_1$, while

$$a^{d_1} \equiv 1 \pmod p.$$

Hence the sum of the members of set K

$$\begin{aligned} &\equiv \mu(d_1) a^{d_1} \pmod p \\ &\equiv \mu(d_1) \pmod p. \end{aligned} \tag{3}$$

Our proof is based on theorem 9 and is significantly simpler than Moller's.

Exercises

1. If S_j denotes the sum of the jth powers of integers $<' n$, use theorem 9 to show that for $n > 1$,

$$S_1 = \tfrac{1}{2} n \phi(n),$$

$$S_2 = \tfrac{1}{3}n^2\phi(n) + \tfrac{1}{6}n \prod_{p \mid n} (1 - p),$$

$$S_3 = \tfrac{1}{4}n^3\phi(n) + \tfrac{1}{4}n^2 \prod_{p \mid n} (1 - p).$$

2. Extend the property given for primitive roots of primes in theorem 10 to powers of primes.

9. An analogy

Another context in which one comes across the term 'primitive roots' is where we talk of the qth roots of unity. The primitive roots, there, are given by

$$\exp(2ir\pi/q), \quad r <' q, \quad (i = \sqrt{-1}).$$

Equivalently, we could define the primitive qth roots of unity as those roots of the equation

$$x^q = 1 \tag{1}$$

which do not satisfy the equation

$$x^k = 1 \tag{2}$$

for any $k < q$.

The primitive roots of $n = 2, 4, p^u, 2p^u, (p \geqslant 3)$, which we have considered in this chapter can be defined similarly as those roots of the congruence

$$x^{\phi(n)} \equiv 1 \pmod{n} \tag{3}$$

which do not satisfy the congruence

$$x^k \equiv 1 \pmod{n} \tag{4}$$

for any $k < \phi(n)$. The analogy between the primitive roots of unity and the primitive roots of an integer is thus clear.

Take, for example, $n = p$, $q = \phi(p)$, $(p \geqslant 3)$. Then we can establish a one–one correspondence between the roots of (3) and those of (1), in which the set of primitive roots of (3) maps onto the set of primitive roots of (1).

Let g be a primitive root of (3) and ξ a primitive root of (1); then the desired mapping is obtained by letting g^k correspond to ξ^k, $1 \leqslant k \leqslant q$.

This analogy is further brought out by the following results (still taking the case $n = p$, $q = \phi(p)$):

The sum of the primitive roots of (1) $= \mu(p - 1)$.

Correspondingly:

The sum of the primitive roots of (3) $\equiv \mu(p - 1) \pmod{p}$.

Again:

The sum of all the roots of (1) $= 0$,

while

The sum of the incongruent roots of (3) $\equiv 0 \pmod p$.

In the aforesaid mapping, the numbers which appertain to d modulo p, correspond to the numbers

$$\xi^{h(p-1)/d}, \quad h <' d; \tag{5}$$

and the sum of the rth powers of numbers which appertain to d modulo p is congruent (mod p) to the sum of the rth powers of the numbers in (5). Notice that the second sum here is Ramanujan's trigonometric sum.

Though the analogy between the primitive roots of (1) and those of (3) appears again and again, it cannot be stretched too far. For example, we cannot say that

$$g \equiv \xi \pmod p.$$

While g is an integer, ξ is a complex number.

Exercises

1. Consider if the primitive roots of p^u, $u > 1$, $p \geqslant 3$, can be mapped onto the primitive qth roots of unity for $q = \phi(p^u)$.
2. Can a one–one correspondence be set up between the numbers which appertain to d modulo n and the numbers

$$\xi^{k\phi(n)/d}, \quad k <' d,$$

ξ being a $\phi(n)$th primitive root of unity?

Appendix

The following table gives one primitive or pseudo-primitive root of each $n \leqslant 50$.

n		n		n		n		n	
1	1	11	2	21	17	31	3	41	6
2	1	12	11	22	13	32	3	42	17
3	2	13	2	23	5	33	2	43	3
4	3	14	3	24	11	34	3	44	35
5	2	15	2	25	2	35	17	45	2
6	5	16	3	26	15	36	11	46	5
7	3	17	3	27	2	37	2	47	5
8	3	18	11	28	3	38	21	48	11
9	2	19	2	29	2	39	2	49	3
10	7	20	7	30	17	40	27	50	27

10. Primitive divisors and primitive factors of $a^n - b^n$

Given two relatively prime positive integers a and b, $b < a$, the expression

$$a^n - b^n$$

represents, for $n \geqslant 1$, a natural number which we shall denote by $v(n)$ for simplicity.

For any given n, a positive integer k will be said to be a *primitive divisor* of $v(n)$, if $k \mid v(n)$ but for each $j < n$, $(k, v(j)) = 1$. The greatest primitive divisor of $v(n)$ will be denoted by $g(n)$. This $g(n)$ could be as small as 1.

Exercises
1. Show that 1 is a primitive divisor of $v(n)$ for each $n \geqslant 1$.
2. Show that 2 is not a primitive divisor of $v(n)$ for any $n > 1$.
3. If k_1 and k_2 are two distinct primitive divisors of $v(n)$, then so also is their l.c.m. $\{k_1, k_2\}$ a primitive divisor of $v(n)$.
4. Every divisor of $g(n)$ is a primitive divisor of $v(n)$.

The expression $a^n - b^n$ can also be looked upon as a polynomial of degree n in the two variables a and b. When we do so, we denote the expression by $V(n)$.

DEFINITION. *If F is a homogeneous polynomial of degree more than 0 in a and b with all coefficients integers (not necessarily positive or non-zero) and the coefficient of the highest degree term in a equal to 1, then F is said to be a primitive factor of $V(n)$ if for a given n*

$$F \mid V(n) \quad \textit{but for each} \quad j < n, (F, V(j)) = 1,$$

that is, the highest common factor of F and $V(j)$ is 1.

Exercises
1. Is 1 a primitive factor of $V(n)$?
2. Show that $a^2 + ab + b^2$ is a primitive factor of $V(3)$.

10.1. Notation

In what follows, capital letters denote polynomials homogeneous in a and b, of degree at least one, coefficients all integral and the coefficient of the highest degree term in a unity. The corresponding small letter is used to denote the value of the polynomial for the given a and b.

The l.c.m. of the polynomials $U_1, U_2, \ldots, U_h$ is itself a polynomial which we denote by $\{U_1, U_2, \ldots, U_h\}$. This polynomial has integral coefficients, is homogeneous, and has unity for the coefficient of the highest degree term in a.

The l.c.m. of $u_1, u_2, \ldots, u_h$ will be denoted by $\{u_1, u_2, \ldots, u_h\}$. It is noteworthy that $(U_1, U_2) = 1$ does not necessarily imply that $(u_1, u_2) = 1$. Thus, for example

$$(a + b, a^2 + b^2) = 1, \quad \text{but for} \quad a = 3, b = 1, (4, 10) = 2.$$

If $W \mid U$, we say W is a *factor* of U; if $w \mid u$, we say w is a *divisor* of u. If $W \mid U$, then $w \mid u$, but the converse is not necessarily true. We denote by (W, U) the highest common factor of W and U, and by (w, u) the greatest common divisor of w and u. The greatest primitive factor of $V(n)$ is denoted by $F(n)$.

10.2. Primitive divisors of $v(n)$

Since 2 cannot be a primitive divisor of $v(n)$ for any $n > 1$, we need consider the primitive divisors of $v(n)$ for only $n \geqslant 2$ and only those that are odd and >2. In the following, p shall denote an odd prime and $n > 2$ unless stated otherwise. If k is a primitive divisor of $v(n)$, then we have

$$a^n \equiv b^n \pmod{k}.$$

Since $(a, b) = 1$, this implies that

$$(a, k) = 1 = (b, k).$$

We can now give some theorems regarding primitive divisors of $v(n)$.

THEOREM 12. *If k is a primitive divisor of $v(n)$, then*

$$k \mid g(n).$$

Proof. Since k and $g(n)$ are primitive divisors of $v(n)$, so also is $\{k, g(n)\}$ a primitive divisor of $v(n)$. This is greater than $g(n)$ unless $k \mid g(n)$. Hence the theorem follows.

THEOREM 13. *If $t \mid v(m_1)$ and $t \mid v(m_2)$, then*

$$t \mid v((m_1, m_2)).$$

Proof. Without loss of generality, we can assume that $m_1 > m_2$. Let $m_1 = qm_2 + m_3$, where $0 \leqslant m_3 < m_2$. Then since

$$(a^{m_1} - b^{m_1}) - a^{m_3}(a^{qm_2} - b^{qm_2}) = b^{qm_2}(a^{m_3} - b^{m_3}) \qquad (1)$$

identically. It follows that $t \mid v(m_3)$. Repeating the process as in Euclid's algorithm for finding the g.c.d. of m_1 and m_2, we get the result stated in the theorem.

As direct consequences of the identical relation (1), we have

THEOREM 14. *If p is a primitive divisor of $v(m)$, then*

$$p \mid v(n) \quad \textit{if and only if} \quad m \mid n.$$

THEOREM 15.

$$V(m) \mid V(n) \quad \textit{if and only if} \quad m \mid n.$$

Moreover, if $(V(m), V(n)) = V(h)$ then $(m, n) = h$.

Alternatively, therefore, k is a primitive divisor of $v(n)$, if $k \mid v(n)$ and $(k, v(j)) = 1$ for every proper divisor j of n.

THEOREM 16. *If p is a primitive divisor of $v(n)$, then*

$$p \equiv 1 \pmod{n}.$$

Proof. Since $p \mid v(n)$, we have $(a, p) = 1 = (b, p)$. Therefore, a natural number j exists, such that

$$a \equiv jb \pmod{p}.$$

Then, since $p \mid v(n)$ but $p \nmid v(m)$ for any $m < n$, $j^n \equiv 1 \pmod{p}$ but $j^m \not\equiv 1 \pmod{p}$ for any $m < n$. This means that j appertains to n modulo p. Hence

$$n \mid \phi(p)$$

and this is the theorem.

COROLLARY. *Every primitive divisor of $v(n)$ is of the form*

$$1 + tn, \quad t \geqslant 0.$$

In particular $g(n)$ is also of this form.

THEOREM 17. *If $p \mid n$, then $p \mid v(n)$ if and only if*

$$p \mid v(n/p).$$

Proof. Let $n = pn_1$, then if $p \mid v(n)$, we must have

$$(a, p) = 1 = (b, p).$$

Hence, by Fermat's Theorem,

$$\begin{aligned} v(n) &= (a^{n_1})^p - (b^{n_1})^p \\ &\equiv a^{n_1} - b^{n_1} \pmod{p}, \end{aligned}$$

so that

$$v(n_1) \equiv 0 \pmod{p}.$$

The converse is proved by retracing the steps, and the theorem is established.

COROLLARY. *If $n = p^j n_1$ where $(p, n_1) = 1$, then $p \mid v(n)$ if and only if $p \mid v(n_1)$.*

THEOREM 18. *If $p \mid n$ and $p \mid v(n/p)$, then*

$$\mathrm{pot}_p\, v(n) = 1 + \mathrm{pot}_p\, v(n/p).$$

Proof. Let $\mathrm{pot}_p\, v(n/p) = t \geqslant 1$, so that

$$a^{n/p} = b^{n/p} + hp^t, \quad (h, p) = 1.$$

Then

$$\begin{aligned} a^n &= (b^{n/p} + hp^t)^p \\ &= b^n + hp^{t+1} b^{n(p-1)/p} \pmod{p^{2t+1}}. \end{aligned}$$

Hence $\mathrm{pot}_p\, v(n) = 1 + t$, and this is the theorem.

COROLLARY. *If* $n = p^j n_1, (p, n_1) = 1$ *and* $p \mid v(n_1)$, *then*

$$\text{pot}_p\, v(n) = j + \text{pot}_p\, v(n_1).$$

THEOREM 19. *If* p *is a non-primitive divisor of* $v(n), (n, p) = 1$, *but it is a primitive divisor of* $v(m)$, *then*

$$\text{pot}_p\, v(n) = \text{pot}_p\, v(m).$$

Proof. Since p is a primitive divisor of $v(m)$ and $p \mid v(n)$, it follows that $m \mid n$. Hence $v(m) \mid v(n)$. By actual division, we have

$$V(n)/V(m) = a^{n-m} + a^{n-2m}b^m + a^{n-3m}b^{2m} + \cdots + b^{n-m}$$

$$= V(m) \cdot \left(a^{n-2m} + 2a^{n-3m}b^m + \cdots + \left(\frac{n}{m} - 1\right)b^{n-2m}\right) + \frac{n}{m}\, b^{n-m}.$$

Since $(p, b) = 1 = \left(p, \dfrac{n}{m}\right)$, the theorem follows.

The results of theorems 18 and 19 can be stated together:

If p *is a primitive divisor of* $v(m)$ *and* $p \mid v(n)$, *then*

$$\text{pot}_p\, v(n) = \text{pot}_p\, n + \text{pot}_p\, v(m).$$

Finally, the p-potency of $v(m)$ of which p is a primitive divisor is given by

THEOREM 20. *If* p *is a primitive divisor of* $v(m)$, *then*

$$\text{pot}_p\, v(m) = \text{pot}_p\, g(m).$$

Proof. Since $g(m)$ is the l.c.m. of the primitive divisors of $v(m)$, the theorem follows.

Exercises

1. Show that $g(m)$ is that largest divisor of $v(m)$ which is prime to $g(m_j)$ for each proper divisor m_j of m.
2. For $a = 5$, $b = 3$, determine $g(m)$ for $m = 1, 2, 3, 4, 5$ and 6.
3. For $a = 2$, $b = 1$ find $g(4)$ and $g(6)$.

10.3. Primitive factors of $V(n)$

As for primitive divisors of $v(n)$, we can define a primitive factor of $V(n)$ as follows.

W is a primitive factor of $V(n)$ if $W \mid V(n)$ and $(W, V(j)) = 1$ for each proper divisor j of n.

The greatest primitive factor $F(n)$ of $V(n)$ is therefore given by

$$F(n) = V(n)/\{V(n_1), V(n_2), \ldots, V(n_h)\}$$

where $n_1, n_2, \ldots, n_h$ are all the proper divisors of n.

Exercises

1. Show that

$$F(1) = a - b; \quad F(2) = a + b; \quad F(3) = a^2 + ab + b^2; \quad F(4) = a^2 + b^2;$$
$$F(5) = a^4 + a^3b + a^2b^2 + ab^3 + b^4; \quad F(6) = a^2 - ab + b^2.$$

2. For $a = 5, b = 3$, evaluate $f(1), f(2), \ldots, f(6)$.

Do you notice any relationship between the values of f's and g's found earlier? Repeat the experiment with $a = 2, b = 1$ and with some other sets of values of a and b.

3. Show that for $j \geqslant 1$,

$$F(p^j) = V(p^j)/V(p^{j-1}).$$

4. Show that

$$F(n) = V(n)/F(n_1)F(n_2)F(n_3)\ldots F(n_h),$$

where $n_1, n_2, n_3, \ldots, n_h$ are all the proper divisors of n.

Another useful expression for $F(n)$ is obtained as follows. Observe that

$$(a - \omega b) \mid F(n)$$

if and only if ω is an nth root of unity which is not at the same time a jth root of unity for some proper divisor j of n. This amounts to asserting that ω is a primitive nth root of unity. Since the coefficient of the highest degree term in a in $F(n)$ is 1, we thus have

$$F(n) = \prod_{t <' n} (a - \omega^t b), \quad \text{with } \omega = \exp(2i\pi/n). \tag{1}$$

For $n \geqslant 3$, this gives

$$F(n) = \prod_{\substack{t < n/2 \\ (t,n)=1}} \{(a - b)^2 + 4ab \sin^2(\pi t/n)\}.$$

Notice that $f(n)$, the value of $F(n)$ for given a and b, is the least when $a = 2, b = 1$. Thus for $n \geqslant 3$,

$$f(n) = \prod_{\substack{t < n/2 \\ (t,n)=1}} (1 + 8 \sin^2(\pi t/n)) > 1.$$

This result will be of use to us a little later.

From (1) it is clear that $F(n)$ is an irreducible polynomial of degree $\phi(n)$ in a and b, with coefficients integral and the coefficient of $a^{\phi(n)}$ equal to 1. This means that $F(n)$ is the unique primitive factor of $V(n)$.

10.4. An inequality for $f(n)$

From the preceding subsection, we have

$$V(n) = \prod_{t \mid n} F(t).$$

Using the Möbius inversion formula, from this we obtain

$$F(n) = \prod_{t \mid n} \{V(n/t)\}^{\mu(t)}. \tag{1}$$

Here t need run through only the square-free divisors of n. The expression is particularly important because it leads to a nice inequality for $f(n)$.

Let q be the number of distinct prime divisors of n. Then, for any $b <' a$, we have

$$a^{(n/t)-1} < v(n/t) < a^{n/t}.$$

Therefore

$$a^{y(n)} < f(n) < a^{x(n)},$$

where

$$\begin{aligned} x(n) &= \sum_{\substack{t \mid n \\ \mu(t)=1}} (n/t) - \sum_{\substack{t \mid n \\ \mu(t)=-1}} ((n/t) - 1) \\ &= \sum_{t \mid n} \mu(t)(n/t) + \sum_{\substack{t \mid n \\ \mu(t)=-1}} 1 \\ &= \phi(n) + \text{number of divisors } t \text{ of } n \text{ with } \mu(t) = -1 \\ &= \phi(n) + \left\{\binom{q}{1} + \binom{q}{3} + \cdots + \binom{q}{2j+1}\right\}, \quad j = (q-1)/2, \\ &= \phi(n) + 2^{q-1}; \end{aligned}$$

and $y(n) = \phi(n) - 2^{q-1}$ similarly.

10.5. A relation between $f(n)$ and $g(n)$

Since $f(n)$ is divisible by every primitive divisor of $v(n)$, it is divisible by $g(n)$. Hence for some natural number $s = s(n)$,

$$f(n) = sg(n).$$

To determine $s(n)$, we need the following

THEOREM 21. *The only primes that can divide $s(n)$ are those which divide both n and $v(n)$.*

Proof. Three cases arise:

(i) If p does not divide $v(n)$, then p cannot divide $v(t)$ for any divisor t of n. Hence p cannot divide $f(n)$ and therefore not $s(n)$ either.

(ii) If $p \mid v(n)$ but $p \nmid n$, let p be a primitive divisor of $v(m)$. Then $m \mid n$ and $(p, m) = 1$. Also $p \mid v(j)$ if and only if $m \mid j$. Hence

$$\begin{aligned}
\operatorname{pot}_p f(n) &= \operatorname{pot}_p \prod_{t \mid n} \{v(t)\}^{\mu(n/t)} \\
&= \sum_{t \mid n} \mu(n/t) \operatorname{pot}_p v(t) \\
&= \sum_{t \mid n/m} \mu\left(\frac{n}{mt}\right) \operatorname{pot}_p v(mt) \\
&= \sum_{t \mid n/m} \mu\left(\frac{n/m}{t}\right) \operatorname{pot}_p v(m) \\
&= \operatorname{pot}_p g(m) \quad \text{or} \quad 0
\end{aligned}$$

according as m is or is not equal to n. In either case, $\operatorname{pot}_p s(n) = 0$, and p does not divide s.

(iii) Let $n = p^j n_1, (p, n_1) = 1, j \geqslant 1$ and let $p \mid v(n)$. If p is a primitive divisor of $v(m)$, then $m \mid n_1$. Also $p \mid v(r)$ if and only if $m \mid r$. Hence

$$\operatorname{pot}_p f(n) = \sum_{t \mid n} \mu(t) \operatorname{pot}_p v(n/t).$$

The only non-zero terms in the sum are those for which $m \mid (n/t)$ and t is a square-free divisor of n. We can, therefore, let t run through the divisors of pn_1/m. This gives

$$\begin{aligned}
\operatorname{pot}_p f(n) &= \sum_{t \mid pn_1/m} \mu(t) \operatorname{pot}_p v(n/t) \\
&= \sum_{t \mid n_1/m} \mu(t) \{\operatorname{pot}_p v(n/t) - \operatorname{pot}_p v(n/pt)\} \\
&= \sum_{t \mid n_1/m} \mu(t) \\
&= 1 \quad \text{or} \quad 0
\end{aligned}$$

according as n_1 is or is not equal to m. Since p is not a primitive divisor of $v(n)$, p does not divide $g(n)$. Hence

$$\operatorname{pot}_p f(n) = \operatorname{pot}_p s(n) = 1 \quad \text{or} \quad 0$$

according as p is or is not a primitive divisor of $v(n_1)$. In the former case

$$n_1 \mid (p-1).$$

This shows that, in this case, p is the largest prime factor of n. We have thus proved that, for $n > 2$,

$$f(n) = \begin{cases} pg(n) & \text{when } n = p^j n_1, j \geqslant 1, \text{ and} \\ & p \text{ is a primitive divisor of } v(n_1); \\ g(n) & \text{otherwise.} \end{cases}$$

Exercise

Verify this result by taking a few examples of your own.

10.6. A problem

What values of n, a, b will give $g(n) = 1$?

Evidently, $g(1) = 1$ if and only if $a - b = 1$. Thus, there are an infinite number of values of a and b for which $g(1) = 1$.

Again $g(2) = 1$ if and only if a and b are odd and $a + b = 2^{j+1}$ with $j \geqslant 1, b <' a$.

Example. For $a = 5, b = 3$, we have $v(1) = 2, v(2) = 16$; so that $g(2) = 1$. The proof of the assertion is left to the reader.

THEOREM 22. *For $n > 2, g(n) = 1$ only when $a = 2, b = 1$ and $n = 6$.*

Proof. We have seen that

$$f(n) = \begin{cases} pg(n) & \text{when } n = p^j n_1 \text{ where } j \geqslant 1, (p, n_1) = 1; \\ & \text{and } p \text{ is a primitive divisor of } v(n_1); \\ g(n) & \text{otherwise.} \end{cases}$$

For $n > 2$ we have $f(n) > 1$. Thus for $g(n)$ to be equal to 1, it is necessary that we have $f(n) = p, n = p^j n_1$ with $(n_1, p) = 1, j \geqslant 1$ and p a primitive divisor of $v(n_1)$. Now, we have shown that

$$\log_a f(n) > \phi(n) - 2^{q-1}, \tag{1}$$

where q is the number of distinct prime divisors of n. Hence

$$\log_a p > \phi(p^j)\phi(n_1) - 2^{q-1}. \tag{2}$$

For $q = 1$, this means that

$$\log_a p > \phi(p^j) - 1 > p - 2. \tag{3}$$

This can hold only for $p = 3$ and that too only for $a = 2, b = 1, j = 1$. For $q = 2$, even if we take $n_1 = 2, j = 1$, we get

$$\log_a p > p - 3. \tag{4}$$

The inequality in (4) can hold only for $p = 3$ or $p = 5$ with $a = 2, b = 1$. Since 5 is not a primitive divisor of $v(2)$, $p = 5$ is ruled out. The only case left to be considered now is $n = 3.2, a = 2, b = 1$. Here 3 is a primitive divisor of $v(2)$ and $f(6) = a^2 - ab + b^2 = 3$. Hence $g(6) = 1$. The theorem is thus proved.

10.7. The infinity of primes of the form 1 (mod n)

From what has been proved in the preceding subsections, it is clear that every two members of the sequence

$$g(n), g(2n), g(3n), \ldots$$

are relatively prime and each (with the possible exception of one of them) has a prime divisor of the form 1 (mod n). This proves that there are an infinity of primes of the said form.

Remark

In this section, our discussion has been primarily based on a paper of G. D. Birkhoff and H. S. Vandiver [*Annals of Mathematics*, second series, vol. 5 (1903–4), 173–180].

11. Congruence properties of $\phi(a^n - b^n)$ and $\phi(a^n + b^n)$

In this section, we present some results which are a little stronger than those of A. Rotkiewicz [*Proc. American Math. Soc.*, **12** (1961), 419–421].

In the following, we denote by $d(n)$ the number of divisors of n and by $d^*(n)$ the number of odd divisors of n.

THEOREM 23. *If a, b, n are natural numbers, $b <' a$, then for $n \geqslant 3$,*

$$\phi(a^n - b^n) \equiv 0 \pmod{n^{d(n)/2}}.$$

Proof. In section 10, we have shown that for $j \geqslant 1$, the $g(j)$'s are relatively prime in pairs. Moreover, for each $j \neq 1$, 2 or 6, $g(j)$ has at least one prime divisor p_j, such that

$$\begin{aligned} p_j &\equiv 1 \pmod{j} \text{ for } j \text{ even,} \\ &\equiv 1 \pmod{2j} \text{ for } j \text{ odd.} \end{aligned}$$

Hence for every j other than 1, 2 or 6, we have

$$\begin{aligned} \phi(g(j)) &\equiv 0 \pmod{j} \text{ for } j \text{ even,} \\ &\equiv 0 \pmod{2j} \text{ for } j \text{ odd.} \end{aligned}$$

If a and b are both odd,

$$a^2 - b^2 \equiv 0 \pmod{8};$$

so that

$$\phi(a^2 - b^2) \equiv 0 \pmod{4}.$$

If a and b are of opposite parity, $a^2 - b^2$ is odd. Therefore

$$\phi(a^2 - b^2) \equiv 0 \pmod{2}, \quad \text{since } a^2 - b^2 \geqslant 3.$$

Finally, except when $a = 2, b = 1, g(6)$ has a prime divisor p_6 such that

$$p_6 \equiv 1 \pmod{6}.$$

In the exceptional case, we have

$$\phi(a^6 - b^6) = \phi(63) = 36.$$

Hence

$$\phi(a^6 - b^6) \equiv 0 \pmod 6.$$

Since $g(j) \mid f(j)$, for all $j \geqslant 1$,

$$\prod_{j \mid n} g(j) \,\Big|\, \prod_{j \mid n} f(j) = a^n - b^n.$$

Hence

$$\phi\left(\prod_{j \mid n} g(j)\right) = \prod_{j \mid n} \phi(g(j)) \mid \phi(a^n - b^n).$$

Now, for odd $n \geqslant 3$,

$$\prod_{\substack{j \mid n \\ j > 1}} (2j) \,\Big|\, \prod_{j \mid n} \phi(g(j)).$$

Therefore, for odd $n \geqslant 3$,

$$\phi(a^n - b^n) \equiv 0 \pmod{2^{d^*(n)-1} n^{\frac{1}{2}d(n)}}. \tag{1}$$

Using the same type of reasoning, it can be shown that (1) holds even when $(n, 6) = 2$ or 3. For $(n, 6) = 6$, the result holds except when $a = 2$, $b = 1$. In the exceptional case, we get

$$\phi(a^n - b^n) \equiv 0 \pmod{2^{d^*(n)-2} n^{\frac{1}{2}d(n)}}. \tag{2}$$

For $\phi(a^n + b^n)$, all that we need to note is that

$$a^n + b^n = v(2n)/v(n).$$

Exercises

1. In our proof of theorem 23, cite at each step the theorem used.
2. Prove (1) when $(n, 6) = 2$ or 3. Examine also the case $(n, 6) = 6$.
3. If $\text{pot}_2\, n = j$, show that

$$\tfrac{1}{2}(2^{j+2} n)^{d^*(n)/2} \mid \phi(a^n + b^n), \qquad \text{if } 3 \mid n;$$

otherwise

$$(2^{j+2} n)^{d^*(n)/2} \mid \phi(a^n + b^n).$$

4. Verify these results for different sets of values of a, b and n.
5. Show that the theorems and results of this section hold even when a and b are not relatively prime.

CHAPTER 3

POWER RESIDUES, QUADRATIC RECIPROCITY AND SUM OF SQUARES

1. Definition

If for a given integer $k > 1$, and n prime to m the congruence

$$x^k \equiv n \pmod{m}$$

has a solution, then n is said to be a k-ic residue of m (or mod m). If it has no solution, n is a k-ic non-residue of m. For $k = 2, 3, 4, 5$ the words quadratic, cubic, biquadratic and quintic are more commonly used. Let

$$m = p_1^{a_1} p_2^{a_2} \dots p_r^{a_r}$$

be the canonical decomposition of m, then for n to be a k-ic residue of m, it is necessary and sufficient that the congruence

$$x^k \equiv n \pmod{p_j^{a_j}}, \quad (n, p_j) = 1,$$

has a solution for each $j \leqslant r$.

In view of this, we need consider the k-ic residues of p^a only, p being a prime $\geqslant 2$. Further, if n is a k-ic residue of m, then so also is any integer n_1 which is congruent to n modulo m. Hence we confine ourselves throughout to k-ic residues of m which are incongruent modulo m.

2. Number of k-ic residues

THEOREM 1. *If p is an odd prime and $a > 0$, then the number of* k-ic *residues of p^a is given by*

$$\phi(p^a)/(k, \phi(p^a)).$$

Proof. Let g be a primitive root of p^a, then taking indices to the base g, the congruence

$$x^k \equiv n \pmod{p^a} \tag{1}$$

takes the form

$$k \operatorname{Ind} x \equiv \operatorname{Ind} n \pmod{\phi(p^a)}. \tag{2}$$

If $(k, \phi(p^a)) = h$, then for (2) to have a solution, it is necessary and sufficient that $h \mid \operatorname{Ind} n$. Hence the k-ic residues of p^a are given by

$$g^h, g^{2h}, g^{3h}, \dots, g^{\phi(p^a)}.$$

Since these integers are all incongruent modulo p^a, the theorem is proved.

Note that if $h \mid \text{Ind } n$, then each of the congruences (1) and (2) has h solutions.

COROLLARY. *If p is an odd prime that divides neither k nor n, then for all $a > 0$, the congruence (1) has exactly $(k, p-1)$ solutions or no solution at all according as n is or is not a k-ic residue of p.*

This follows because $h = (k, \phi(p^a)) = (k, p^{a-1}(p-1)) = (k, p-1)$ so that h is independent of a.

THEOREM 2. *The number of k-ic residues of 2^a, $a \geqslant 3$, is 2^{a-1} or $2^{a-2}/(k, 2^{a-2})$ according as k is odd or even.*

Proof. If k is odd, then the congruence

$$x^k \equiv n \pmod{2^a} \tag{3}$$

where $(n, 2) = 1$, can be written, using generalized indices, as

$$k \text{ Ind } x \equiv \text{Ind } n \pmod{[2, 2^{a-2}]}. \tag{4}$$

Since $(k, 2) = 1$, (4) has a solution for every $n <' 2^a$. Hence the number of k-ic residues of 2^a for odd values of k is 2^{a-1}.

If k is even, (4) has a solution if and only if

$$n \equiv 3^j \pmod{2^a}$$

for some j in the range 1 to 2^{a-2} with $(k, 2^{a-2}) \mid j$. Hence, the number of k-ic residues of 2^a for k even, is

$$2^{a-2}/(k, 2^{a-2}).$$

Remark

When k is odd, the k-ic residues of 2^a, $a \geqslant 3$, are all the odd numbers $< 2^a$. When k is even, those are the numbers

$$3^h, 3^{2h}, 3^{3h}, \ldots, 3^{uh}$$

where

$$h = (k, 2^{a-2}), \quad u = 2^{a-2}/h.$$

The following theorem suggested by S. Chowla will be found to be of interest.

THEOREM 3. *If the set*

$$A = \{a_1, a_2, \ldots, a_n\}, \quad (a_j, p) = 1$$

is closed modulo p^u under multiplication, p being an odd prime, then the a's are k-ic residues of p^u for

$$k = \phi(p^u)/n.$$

Proof. Let g be a primitive root of p^u, then corresponding to the given set of integers A, there is another set of n distinct positive integers

$$B = \{b_1, b_2, b_3, \ldots, b_n\}$$

given by the relation

$$b_j = \text{Ind}_g\, a_j, \quad j = 1, 2, 3, \ldots, n;$$

with

$$0 < b_j \leqslant \phi(p^u).$$

Since every a is prime to p^u, we have

$$a_j^{\phi(p^u)} \equiv 1 \pmod{p^u}.$$

Hence, 1 must be one of the a's. Consequently, $\phi(p^u)$ must be one of the b's. Again, for a suitable t and $r, s, t \leqslant n$

$$a_r a_s \equiv a_t \pmod{p^u},$$

so that

$$b_r + b_s \equiv b_t \pmod{\phi(p^u)}.$$

This implies that the set B is closed under addition modulo $\phi(p^u)$. If k is the greatest common divisor of the integers $b_1, b_2, \ldots, b_n$ then the numbers $k, 2k, 3k, \ldots, \phi(p^u)$ can be represented modulo $\phi(p^u)$ as the sums, differences and multiples of b's and no other numbers can be so represented. Hence, the b's must be the numbers $k, 2k, 3k, \ldots, \phi(p^u)$ in some order. Arrange the b's so that

$$b_j = jk, \quad j = 1, 2, 3, \ldots, n;$$

then

$$a_j \equiv g^{jk} \pmod{p^u}.$$

Hence the a's are k-ic residues of p^u and k is given by

$$k = \phi(p^u)/n.$$

3. Proper and improper k-ic residues

If $(k, \phi(p^{u-v})) = h$ where p is a prime $\geqslant 2$ and

$$v = \begin{cases} 1 & \text{if } p = 2, u \geqslant 3 \text{ and } k \text{ is even,} \\ 0 & \text{if } p \geqslant 3, \end{cases}$$

then the k-ic residues of p^u are the same as the h-ic residues. We call these improper k-ic residues of p^u when $h < k$ and proper k-ic residues when $h = k$. There is no loss of generality if we consider only the proper k-ic residues of p^u and this we do. Thus, in what follows, we assume that p is an odd prime and

$$k \mid \phi(p^u). \tag{1}$$

We further denote by $R_k(m)$ the set* of k-ic residues of m.

Evidently, if $d \mid k$, then for all $m \geqslant 2$,

$$R_k(m) \subseteq R_d(m). \tag{2}$$

3.1. Reduction of the problem

DEFINITION. *If C and D are two sets of integers, then*

$$C \oplus D$$

denotes the set of all numbers obtained by adding each element of C to each element of D.

The following theorems enable us to limit the discussion on k-ic residues.

THEOREM 4. *If p is an odd prime and $(k, p) = 1$, then for $a > 1$,*

$$R_k(p^a) = R_k(p) \oplus \{tp : 0 \leqslant t \leqslant p^{a-1} - 1\}.$$

Proof. Since every k-ic residue of p^a is also a k-ic residue of p, the k-ic residues of p^a must be found among the elements of the set

$$R_k(p) \oplus \{tp : 0 \leqslant t \leqslant p^{a-1} - 1\}.$$

Since the elements of $R_k(p)$ can be assumed to be $<p$ and incongruent modulo p, the elements of this set are $<p^a$ and incongruent modulo p^a. Moreover, their number is $\phi(p^a)/k$. Since p^a has exactly $\phi(p^a)/k$ k-ic residues incongruent modulo p^a, the elements of the set are nothing but the k-ic residues of p^a.

From theorem 4, it follows that when p is an odd prime and $(k, p) = 1$, it is enough to study the k-ic residues of p itself in place of those of p^a. When $(k, p) > 1$, we have the

THEOREM 5. *If $k = p^{a-1}q$, $a > 1$, $(q, p) = 1$, $q \mid (p - 1)$, then for $u > a$,*

$$R_k(p^u) = R_k(p^a) \oplus \{tp^a : 0 \leqslant t \leqslant p^{u-a} - 1\}.$$

This is a generalization of theorem 4. The proof is left to the reader.

This shows that when $k = p^{a-1}q$ with $q \mid (p - 1)$, it is enough to study the k-ic residues of p^a in place of those of p^u, $u > a$. The next theorem further reduces the problem of k-ic residues.

THEOREM 6. *If $k = p^{a-1}q$ and $q \mid (p - 1)$, p being an odd prime as before, then the k-ic residues of p^a are given by the p^{a-1}-th powers of the q-ic residues of p.*

Proof. Let g be a primitive root of p^a, then the k-ic residues of p^a are

$$g^k, g^{2k}, g^{3k}, \ldots, g^{\phi(p^a)};$$

*We can take the elements of this set to be $<' m$.

while the q-ic residues of p are

$$g^q, g^{2q}, g^{3q}, \ldots, g^{\phi(p)}.$$

The theorem follows readily now.

We can write the result of theorem 6 in the elegant form:

$$R_{p^{a-1}q}(p^a) = \{R_q(p)\}^{p^{a-1}}, \quad q \mid (p-1).$$

THEOREM 7. *If $k = 2^b$, then the k-ic residues of 2^a, $a \geqslant 3$, are given by*

$$R_{2^b}(2^a) = \begin{cases} \{1\} & \text{when } b \geqslant a-2 \geqslant 1, \\ \{1 + j2^{b+2} : 0 \leqslant j \leqslant 2^{a-b-2} - 1\} & \text{when } 1 \leqslant b < a-2. \end{cases}$$

The proof is again left to the reader.

The study of k-ic residues may thus be confined to that of the proper k-ic residues of odd primes p alone.

4. Characterization of proper k-ic residues and non-residues

Throughout this section p denotes an odd prime, $k \mid (p-1)$, and we write q for $(p-1)/k$.

If g is a primitive root of p, then the numbers

$$g, g^2, g^3, \ldots, g^{p-1} \tag{1}$$

form a reduced residue set modulo p. Classify the numbers prime to p by placing all those a's for which

$$\operatorname{Ind}_g a \equiv r \pmod{k}, \quad 0 \leqslant r < k, \tag{2}$$

in one class. Call this the class r. For any a belonging to the class r, we define

$$\left(\frac{a}{p}\right)_k = \left(\frac{a}{p}\right) = w^r \tag{3}$$

where

$$w = \exp(2i\pi/k), \quad (i = \sqrt{-1}). \tag{4}$$

Thus w is a primitive kth root of unity. We take

$$\left(\frac{a}{p}\right) = 0 \quad \text{when} \quad (a, p) = p. \tag{5}$$

From the definition of $\left(\frac{a}{p}\right)$ it readily follows that

$$\left(\frac{a}{p}\right) = \left(\frac{a'}{p}\right), \quad \text{if } a \equiv a' \pmod{p}, \tag{6}$$

and

$$\left(\frac{ab}{p}\right) = \left(\frac{a}{p}\right)\left(\frac{b}{p}\right), \quad \text{for all } a \text{ and } b. \tag{7}$$

Notice that

$$\left(\frac{a}{p}\right) = 1 \tag{8}$$

if and only if a is a k-ic residue of p. A criterion for k-ic residues is provided by the

THEOREM 8. *a is a k-ic residue of p if and only if*

$$a^q \equiv 1 \pmod{p}.$$

Proof. If a is a k-ic residue of p, then for some integer x prime to p, we have

$$x^k \equiv a \pmod{p}.$$

Hence

$$a^q \equiv x^{p-1} \equiv 1 \pmod{p}.$$

If, on the other hand,

$$a^q \equiv 1 \pmod{p},$$

then for g a primitive root of p, we must have

$$q \operatorname{Ind}_g a \equiv 0 \pmod{(p-1)}$$

so that

$$\operatorname{Ind}_g a \equiv 0 \pmod{(p-1)/q}.$$

Hence

$$k \mid \operatorname{Ind}_g a,$$

which implies that a is a k-ic residue of p.

As a direct consequence of theorem 8, the k-ic residue character of a number is now given by the

THEOREM 9. *For $(a, p) = 1$,*

$$1 + a^q + a^{2q} + \cdots + a^{(k-1)q} \equiv k \text{ or } 0 \pmod{p}$$

according as a is or is not a k-ic residue of p.

Proof. Since $(a, p) = 1$, we have

$$a^{p-1} - 1 \equiv 0 \pmod{p},$$

that is

$$(a^q - 1)(1 + a^q + a^{2q} + \cdots + a^{(k-1)q}) \equiv 0 \pmod{p}.$$

If a is a k-ic residue of p

$$a^q - 1 \equiv 0 \pmod{p}$$

so that

$$1 + a^q + a^{2q} + \cdots + a^{(k-1)q} \equiv k \pmod{p}.$$

If a is a k-ic non-residue of p, we have

$$1 + a^q + a^{2q} + \cdots + a^{(k-1)q} \equiv 0 \pmod{p}.$$

Hence the criterion.

This leads us to a

Second definition of $\left(\dfrac{a}{p}\right)$. We define

$$\left(\frac{a}{p}\right) \equiv a^q \pmod{p}.$$

This definition of $\left(\dfrac{a}{p}\right)$ is of course different from the one given at the beginning of this section. Notice that as a takes the values

$$g, g^2, g^3, \ldots, g^{p-1},$$

where g is a primitive root of p, a^q takes k incongruent values modulo p, each q times. These values are

$$g^q, g^{2q}, g^{3q}, \ldots, g^{kq}.$$

Thus there are q a's for which

$$\left(\frac{a}{p}\right) \equiv g^{jq} \pmod{p}, \quad j = 1, 2, 3, \ldots, k.$$

With this definition of $\left(\dfrac{a}{p}\right)$, a is a k-ic residue of p if and only if

$$\left(\frac{a}{p}\right) \equiv 1 \pmod{p}. \tag{9}$$

Relations (5)–(7) now become congruences modulo p in place of equalities.

Whereas the absolute value of each character in the first definition is 1, this is not necessarily so in the second – even if we slightly alter the definition by saying that $\left(\dfrac{a}{p}\right)$ is the least absolute residue of a^q modulo p. Each of the two definitions has, however, its own advantages and disadvantages.

We illustrate the use of the second definition below.

4.1. Evaluation of $\left(\dfrac{a}{p}\right)$

For $a = 1, -1$ or 2 the value of $\left(\dfrac{a}{p}\right)$ is readily given by the definition itself.

For $a > 2$, the value of $\left(\dfrac{a}{p}\right)$ will be known if we know for each prime $p_j < p$, the value of $\left(\dfrac{p_j}{p}\right)$. In fact if

$$a = p_1^{u_1} p_2^{u_2} p_3^{u_3} \dots p_t^{u_t}$$

is the canonical decomposition of a. Then since a can be assumed to be less than p, each p_j in the decomposition is also $< p$. Moreover

$$\left(\frac{a}{p}\right) \equiv \left(\frac{p_1}{p}\right)^{u_1} \left(\frac{p_2}{p}\right)^{u_2} \left(\frac{p_3}{p}\right)^{u_3} \cdots \left(\frac{p_t}{p}\right)^{u_t} \pmod{p}.$$

Hence the assertion.

For the evaluation of $\left(\dfrac{p_j}{p}\right)$ we make a judicious use of the relations:

$$\left(\frac{a}{p}\right) \equiv \left(\frac{a'}{p}\right) \pmod{p}, \quad \text{if } a \equiv a' \pmod{p}, \tag{10}$$

$$\left(\frac{ab}{p}\right) \equiv \left(\frac{a}{p}\right)\left(\frac{b}{p}\right) \pmod{p} \text{ for all } a \text{ and } b. \tag{11}$$

We illustrate how this is done by an example.

Example. Let $p = 19, k = 3$, so that $q = 6$. Then taking all congruences modulo 19, we have

$$\left(\frac{-1}{19}\right) \equiv 1 \equiv \left(\frac{1}{19}\right)$$

$$\left(\frac{2}{19}\right) \equiv 2^6 \equiv 7.$$

Now

$$\left(\frac{3}{19}\right) \equiv \left(\frac{-16}{19}\right) \equiv \left(\frac{-1}{19}\right)\left(\frac{2}{19}\right)^4 \equiv \left(\frac{2}{19}\right) \equiv 7$$

$$\left(\frac{5}{19}\right) \equiv \left(\frac{24}{19}\right) \equiv \left(\frac{2}{19}\right)^3\left(\frac{3}{19}\right) \equiv \left(\frac{3}{19}\right) \equiv 7$$

$$\left(\frac{7}{19}\right) \equiv \left(\frac{45}{19}\right) \equiv \left(\frac{3}{19}\right)^2\left(\frac{5}{19}\right) \equiv 7^3 \equiv 1$$

$$\left(\frac{11}{19}\right) \equiv \left(\frac{30}{19}\right) \equiv \left(\frac{2}{19}\right)\left(\frac{3}{19}\right)\left(\frac{5}{19}\right) \equiv 7^3 \equiv 1$$

$$\left(\frac{13}{19}\right) \equiv \left(\frac{32}{19}\right) \equiv \left(\frac{2}{19}\right)^5 \equiv 11$$

and finally

$$\left(\frac{17}{19}\right) \equiv \left(\frac{-2}{19}\right) \equiv \left(\frac{-1}{19}\right)\left(\frac{2}{19}\right) \equiv 7.$$

Thus, the only primes less than 19 which are cubic residues of 19 are 7 and 11.

It may be observed that it is not difficult to evaluate $\left(\frac{a}{p}\right)$ for given a and p. It is, however, very difficult to answer the question: 'For what type of primes p, is a given number a a proper k-ic residue of p?', except when $k = 2$, as we shall see in the next few sections.

5. Quadratic residues: Legendre's symbol and Euler's criterion

Much more is known about the quadratic residues of primes than about their cubic and higher power residues. In this and some of the following sections we, therefore, study quadratic residues only.

We first recount some properties of quadratic residues which follow directly from the results proved in the foregoing sections. We continue to denote odd primes by p.

(i) p has exactly $\frac{1}{2}(p-1)$ quadratic residues and as many quadratic non-residues;

(ii) The Legendre symbol $\left(\frac{a}{p}\right)$, which shall now be written (a/p), can be defined as

$$(a/p) = \begin{cases} 0 & \text{if } (a,p) = p. \\ 1 & \text{if } a \text{ is a quadratic residue of } p. \\ -1 & \text{if } a \text{ is a quadratic non-residue of } p. \end{cases}$$

Sometimes we shall use Gauss' notations

aRp to mean that a is a quadratic residue of p,

and

bNp to indicate that b is a quadratic non-residue of p;

(iii) $(a/p) = (a'/p)$ if $a \equiv a' \pmod p$,
$(ab/p) = (a/p)(b/p)$ for all a and b;

(iv) $(a^2/p) = 1$ for $(a,p) = 1$;

(v) $(-1/p) = (-1)^{\frac{1}{2}(p-1)}$.

This implies that -1 is a quadratic residue of every prime p of the form $4t + 1$ and a quadratic non-residue of every prime of the form $4t + 3$.

(vi) We have

$$(a/p) \equiv a^{\frac{1}{2}(p-1)} \pmod p;$$

(this is known as Euler's criterion).

Euler's criterion can also be deduced from the

THEOREM 10. *For* $(a, p) = 1$,

$$(p-1)! \equiv -(a/p)a^{\frac{1}{2}(p-1)} \pmod p.$$

Proof. Let

$$S = \{1, 2, 3, \ldots, p-1\}.$$

Then it is easy to show that for every $x \in S$ there is a unique $x' \in S$, such that

$$xx' \equiv a \pmod p. \tag{1}$$

(i) When $(a/p) = -1$, the congruence

$$x^2 \equiv a \pmod p \tag{2}$$

has no solution. Hence x and x' in (1) are distinct. The members of the set S can, therefore, be paired off into $\frac{1}{2}(p-1)$ pairs, such that for each pair, relation (1) holds good. Thus, we have on multiplication

$$\begin{aligned} 1 \,.\, 2 \,.\, 3 \ldots (p-1) &\equiv a^{\frac{1}{2}(p-1)} \pmod p \\ &\equiv -(a/p)a^{\frac{1}{2}(p-1)} \pmod p. \end{aligned}$$

(ii) When $(a/p) = 1$, the congruence (2) has two solutions $x_0, p - x_0$. Evidently x_0 and $p - x_0$ are distinct. The members of S other than x_0 and $p - x_0$ can be paired off into $\frac{1}{2}(p-3)$ pairs, each pair satisfying relation (1). Hence

$$\begin{aligned} 1 \,.\, 2 \,.\, 3 \ldots (p-1) &\equiv a^{\frac{1}{2}(p-3)} x_0(p - x_0) \pmod p \\ &\equiv -a^{\frac{1}{2}(p-3)} x_0^2 \pmod p \\ &\equiv -a^{\frac{1}{2}(p-1)} \pmod p \\ &\equiv -(a/p)a^{\frac{1}{2}(p-1)} \pmod p. \end{aligned}$$

Now for $a = 1$, theorem 10 gives

$$(p-1)! \equiv -1 \pmod p \quad \text{[Wilson's Theorem]}. \tag{3}$$

This in turn gives

$$(a/p) \equiv a^{\frac{1}{2}(p-1)} \pmod p.$$

THEOREM 11. *For* $p \equiv 1 \pmod 4$, $m = \frac{1}{2}(p-1)$,

$$(m!)^2 \equiv -1 \pmod p.$$

Proof. We have

$$(p-1)! = 1 \,.\, 2 \,.\, 3 \ldots m \,.\, (p-m)(p-m+1) \ldots (p-1)$$
$$\equiv (1 \,.\, 2 \,.\, 3 \ldots m)^2 \pmod{p}.$$

Since by Wilson's Theorem

$$(p-1)! \equiv -1 \pmod{p},$$

the theorem follows immediately.

For primes of the form $4t + 1$, the theorem furnishes a solution of the congruence

$$x^2 \equiv -1 \pmod{p}.$$

THEOREM 12. *For* $p \equiv 3 \pmod 4$, $m = \frac{1}{2}(p-1)$,

$$m! \equiv (-1)^n \pmod{p}$$

where n is the number of quadratic non-residues b of p in the interval $(0, \frac{1}{2}p)$.

Proof. We have

$$(p-1)! = 1 \,.\, 2 \,.\, 3 \ldots m \,.\, (p-m)(p-m+1) \ldots (p-1)$$
$$\equiv (-1)^m (m!)^2 \pmod{p}.$$

Since m is odd, using Wilson's Theorem, we get

$$m! \equiv 1 \quad \text{or} \quad -1 \pmod{p}.$$

Let

$$m! \equiv (-1)^u \pmod{p}.$$

By Euler's criterion, we then have

$$(m!/p) \equiv (m!)^m \equiv (-1)^u \pmod{p}.$$

Again

$$(m!/p) = \prod_{j=1}^{m} (j/p) = \prod_{\substack{aRp \\ a \leqslant m}} (a/p) \prod_{\substack{bNp \\ b \leqslant m}} (b/p) = (-1)^n.$$

Hence

$$u \equiv n \pmod{2}.$$

This proves the theorem.

THEOREM 13. *Let r denote the number of residues and n the number of non-residues of p in the interval* $(0, \frac{1}{2}p)$, *then r is or is not equal to n according as* $p \equiv 1$ *or* $3 \pmod 4$.

Proof. We have

$$\begin{aligned}((p-a)/p) &= (-a/p) = (-1/p)(a/p)\\ &= (a/p) \quad \text{if} \quad p \equiv 1 \pmod 4\\ &= -(a/p) \quad \text{if} \quad p \equiv 3 \pmod 4.\end{aligned}$$

Hence, when $p \equiv 1 \pmod 4$, there are as many residues of p in the interval $(0, \frac{1}{2}p)$ as in the interval $(\frac{1}{2}p, p)$, and the same is true of the non-residues. Since the total number of residues of p in the interval $(0, p)$ is $\frac{1}{2}(p-1)$ and similarly for the non-residues, we have

$$r = \tfrac{1}{4}(p-1) = n.$$

When $p \equiv 3 \pmod 4$, we have

$$r + n = \tfrac{1}{2}(p-1).$$

Since $\frac{1}{2}(p-1)$ is odd, r and n cannot be equal.

Remark

Actually it is known that for primes of the form $4t + 3$ there are more quadratic residues than non-residues in the interval $(0, \frac{1}{2}p)$, a result due to Dirichlet, but no really elementary proof is available.

6. Gauss' lemma

The study of quadratic residues is greatly facilitated by the following very important lemma due to Gauss, nothing like which is available for higher power residues.

Consider the set of $\frac{1}{2}(p-1)$ numbers

$$a, 2a, 3a, \ldots, \tfrac{1}{2}(p-1)a, \tag{1}$$

where $(a, p) = 1$. Let

$$ja \equiv r_j \pmod p, \quad 0 < r_j < p, \quad 1 \leqslant j < \tfrac{1}{2}p.$$

If in the set of numbers

$$r_1, r_2, r_3, \ldots, r_{(p-1)/2}, \tag{2}$$

which are surely incongruent modulo p, m of the r's are each $> \frac{1}{2}p$, where $m \geqslant 0$, then Gauss' lemma states that

$$(a/p) = (-1)^m. \tag{3}$$

Proof. We notice that $m = 0$ for

$$a \equiv 1 \pmod p,$$

and we have

$$(a/p) = (1/p) = 1 = (-1)^0$$

so that the lemma holds in this case.

Again $m = \frac{1}{2}(p-1)$ for

$$a \equiv -1 \pmod{p},$$

and the lemma holds by Euler's criterion.

When $0 < m < \frac{1}{2}(p-1)$, we consider the three sets of numbers:

(i) $A = \{a_1, a_2, a_3, \ldots, a_m\}$

consisting of the r's $> \frac{1}{2}p$;

(ii) $B = \{b_1, b_2, b_3, \ldots, b_n\}, \quad n = [\frac{1}{2}p] - m,$

consisting of the r's $< \frac{1}{2}p$; and

(iii) $C = \{p - a_1, p - a_2, p - a_3, \ldots, p - a_m\}.$

We assert that no member of C is congruent to any member of B, though each member of C is $< \frac{1}{2}p$.

If possible, let

$$p - a_t \equiv b_s \pmod{p}.$$

Since a_t and b_s are elements of (1), positive integers x and y not exceeding $\frac{1}{2}(p-1)$ exist, such that

$$p - ax \equiv ay \pmod{p}.$$

Since $(a, p) = 1$, this would imply that

$$x + y \equiv 0 \pmod{p}.$$

But this is impossible, because

$$0 < x + y < p.$$

Hence the assertion.

Evidently, therefore, the union of sets B and C is the set

$$D = \{1, 2, 3, \ldots, \tfrac{1}{2}(p-1)\}.$$

Now

$$\begin{aligned}\left(\frac{p-1}{2}\right)! &= \prod_{t=1}^{m} (p - a_t) \cdot \prod_{s=1}^{n} b_s \\ &\equiv (-1)^m \prod_{t=1}^{m} a_t \cdot \prod_{s=1}^{n} b_s \pmod{p} \\ &\equiv (-1)^m \prod_{j=1}^{\frac{1}{2}(p-1)} r_j \pmod{p}\end{aligned}$$

$$\equiv (-1)^m \prod_{j=1}^{\frac{1}{2}(p-1)} (ja) \pmod{p}$$

$$\equiv (-1)^m \left(\frac{p-1}{2}\right)!\, a^{\frac{1}{2}(p-1)} \pmod{p}.$$

Hence

$$a^{\frac{1}{2}(p-1)} \equiv (-1)^m \pmod{p}.$$

Using Euler's criterion, we get

$$(a/p) \equiv (-1)^m \pmod{p}.$$

Since p is an odd prime $\geqslant 3$ and $(a/p) = 1$ or -1, we must have

$$(a/p) = (-1)^m.$$

This proves Gauss' lemma.

6.1. Evaluation of $(2/p)$

Gauss' lemma can be stated in the following alternative form:

If $(a, p) = 1$, then

$$(a/p) = (-1)^m,$$

where m is the number of negative least absolute residues modulo p of the products $ja, j = 1, 2, 3, \ldots, \frac{1}{2}(p-1)$.

As a first application of Gauss' lemma, we prove

THEOREM 14. *For any odd prime p,*

$$(2/p) = (-1)^u,$$

where $u = (p^2 - 1)/8$.

Proof. Consider the numbers

$$2.1, 2.2, 2.3, \ldots, 2.\tfrac{1}{2}(p-1)$$

None of these exceeds p. Those that exceed $\frac{1}{2}p$ are

$$2\left(\left[\frac{p}{4}\right]+1\right),\quad 2\left(\left[\frac{p}{4}\right]+2\right), \ldots, 2\left[\frac{p}{2}\right].$$

Hence, m in Gauss' lemma is given by

$$m = \left[\frac{p}{2}\right] - \left[\frac{p}{4}\right].$$

Now for $p = 4t + 1$,

$$\left[\frac{p}{2}\right] - \left[\frac{p}{4}\right] = 2t - t = t,$$

while

$$\frac{p^2 - 1}{8} = 2t^2 + t \equiv t \pmod 2.$$

Similarly, for $p = 4t + 3$,

$$\left[\frac{p}{2}\right] - \left[\frac{p}{4}\right] = (2t + 1) - t = t + 1,$$

while

$$\frac{p^2 - 1}{8} = 2t^2 + 3t + 1 \equiv t + 1 \pmod 2.$$

Hence

$$u \equiv m \pmod 2$$

and the theorem follows immediately.

Observe that the theorem implies that 2 is a quadratic residue of all primes of the form $8t + 1$ and $8t + 7$, and a quadratic non-residue of primes of the form $8t + 3$ and $8t + 5$.

The main point in the application of Gauss' lemma is the determination of m modulo 2.

7. The quadratic reciprocity law

This celebrated law is one of the most beautiful in the literature. It was first conjectured by Euler. Gauss discovered it for himself, reportedly at the age of 19, and gave several proofs. His third proof, based on the lemma of section 6 and a lemma of Eisenstein, will be found in most of the text books. We give here an elegant version due to L. Carlitz† of a proof given by Frobenius. About fifty proofs of the law are known. Among these there is also the interesting 'low energy proof' of D. H. Lehmer.‡ We leave it to the reader to study it for himself.

THEOREM 15 (THE QUADRATIC RECIPROCITY LAW). *If p_1 and p_2 are two distinct odd primes, then*

$$(p_2/p_1)(p_1/p_2) = (-1)^{q_1 q_2},$$

† *Proc. American Math. Soc.*, **11** (1960), 563–565.
‡ *American Math. Monthly*, **64** (1957), 103–106.

where $q_1 = (p_1 - 1)/2, \quad q_2 = (p_2 - 1)/2.$

Proof. By Gauss' lemma, we have

$$(p_2/p_1) = (-1)^{m_1},$$

where m_1 is the number of products $hp_2, h = 1, 2, \ldots, q_1$, for which the least absolute residue modulo p_1 is negative, i.e.

$$hp_2 = kp_1 - r_1, \quad 1 \leqslant r_1 \leqslant (p_1 - 1)/2.$$

In other words, m_1 is the number of h, $1 \leqslant h \leqslant q_1$, for which

$$-(p_1 - 1)/2 \leqslant hp_2 - kp_1 \leqslant -1 \tag{1}$$

has an integral solution k. The right hand inequality in (1) cannot hold for any $k < 1$. The left hand inequality implies that we have

$$kp_1 \leqslant hp_2 + q_1 \leqslant q_1(p_2 + 1). \tag{2}$$

Hence

$$k \leqslant (p_2 - 1)/2 \quad \text{so that} \quad 1 \leqslant k \leqslant q_2. \tag{3}$$

Similarly

$$(p_1/p_2) = (-1)^{m_2},$$

where m_2 is the number of $k, k = 1, 2, \ldots, q_2$, for which

$$-q_2 \leqslant kp_1 - hp_2 \leqslant -1 \tag{4}$$

has an integral solution h with $1 \leqslant h \leqslant q_1$. Since (4) is equivalent to

$$1 \leqslant hp_2 - kp_1 \leqslant q_2, \tag{5}$$

the number-pairs (h, k) satisfying (1) are distinct from those satisfying (4). Moreover, $hp_2 - kp_1$ cannot be zero. Hence of the q_1q_2 number-pairs (h, k) with $1 \leqslant h \leqslant q_1, 1 \leqslant k \leqslant q_2$, there are exactly $m_1 + m_2$ which satisfy the inequality

$$-\tfrac{1}{2}(p_1 - 1) \leqslant hp_2 - kp_1 \leqslant \tfrac{1}{2}(p_2 - 1). \tag{6}$$

Of the rest, that is those that do not satisfy (6), let u be the number of those for which

$$hp_2 - kp_1 < -\tfrac{1}{2}(p_1 - 1). \tag{7}$$

Then, for the number-pair (h', k') given by

$$h' = \tfrac{1}{2}(p_1 + 1) - h, \quad k' = \tfrac{1}{2}(p_2 + 1) - k, \tag{8}$$

we have

$$\begin{aligned} h'p_2 - k'p_1 &= \tfrac{1}{2}(p_1 + 1)p_2 - hp_2 - \tfrac{1}{2}(p_2 + 1)p_1 + kp_1 \\ &= \tfrac{1}{2}(p_2 - p_1) - (hp_2 - kp_1) > \tfrac{1}{2}(p_2 - 1). \end{aligned}$$

and conversely.

Hence the number of pairs (h, k) for which

$$hp_2 - kp_1 > \tfrac{1}{2}(p_2 - 1) \tag{9}$$

is also u. We thus have

$$q_1 q_2 = m_1 + m_2 + 2u \tag{10}$$

and the quadratic reciprocity law follows immediately.

We can give a nice geometrical interpretation of the results proved above. In the following figure drawn with $p_1 = 19, p_2 = 11$, XOY is the frame of reference. The rectangle $ABCD$ is bounded by the lines $x = \frac{1}{2}, y = \frac{1}{2}, x = \frac{1}{2}p_1, y = \frac{1}{2}p_2$. OEC, PQ, RS are respectively the lines:

$$p_2 x - p_1 y = 0, \quad p_2 x - p_1 y + \tfrac{1}{2}p_1 = 0,$$

and

$$p_2 x - p_1 y - \tfrac{1}{2}p_2 = 0.$$

Evidently then m_1 is the number of lattice points in the area enclosed by $PAECQ$; m_2 the number of lattice points in the area enclosed by $ERSC$; and the triangle RBS can be cut and placed so as to fit on the triangle QDP, the lattice points in one fitting upon the lattice points in the other, showing that the two triangles have the same number of lattice points each, say u. Since the total

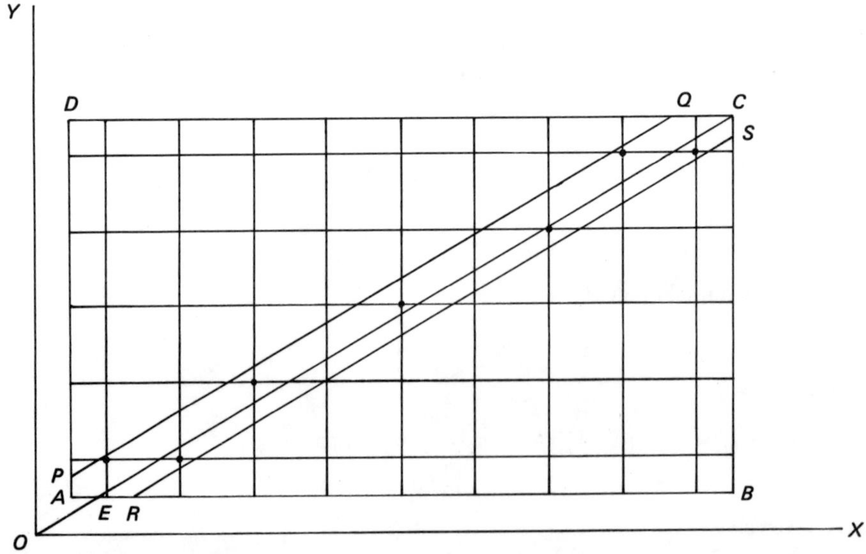

number of lattice points in the rectangle $ABCD$ is q_1q_2 and there are no lattice points on the lines OEC, PQ, RS within the rectangle, we get

$$q_1q_2 = m_1 + m_2 + 2u$$

as stated.

7.1. Evaluation of (a/p)

The quadratic reciprocity law and the results of **section 5** enable us to find the value of (a/p) for all a. The following example will explain the procedure.

Example. Evaluate (17/23).
We have

$$(17/23)(23/17) = (-1)^{8 \cdot 11} = 1.$$

Now

$$(23/17) = (6/17) = (2/17)(3/17).$$

But

$$(2/17) = 1$$

and

$$(3/17)(17/3) = (-1)^{1 \cdot 8} = 1.$$

Again

$$(17/3) = (2/3) = -1.$$

Hence

$$(23/17) = -1$$

and consequently

$$(17/23) = -1.$$

This means that 17 is a quadratic non-residue of 23.

8. Jacobi's symbol

Let m be an odd natural number and $(n, m) = 1$. Further let

$$m = p_1^{c_1} p_2^{c_2} p_3^{c_3} \dots p_j^{c_j}$$

be the canonical decomposition of m. Then Jacobi's symbol (n/m) is defined by the relation

$$(n/m) = \prod_{t=1}^{j} (n/p_t)^{c_t},$$

where the symbols on the right are those of Legendre. For $m = 1$, we take

$$(n/m) = 1.$$

Jacobi's symbol is evidently a generalization of Legendre's. It must however be clearly understood that if $(n/m) = 1$, it does not necessarily follow that n is a quadratic residue of m. On the other hand, if $(n/m) = -1$, then it is certain that n is a quadratic non-residue of m. Thus, we have

$$(3/35) = (3/5)(3/7) = (-1)(-1) = 1.$$

But 3 is not a quadratic residue of 35 because it is not a quadratic residue of 5 or 7. On the other hand,

$$(3/245) = (3/5)(3/7)^2 = (-1)(-1)^2 = -1,$$

and we can definitely say that 3 is not a quadratic residue of 245.

8.1. Properties of Jacobi's symbol

Jacobi's symbol has many of the properties of Legendre's symbol. We state these and give proofs for some of them leaving the others for the reader to establish.

(i) If m is an odd positive integer, and

$$n_1 \equiv n_2 \pmod{m}, \quad (n_1, m) = 1,$$

then

$$(n_1/m) = (n_2/m).$$

(ii) If m is an odd positive integer and $(n_1 n_2, m) = 1$, then

$$(n_1 n_2/m) = (n_1/m)(n_2/m).$$

(iii) If m_1 and m_2 are two odd positive integers and $(n, m_1 m_2) = 1$, then

$$(n/m_1)(n/m_2) = (n/m_1 m_2).$$

If m's are odd positive integers, then

(iv) $(-1/m) = (-1)^{\frac{1}{2}(m-1)}$,

(v) $(2/m) = (-1)^{(m^2-1)/8}$,

(vi) $(m_1/m_2)(m_2/m_1) = (-1)^{\frac{1}{4}(m_1-1)(m_2-1)}, \quad (m_1, m_2) = 1.$

We shall prove the last three results. The proofs need the following

LEMMA. *If $n_1, n_2, \ldots, n_j$ are odd integers, then*

$$n_1 n_2 \ldots n_j - 1 \equiv (n_1 - 1) + (n_2 - 1) + \cdots + (n_j - 1) \pmod 4$$

and

$$n_1^2 n_2^2 \ldots n_j^2 - 1 \equiv (n_1^2 - 1) + (n_2^2 - 1) + \cdots + (n_j^2 - 1) \pmod{16}.$$

Proof. We have

$$(n_1 - 1)(n_2 - 1) = n_1 n_2 - n_1 - n_2 + 1.$$

Therefore if n_1 and n_2 are odd, then

$$n_1 n_2 - 1 \equiv (n_1 - 1) + (n_2 - 1) \pmod 4. \tag{1}$$

Similarly, we have

$$(n_1^2 - 1)(n_2^2 - 1) = n_1^2 n_2^2 - n_1^2 - n_2^2 + 1,$$

whence for n_1 and n_2 odd, it follows that

$$n_1^2 n_2^2 - 1 \equiv (n_1^2 - 1) + (n_2^2 - 1) \pmod{16}. \tag{2}$$

The results stated in the lemma follow readily from (1) and (2).

Proof of (iv). Let $m = p_1^{c_1} p_2^{c_2} \ldots p_j^{c_j}$ be the canonical decomposition of m. Then

$$\begin{aligned}(-1/m) &= \prod_{t=1}^{j} (-1/p_t)^{c_t} \\ &= \prod_{t=1}^{j} (-1)^{c_t(p_t - 1)/2} \\ &= (-1)^u,\end{aligned}$$

where

$$\begin{aligned}2u &= c_1(p_1 - 1) + c_2(p_2 - 1) + \cdots + c_j(p_j - 1) \\ &\equiv p_1^{c_1} p_2^{c_2} \ldots p_j^{c_j} - 1 \pmod 4\end{aligned}$$

that is

$$u \equiv (m - 1)/2 \pmod 2.$$

Hence

$$(-1/m) = (-1)^{\frac{1}{2}(m-1)}.$$

Proof of (v). Using the same expression for m as in the proof of (iv), we have

$$(2/m) = \prod_{t=1}^{j} (2/p_t)^{c_t}$$

$$= \prod_{t=1}^{j} (-1)^{c_t(p_t^2-1)/8}$$

$$= (-1)^v,$$

where

$$8v = c_1(p_1^2 - 1) + c_2(p_2^2 - 1) + \cdots + c_j(p_j^2 - 1)$$
$$\equiv (p_1^2)^{c_1}(p_2^2)^{c_2} \ldots (p_j^2)^{c_j} - 1 \pmod{16},$$

that is

$$v \equiv (m^2 - 1)/8 \pmod 2.$$

Hence

$$(2/m) = (-1)^{(m^2-1)/8}.$$

Proof of (vi). Let

$$m_1 = \prod_{t=1}^{j} p_{1t}^{c_{1t}}$$

and

$$m_2 = \prod_{s=1}^{h} p_{2s}^{c_{2s}}$$

be the canonical decompositions of m_1 and m_2. Then since m_1 and m_2 are odd and prime to each other, the p's are distinct odd primes. Now

$$(m_1/m_2) = \prod_{s=1}^{h} (m_1/p_{2s})^{c_{2s}}$$

$$= \prod_{s=1}^{h} \prod_{t=1}^{j} (p_{1t}/p_{2s})^{c_{1t}c_{2s}}$$

and

$$(m_2/m_1) = \prod_{t=1}^{j} \prod_{s=1}^{h} (p_{2s}/p_{1t})^{c_{2s}c_{1t}}.$$

Hence

$$(m_1/m_2)(m_2/m_1) = \prod_{t=1}^{j} \prod_{s=1}^{h} \{(p_{1t}/p_{2s})(p_{2s}/p_{1t})\}^{c_{1t}c_{2s}}$$

$$= (-1)^w,$$

where, for certain integers d_1, d_2,

$$\begin{aligned}4w &= \sum_{t=1}^{j} \sum_{s=1}^{h} c_{1t}c_{2s}(p_{1t}-1)(p_{2s}-1)\\ &= \sum_{t=1}^{j} c_{1t}(p_{1t}-1) \cdot \sum_{s=1}^{h} c_{2s}(p_{2s}-1)\\ &= (m_1-1+4d_1)(m_2-1+4d_2),\\ &\equiv (m_1-1)(m_2-1) \pmod 8,\end{aligned}$$

since m_1, m_2 are odd. Thus

$$w \equiv \tfrac{1}{4}(m_1-1)(m_2-1) \pmod 2.$$

The result follows directly.

8.2. Application of Jacobi's symbol

How the Jacobi symbol can be utilized to determine the quadratic character of a modulo a prime $p \geqslant 3$, will be clear from the following example.

Example. Evaluate (33/73).
We have

$$\begin{aligned}(33/73)(73/33) &= (-1)^{16.36} = 1\\ (73/33) &= (7/33)\\ (7/33)(33/7) &= (-1)^{3.16} = 1\\ (33/7) &= (5/7)\\ (5/7)(7/5) &= (-1)^{2.3} = 1\\ (7/5) &= (2/5) = -1.\end{aligned}$$

Hence $(33/73) = -1.$

Note that the use of Jacobi's symbol in place of Legendre's leads to a lot of saving of steps.

9. An extension of the Lehmer property of the Jacobi symbol

A remarkable property, which could itself serve as a definition of the Jacobi symbol, was obtained by D. H. Lehmer [*Australian J. Math.* **1** (1959), 385–395]. Two proofs of the property were given by J. L. Brenner [*Duke Math. J.* **29** (1962), 13–31]. We give here a short and simple proof of an extension of the Lehmer property to k-ic residues, due to the author.

Let m and n be positive integers with $(m, n) = 1$ and n odd. We define an equivalence relation among the residue classes

$$1, 2, 3, \ldots, (n-1) \pmod{n}$$

by

$$t_1 \sim t_2$$

if and only if there exists an integer h such that

$$m^h t_1 \equiv t_2 \pmod{n}. \tag{1}$$

Then what Lehmer proved was that the Jacobi symbol (m/n) is given by

$$(m/n) = (-1)^N, \tag{2}$$

where N is the number of equivalence classes defined above. We extend this property here to k-ic residues for $k > 1$.

In what follows,

(i) p's are primes of the form $kq + 1$.
(ii) We take

$$n = p_1^{a_1} p_2^{a_2} p_3^{a_3} \ldots p_r^{a_r},$$

as the canonical decomposition of n.
(iii) $(m, n) = 1$.
(iv) g is a common primitive root of $p_v^{a_v}$, $v = 1, 2, 3, \ldots, r$.
(v) We define the generalized Jacobi symbol by the relation

$$\left(\frac{m}{n}\right)_k = w^N,$$

where

$$w = \exp(2\pi i/k), \quad (i = \sqrt{-1}).$$

In particular, for $n = p$, we get the analogue of Legendre's symbol.

9.1. An expression for N

The numbers

$$1, 2, 3, \ldots, (n-1)$$

can be placed in classes $C(d)$, where $d \mid n, d > 1$, and

$$\begin{aligned} C(d) &= \{b : (b, n) = n/d, \quad b \leqslant (n-1)\} \\ &= \{jn/d : j <' d\}. \end{aligned} \tag{3}$$

If m appertains to $h(d)$ modulo d, then the numbers $j <' d$ can be further divided into subclasses as follows. We place two numbers j_1 and j_2 (each $<' d$) in one subclass if for some integer h

$$m^h j_1 \equiv j_2 \pmod{d}. \tag{4}$$

Since m appertains to $h(d)$ modulo d, each such subclass has exactly $h(d)$ members. Hence the number of subclasses corresponding to $C(d)$ is

$$\phi(d)/h(d). \tag{5}$$

Thus the total number of subclasses N to which this subdivision leads is given by

$$N = \sum_{\substack{d|n \\ d>1}} \phi(d)/h(d). \tag{6}$$

9.2. Proof of Lehmer's Theorem

THEOREM 16. *The generalized Jacobi symbol is multiplicative in n.*

Proof. Let

$$m = g^{q_v} \pmod{p_v},$$

and

$$(q_v, \phi(p_v)) = u_v.$$

Then, we have

$$h(p_v) = \phi(p_v)/u_v, \tag{7}$$

and

$$h(p_v^{c_v}) = p_v^{c_v - 1} h(p_v), \quad c_v > 0. \tag{8}$$

So that

$$\phi(p_v^{c_v})/h(p_v^{c_v}) = u_v, \quad \text{for each } c_v \geqslant 1. \tag{9}$$

Now, let

$$d = \prod_{v=1}^{r} p_v^{c_v}, \quad 0 \leqslant c_v \leqslant a_v.$$

Then,

$$h(d) = \text{l.c.m. of } h(p_v^{c_v})\text{'s}. \tag{10}$$

Since every prime factor of n, and therefore of d, is of the form $kx + 1$, for $d > 1$, we have

$$\phi(d)/h(d) \equiv 0 \pmod{k}$$

except possibly when d is a power of a single prime. Therefore, modulo k, we have

$$N = \sum_{\substack{d|n \\ d>1}} \phi(d)/h(d),$$

$$\equiv \sum_{v=1}^{r} \sum_{c_v=1}^{a_v} \phi(p_v^{c_v})/h(p_v^{c_v}),$$

$$\equiv \sum_{v=1}^{r} a_v u_v.$$

Hence

$$\left(\frac{m}{n}\right)_k = \prod_{v=1}^{r} \left(\frac{m}{p_v}\right)_k^{a_v}. \tag{11}$$

In particular,

$$\left(\frac{m}{p_v}\right)_k = \exp(2\pi i u_v/k), \tag{12}$$

which is equal to 1, if and only if $k \mid u_v$, i.e. if and only if m is a k-ic residue of p_v. Evidently $\left(\frac{m}{k}\right)_k$ can be equal to 1, even when m is not a k-ic residue of n.

Remark

In practice, the N equivalence classes are easy to write by the following sieving process.

Take the set of residue classes modulo n:

$$1, 2, 3, \ldots, (n-1). \tag{13}$$

Start with 1, and remove from (13) the residue classes

$$m, m^2, m^3, \ldots \text{ (modulo } n).$$

Next take the first residue class after 1 which has survived the sieving. Suppose it is j. Now remove from (13) the residue classes

$$jm, jm^2, jm^3, \ldots \text{ (modulo } n).$$

Repeat the process with the first residue class after j which has not been sieved out yet, and so on. At the end what will be left will be the representatives of the N classes. The procedure is suggested directly by the definition of equivalence itself as in (1).

Example. Take $m = 2, n = 45$, and $k = 2$. The residue classes to be considered modulo 45 are

$$1, 2, 3, \ldots, 44.$$

The sieving process can be presented in the following tabular form:

Starting residue	Residue classes sieved out
1	2, 4, 8, 16, 32, 19, 38, 31, 17, 34, 23
3	6, 12, 24
5	10, 20, 40, 35, 25
7	14, 28, 11, 22, 44, 43, 41, 37, 29, 13, 26
9	18, 36, 27
15	30
21	42, 39, 33

Hence

$$N = 7.$$

From the formula for N also, we have

$$h(3) = 2, \quad h(5) = 4, \quad h(9) = 6, \quad h(15) = 4, \quad h(45) = 12$$
$$\phi(3) = 2, \quad \phi(5) = 4, \quad \phi(9) = 6, \quad \phi(15) = 8, \quad \phi(45) = 24$$

so that $N = 7$. This gives

$$\left(\frac{2}{45}\right)_2 = (-1)^7 = -1.$$

Exercises

1. Find the number of cubic residues of 13 and 17.
2. Given that 2 is a primitive root of 13, find the cubic and biquadratic residues of 13.
3. Find the biquadratic residues of 17. Hence list the biquadratic residues of 17^2.
4. Obtain the 26th power residues of 53^2 and of 13^2.
5. List the biquadratic residues of 64.
6. Use the criterion of theorem 9 to find if 5 is a cubic residue of 19.
7. Find the prime quintic residues of 31 and 41.
8. Use Gauss' lemma to determine the quadratic characters of 7 and 17 modulo each of the two primes 23 and 41.
9. If p is a prime of the form $kq + 1$, and $(m, p) = 1$, then show that m is a k-ic residue of p if and only if N the number of Lehmer classes modulo p is a multiple of k. Is 7 a quintic residue of 31?
10. Taking $m = 3, k = 6$ and $n = 13^2 \,.\, \underline{19} \,.\, \underline{37}$ rewrite the proof of theorem 16 to verify each step in the proof. Evaluate $\left(\dfrac{m}{n}\right)_6$ in this case.

10. Integers representable as sums of two squares

A non-negative integer n is said to be representable as a sum of two squares when it can be written in the form

$$n = u + v, \tag{1}$$

where u and v are squares of integers, x and y say.

The representation (1) is said to be unique when the only other representation of n in that form is

$$n = v + u.$$

It is primitive when $(u, v) = 1$, with u positive and $v \geqslant 0$.

Not all natural numbers can be represented as sums of two squares and there are those that can be so represented but not primitively. We have, for example:

THEOREM 17. *No number of the form $4t + 3$ can be represented as a sum of two squares.*

Proof. If possible, let

$$4t + 3 = x^2 + y^2. \tag{2}$$

Evidently, the integers x and y are of opposite parity. Suppose x is odd and y is even. Then

$$x^2 \equiv 1 \pmod 4, \quad y^2 \equiv 0 \pmod 4.$$

Thus the right side of (2) is $\equiv 1 \pmod 4$, while its left side is $\equiv 3 \pmod 4$. Hence the theorem.

In particular, it follows that no prime of the form $4t + 3$ can be represented as a sum of two squares.

THEOREM 18. *No multiple of 4 can be represented primitively as a sum of two squares.*

Proof. If possible, let

$$4m = x^2 + y^2, \quad (x, y) = 1. \tag{3}$$

Then x and y must both be odd and the right side of (3) is $\equiv 2 \pmod 4$, while the left side is $\equiv 0 \pmod 4$.

Our main tool in dealing with the problem of representation of integers as sums of two squares is Brahmagupta's identity:

$$\begin{aligned}(a^2 + b^2)(c^2 + d^2) &= (ac - bd)^2 + (ad + bc)^2 \\ &= (ac + bd)^2 + (ad - bc)^2.\end{aligned}$$

Unless $a = b$ or $c = d$, the two representations are distinct.

Brahmagupta's identity implies

THEOREM 19. *The product of two representable numbers is representable.*

10.1. Some basic results and theorems.

THEOREM 20. *Every prime of the form $4t + 1$ can be represented as a sum of two squares and the representation is primitive and unique.*

Proof. We use what is known as the 'method of descent'. Let p denote any prime of the form $4t + 1$. Then since -1 is a quadratic residue of p, there is a positive integer $u \leqslant \frac{1}{2}(p-1)$, such that

$$u^2 + 1 \equiv 0 \pmod{p}. \tag{4}$$

For some natural number $m < p$, therefore

$$u^2 + 1 = mp. \tag{5}$$

Let S be the set of all natural numbers m for which mp can be represented as a sum of two squares. From (5) it follows that S is not empty. Hence, it has a smallest member m_0, say.

If $m_0 = 1$, there is nothing to prove. If $m_0 > 1$, then for certain integers x_0, y_0, we have

$$m_0 p = x_0^2 + y_0^2, \quad 1 < m_0 < p. \tag{6}$$

Evidently m_0 cannot divide both x_0 and y_0, for then it will have to divide p, a prime. Hence we can write

$$x_0 \equiv x_1 \pmod{m_0}, \quad y_0 \equiv y_1 \pmod{m_0} \tag{7}$$

with x_1, y_1 not both zero, and

$$|x_1| \leqslant \tfrac{1}{2}m_0, \quad |y_1| \leqslant \tfrac{1}{2}m_0,$$

whence

$$0 < x_1^2 + y_1^2 \leqslant \tfrac{1}{2}m_0^2 < m_0^2.$$

Now, from (6) and (7), we have

$$x_1^2 + y_1^2 \equiv x_0^2 + y_0^2 \equiv 0 \pmod{m_0}.$$

We can, therefore, write

$$x_1^2 + y_1^2 = m_1 m_0, \quad \text{with} \quad 0 < m_1 < m_0. \tag{8}$$

From (6) and (8), using Brahmagupta's identity, we get

$$m_0^2 m_1 p = (x_0 x_1 + y_0 y_1)^2 + (x_0 y_1 - x_1 y_0)^2.$$

Now

$$x_0 x_1 + y_0 y_1 \equiv x_1^2 + y_1^2 \equiv 0 \pmod{m_0},$$
$$x_0 y_1 - x_1 y_0 \equiv x_1 y_1 - x_1 y_1 \equiv 0 \pmod{m_0}.$$

Therefore

$$m_1 p = a^2 + b^2, \tag{9}$$

where

$$a = (x_0 x_1 + y_0 y_1)/m_0, \quad b = (x_0 y_1 - x_1 y_0)/m_0.$$

This contradicts the minimality of m_0. Hence $m_0 = 1$, and the first part of the theorem is proved.

We now have

$$p = x_0^2 + y_0^2. \tag{10}$$

We assert that $(x_0, y_0) = 1$, for if $(x_0, y_0) = g > 1$, then

$$g^2 \mid p \text{ (a prime).}$$

To establish uniqueness of representation (10) we give the elegant proof due to Murad Ismail. Besides (10), assume that p has another essentially distinct representation:

$$p = c^2 + d^2.$$

Then, since x_0, y_0 are of opposite parity, as are c, d, we can assume, without loss of generality, that x_0 and c are even and y_0 and d are odd; further that $x_0 > c$, $y_0 < d$. Let

$$(x_0 - c, y_0 + d) = 2r, \quad (x_0 + c, d - y_0) = 2t.$$

Write

$$(x_0 - c)/2r = A, \quad (d + y_0)/2r = B,$$
$$(x_0 + c)/2t = C, \quad (d - y_0)/2t = D.$$

Then, since

$$x_0^2 - c^2 = d^2 - y_0^2,$$

we have

$$AC = BD, \quad \text{with } (A, B) = 1 = (C, D).$$

Hence

$$A = D, \quad B = C.$$

Also

$$p = c^2 + d^2 = (tC - rA)^2 + (rB + tD)^2$$
$$= (t^2 + r^2)(C^2 + D^2).$$

This factorization of p contradicts the fact that p is a prime. Hence follows the uniqueness of representation and the theorem is proved.

THEOREM 21. *If p, an odd prime divisor of m, and m are both representable primitively as sums of two squares each, then so are mp and m/p.*

Proof. Let

$$m = x^2 + y^2, \quad (x, y) = 1,$$

$$p = a^2 + b^2, \quad (a, b) = 1.$$

Then, since $p \mid m$, these relations imply that

$$(p, xy) = 1, \quad (p, ab) = 1.$$

Now

$$\begin{aligned}(ax - by)(ax + by) &= a^2x^2 - b^2y^2 \\ &= a^2(x^2 + y^2) - y^2(a^2 + b^2) \\ &\equiv 0 \pmod{p}.\end{aligned}$$

Hence $p \mid (ax - by)$ or $p \mid (ax + by)$. The number p cannot divide both because then it will have to divide $2ax$. Assume that $p \mid (ax - by)$ and not $(ax + by)$. Then, since from Brahmagupta's identity,

$$mp = (ax - by)^2 + (ay + bx)^2 \tag{11}$$

$$= (ax + by)^2 + (ay - bx)^2, \tag{12}$$

it follows that $p \mid (ay + bx)$ as well. Let $(ax - by, ay + bx) = pg$. Then

$$pg \mid a(ax - by) + b(ay + bx) = x(a^2 + b^2) = xp$$

$$pg \mid a(ay + bx) - b(ax - by) = y(a^2 + b^2) = yp.$$

Since $(x, y) = 1$, we must have $g = 1$. Now (11) readily gives

$$m/p = u^2 + v^2$$

with $u = (ax - by)/p$, $v = (ay + bx)/p$ and $(u, v) = 1$. Again since $p \nmid (ax + by)$, we have $p \nmid (ay - bx)$ too. Moreover, it can be readily shown that

$$(ax + by, ay - bx) = 1.$$

Hence (12) itself provides the primitive representation for mp. This proves the theorem.

Remark

When p is not a divisor of m, both (11) and (12) give primitive representations for mp.

THEOREM 22. *If p is a prime of the form $4t + 1$, then p^k, $k \geqslant 1$, has a unique primitive representation as a sum of two squares.*

Proof. The existence of a primitive representation for p^k is a direct consequence of theorem 21. That this representation is unique, is proved as follows:

Let

$$p^k = a^2 + b^2 = c^2 + d^2,$$

with $(a, b) = 1 = (c, d)$, $(p, ab) = 1 = (p, cd)$.

Then, Brahmagupta's identity gives

$$\begin{aligned} p^{2k} &= (ac - bd)^2 + (ad + bc)^2 \\ &= (ac + bd)^2 + (ad - bc)^2. \end{aligned} \tag{13}$$

Now

$$\begin{aligned} (ac - bd)(ac + bd) &= a^2(c^2 + d^2) - d^2(a^2 + b^2) \\ &\equiv 0 \pmod{p^k}. \end{aligned}$$

Since p cannot divide both $(ac - bd)$ and $(ac + bd)$, p^k divides one and only one of them. Assume that p^k divides $(ac - bd)$. Then from (13) it follows that p^k divides $(ad + bc)$ also. Taking a, b, c and d as positive integers, we readily infer that

$$ad + bc = p^k, \quad ac - bd = 0.$$

Thus, we get

$$a/b = d/c.$$

Since the two fractions are in their lowest terms, we must have

$$a = d, \quad b = c,$$

and the uniqueness of representation is established.

THEOREM 23. *If p is a prime of the form $4t + 3$, then no multiple of p can be represented primitively as a sum of two squares.*

Proof. If possible, let for some positive integer m,

$$mp = u^2 + v^2, \quad (u, v) = 1, \quad (p, uv) = 1.$$

Then, we have

$$u^2 \equiv -v^2 \pmod{p}.$$

This means that $-v^2$, and therefore -1, is a quadratic residue of p. But this cannot be so because p is a prime of the form $4t + 3$. Hence the theorem.

THEOREM 24. *If an even number m is primitively representable as a sum of two squares, then so also is $m/2$ but this is not true of $2m$.*

The proof is simple and is left to the reader.

THEOREM 25. *Every prime divisor of $n^2 + 1$ can be expressed as a sum of two squares.*

Proof. The theorem holds for $n = 1$. Assume that it holds for each $n \leqslant (m - 1)$, that is, assume that the prime divisors of

$$1^2 + 1, 2^2 + 1, 3^2 + 1, \ldots, (m - 1)^2 + 1$$

can all be expressed as sums of two squares each. Then we assert that the same is true of every prime divisor p of $m^2 + 1$. If $p \mid (m^2 + 1)$ and $p < m$, then $p \mid (m - p)^2 + 1$. Hence by the induction hypothesis, p is representable as a sum of two squares.

If $p \mid (m^2 + 1)$ and $p > m$, let

$$m^2 + 1 = jp, \quad j < m.$$

Every prime factor of j is less than m and can therefore be expressed as a sum of two squares. Removing each one of these primes from $(m^2 + 1)$ as in the proof of theorem 21, we get an expression for p as a sum of two squares. Primes which appear repeatedly in the canonical decomposition of j can be removed in several steps or in one. This proof is due to S. Chowla.

COROLLARY. *Every prime of the form* $4t + 1$ *can be represented as a sum of two squares.*

Proof. Use theorem 11.

10.2. Summary

To sum up, the only numbers that can be expressed primitively as sums of two squares are those which are not divisible either by 4 or by any prime of the form $4t + 3$.

11. Jacobsthal's explicit formula for the representation of a prime $p \equiv 1 \pmod 4$ as a sum of two squares

In the preceding section, we have shown that every prime p of the form $4t + 1$ can be represented as a sum of two squares uniquely. Jacobsthal's formula gives this representation explicitly. The exposition is due to S. Chowla.

THEOREM 26. *If r stands for an arbitrary quadratic residue and n for an arbitrary quadratic non-residue of p, and*

$$F(m) = \sum_{j=1}^{p} \left(\frac{j}{p}\right)\left(\frac{j^2 - m}{p}\right),$$

where the symbols on the right are Legendre's, then

$$p = (\tfrac{1}{2}F(r))^2 + (\tfrac{1}{2}F(n))^2.$$

The proof depends on some lemmas which we first prove.

LEMMA 1. *For every integer $x \not\equiv 0 \pmod p$,*

$$F(m) = \left(\frac{x}{p}\right) F(mx^2).$$

Proof. We have

$$F(m) = \left(\frac{x}{p}\right)^4 \sum_{j=1}^{p} \left(\frac{j}{p}\right)\left(\frac{j^2 - m}{p}\right)$$

$$= \left(\frac{x}{p}\right) \sum_{j=1}^{p} \left(\frac{jx}{p}\right)\left(\frac{j^2x^2 - mx^2}{p}\right). \tag{1}$$

Since $(x, p) = 1$, jx runs through a complete residue set modulo p as j takes the values $1, 2, 3, \ldots, p$. Hence replacing jx by j in (1), we get

$$F(m) = \left(\frac{x}{p}\right) \sum_{j=1}^{p} \left(\frac{j}{p}\right)\left(\frac{j^2 - mx^2}{p}\right),$$

which proves the lemma.

LEMMA 2.

$$\sum_{m=1}^{p} (F(m))^2 = \tfrac{1}{2}(p-1)\{(F(r)\}^2 + (F(n))^2\}.$$

Proof. We can always select integers x and y prime to p, such that if

$$\left(\frac{m}{p}\right) = 1 \qquad \text{then } mx^2 \equiv r \pmod{p},$$

and if

$$\left(\frac{m}{p}\right) = -1 \qquad \text{then } my^2 \equiv n \pmod{p}.$$

Thus, by lemma 1, we have

$$(F(m))^2 = (F(r))^2 \quad \text{or} \quad (F(n))^2$$

according as m is or is not a quadratic residue of p. Now, there are exactly $\frac{1}{2}(p-1)$ values of m for which m is a quadratic residue of p and exactly as many for which it is a quadratic non-residue. Thus, since $F(p) = 0$,

$$\sum_{m=1}^{p} (F(m))^2 = \tfrac{1}{2}(p-1)\{(F(r))^2 + (F(n))^2\},$$

and this is the lemma.

LEMMA 3. *For any integer t,*

$$G(t) = \sum_{j=1}^{p} \left(\frac{j^2 - t}{p}\right) = (p-1) \quad \textit{or} \quad -1$$

according as p does or does not divide t.

Proof. The case $p \mid t$ is trivial. When $(t, p) = 1$, we observe that for $(x, p) = 1$

$$G(t) = \left(\frac{x^2}{p}\right) \sum_{j=1}^{p} \left(\frac{j^2 - t}{p}\right) = \sum_{j=1}^{p} \left(\frac{(jx)^2 - tx^2}{p}\right)$$

$$= \sum_{j=1}^{p} \left(\frac{j^2 - tx^2}{p}\right) = G(x^2 t).$$

Hence, arguing as before, we have

$$G(t) = G(r) \quad \text{or} \quad G(n)$$

according as t is or is not a quadratic residue of p; thus

$$\sum_{t=1}^{p} G(t) = \tfrac{1}{2}(p-1)\{G(r) + G(n)\} + (p-1). \tag{2}$$

Again since for any fixed j, $(j^2 - t)$ runs through a complete residue set modulo p as t does, we have

$$\sum_{t=1}^{p} G(t) = \sum_{t=1}^{p} \sum_{j=1}^{p} \left(\frac{j^2 - t}{p}\right) = \sum_{j=1}^{p} \sum_{t=1}^{p} \left(\frac{j^2 - t}{p}\right)$$

$$= \sum_{j=1}^{p} \sum_{t=1}^{p} \left(\frac{t}{p}\right) = 0. \tag{3}$$

From (2) and (3), we now get

$$G(r) + G(n) = -2. \tag{4}$$

Again

$$G(r) = G(1) = \sum_{j=1}^{p} \left(\frac{j^2 - 1}{p}\right) = \sum_{j=2}^{p} \left(\frac{j-1}{p}\right)\left(\frac{j+1}{p}\right)$$

$$= \sum_{j=1}^{p-1} \left(\frac{j}{p}\right)\left(\frac{j+2}{p}\right). \tag{5}$$

Since the set of integers $1, 2, 3, \ldots, (p-1)$ can be mapped onto itself so that the integer j is transformed into j', where

$$jj' \equiv 1 \pmod{p},$$

we have

$$G(r) = \sum_{j=1}^{p-1} \left(\frac{j'^2}{p}\right)\left(\frac{j}{p}\right)\left(\frac{j+2}{p}\right)$$

$$= \sum_{j=1}^{p-1} \left(\frac{jj'}{p}\right)\left(\frac{jj' + 2j'}{p}\right)$$

$$= \sum_{j'=1}^{p-1} \left(\frac{2j' + 1}{p}\right). \tag{6}$$

Now as j' runs through a complete residue set modulo p, $(2j' + 1)$ also does the same. Hence

$$\sum_{j'=1}^{p} \left(\frac{2j' + 1}{p}\right) = \sum_{j'=1}^{p} \left(\frac{j'}{p}\right) = 0. \tag{7}$$

From (6) and (7), we get

$$G(r) = \sum_{j'=1}^{p-1} \left(\frac{2j' + 1}{p}\right) - \sum_{j'=1}^{p} \left(\frac{2j' + 1}{p}\right) = -1. \tag{8}$$

Finally, (4) and (8) give

$$G(n) = -1 = G(r),$$

which proves the lemma.

LEMMA 4. *For $(a, p) = 1$ and $t = b^2 - 4ac$,*

$$\sum_{j=1}^{p} \left(\frac{aj^2 + bj + c}{p}\right) = \left(\frac{a}{p}\right) G(t).$$

Proof. We have

$$\sum_{j=1}^{p} \left(\frac{aj^2 + bj + c}{p}\right) = \left(\frac{4a^2}{p}\right) \sum_{j=1}^{p} \left(\frac{aj^2 + bj + c}{p}\right)$$

$$= \left(\frac{a}{p}\right) \sum_{j=1}^{p} \left(\frac{(2aj + b)^2 - t}{p}\right)$$

$$= \left(\frac{a}{p}\right) \sum_{j=1}^{p} \left(\frac{j^2 - t}{p}\right)$$

$$= \left(\frac{a}{p}\right) G(t),$$

which is the lemma.

Proof of the theorem. We have

$$(F(m))^2 = \sum_{u=1}^{p} \left(\frac{u}{p}\right)\left(\frac{u^2 - m}{p}\right) \cdot \sum_{v=1}^{p} \left(\frac{v}{p}\right)\left(\frac{v^2 - m}{p}\right)$$

$$= \sum_{u=1}^{p-1} \sum_{v=1}^{p-1} \left(\frac{u}{p}\right)\left(\frac{v}{p}\right)\left(\frac{m^2 - (u^2 + v^2)m + u^2 v^2}{p}\right).$$

Hence

$$\sum_{m=1}^{p} (F(m))^2 = \sum_{m=1}^{p} \sum_{u=1}^{p-1} \sum_{v=1}^{p-1} \left(\frac{u}{p}\right)\left(\frac{v}{p}\right)\left(\frac{m^2 - (u^2 + v^2)m + u^2 v^2}{p}\right)$$

$$= \sum_{u=1}^{p-1} \sum_{v=1}^{p-1} \left(\frac{u}{p}\right)\left(\frac{v}{p}\right) G((u^2 - v^2)^2).$$

In view of lemma 3, for every x prime to p, this gives

$$\sum_{m=1}^{p} (F(m))^2 = \sum_{u=1}^{p-1} \sum_{v=1}^{p-1} \left(\frac{ux}{p}\right)\left(\frac{vx}{p}\right) G(((ux)^2 - (vx)^2)^2).$$

Select x so that $ux \equiv 1 \pmod p$. As u runs through a reduced residue set modulo p, so does x, and as x runs through a reduced residue set modulo p, so does vx. Hence

$$\sum_{m=1}^{p} (F(m))^2 = \sum_{u=1}^{p-1} \sum_{v=1}^{p-1} \left(\frac{v}{p}\right) G((1 - v^2)^2)$$

$$= (p-1) \sum_{v=1}^{p-1} \left(\frac{v}{p}\right) G((1 - v^2)^2).$$

When $v = 1$ or $(p - 1)$, we have $\left(\frac{v}{p}\right) = 1$, and

$$G((1 - v^2)^2) = G(0) = p - 1.$$

For the remaining $(p - 3)$ values of v, we have

$$G((1 - v^2)^2) = -1$$

in each case, while

$$\left(\frac{v}{p}\right) = -1 \quad \text{for} \quad \tfrac{1}{2}(p-1) \text{ of the values,}$$

$$\left(\frac{v}{p}\right) = 1 \quad \text{for the remaining } \tfrac{1}{2}(p-5) \text{ values of } v.$$

Hence

$$\sum_{m=1}^{p} (F(m))^2 = (p-1)(2(p-1)+2)$$
$$= 2p(p-1).$$

Equating this evaluation with the one obtained in lemma 2, we get the result stated in the theorem.

Observe further that

$$F(m) = \sum_{u=1}^{p-1} \left(\frac{u}{p}\right)\left(\frac{u^2-m}{p}\right)$$
$$= 2\sum_{u=1}^{\frac{1}{2}(p-1)} \left(\frac{u}{p}\right)\left(\frac{u^2-m}{p}\right)$$

so that $F(m)$ is even for all m. Hence the theorem gives a representation of p as a sum of two squares of integers.

Theorem 26 can also be stated in the form:

If r is any quadratic residue and n any quadratic non-residue of p, a prime of the form $4t+1$, then

$$p = (H(r))^2 + (H(n))^2,$$

where

$$H(m) = \sum_{\substack{u=1 \\ uv\equiv 1 \pmod p}}^{\frac{1}{2}(p-1)} \left(\frac{|u-mv|}{p}\right).$$

The proof is left to the reader.

12. The number of representations of n as a sum of two squares

We can now take up the problem of finding the number of representations of a natural number n, as a sum of two squares of integers, not necessarily positive or primitive.

The problem reduces to that of determining the number of lattice points on the circle given by the equation

$$x^2 + y^2 = n. \qquad (1)$$

Evidently, if any point (u, v) lies on the circle, so do the points

$$(u, -v), \quad (-u, v), \quad (-u, -v).$$

Unless $uv = 0$, these points are distinct. The points of intersection of (1) and the axes of x and y are lattice points if and only if n is a square. Let $n = c^2$, where c is a positive integer. Then

$$(c, 0), \quad (0, c), \quad (-c, 0), \quad (0, -c)$$

are all lattice-points on (1). Let $C(n)$ denote that portion of the arc of (1) which lies in the first quadrant, in which we include the point of intersection of (1) and OX but not the point of intersection of (1) and OY. If $r_2(n)$ denotes the number of lattice points on (1) and $r(n)$ the number of lattice points on $C(n)$, evidently then

$$r_2(n) = 4\,r(n).$$

Now, let $P(x_1, y_1)$ be any lattice point on $C(n)$ and suppose $(x_1, y_1) = g$. Then $Q(x_1/g, y_1/g)$ is a lattice point on $C(n/g^2)$ and conversely. We call Q a primitive lattice point of $C(n/g^2)$ because its coordinates are relatively prime positive integers. Taking $(x, 0) = x$ we will regard $(1, 0)$ also as a primitive lattice point whenever necessary. Denoting by $N(m)$ the number of primitive lattice points on $C(m)$, it is clear that

$$r(n) = \sum_{g^2 \mid n} N(n/g^2). \tag{2}$$

Our problem has thus been reduced to that of finding a formula for $N(m)$ for any given positive integer m.

12.1. A formula for $N(m)$

Since $1 = 1^2 + 0^2$, we have $N(1) = 1$. Again since $2 = 1^2 + 1^2$, we have $N(2) = 1$. The theorems of sections 10 further show that for $k \geqslant 1$,

$$N(p^k) = N(2p^k) = 2 \quad \text{or} \quad 0$$

according as $p \equiv 1$ or $3 \pmod 4$. Also, for $k \geqslant 2$,

$$N(2^k) = 0.$$

These results will completely determine $N(m)$ if we can prove that $N(m)$ is a multiplicative function. For this purpose, it will be enough to prove

THEOREM 27. *If $m = p^k t$, where p is a prime $\geqslant 2$, $(p, t) = 1$ and $t > 2$, then*

$$N(m) = N(p^k)\,.\,N(t).$$

Proof. If m has a prime divisor of the form $4j + 3$ or it is divisible by 4, then there is nothing to prove, Since $N(m) = 0$ and so also is at least one of the numbers $N(p^k)$ or $N(t)$. We can, therefore, assume that p is a prime of the form $4j + 1$ and t has no prime divisor of the form $4j + 3$, nor is it divisible by 4, and $t > 2$. Hence t and p^k can both be represented as sums of two squares. Let

$$t = a^2 + b^2, \quad (a, b) = 1, \quad a, b > 0,$$

$$p^k = c^2 + d^2, \quad (c, d) = 1, \quad c, d > 0.$$

These assumptions are justified by the fact that neither of the circles

$$x^2 + y^2 = t, \tag{3}$$

$$x^2 + y^2 = p^k, \tag{4}$$

has a primitive lattice point on the axes OX or OY.

We proceed to show that with the point-pair $A(a, b)$, $B(c, d)$ we can associate a unique primitive point $P(u, v)$, $u > 0$, $v > 0$, on the circle

$$x^2 + y^2 = m. \tag{5}$$

It is easy to see that point $Q(ac - bd, ad + bc)$ is a primitive point on (5). If $ac - bd$ is positive, we take P at Q, so that

$$u = ac - bd, \quad v = ad + bc. \tag{6}$$

If $ac - bd$ is negative, we let Q move along (5) in the clockwise direction through a right angle round O to get at P. Thus, we have

$$u = ad + bc, \quad v = bd - ac. \tag{7}$$

We now show that with any given primitive point $P(u, v)$, $u, v > 0$, on (5), is associated a unique primitive point-pair A, B with A on (3) and B on (4) and each in the first quadrant of the relevant circle.

We note that in the first quadrant of (4), there are just two primitive points (c, d) and (d, c) say. Thus we have

$$m = u^2 + v^2, \quad (u, v) = 1, \quad u, v > 0, \tag{8}$$

$$p^k = c^2 + d^2, \quad (c, d) = 1, \quad c, d > 0. \tag{9}$$

Hence

$$m \,.\, p^k = (uc - vd)^2 + (ud + vc)^2 \tag{10}$$

$$= (ud - vc)^2 + (uc + vd)^2. \tag{11}$$

Now

$$(uc - vd)(uc + vd) = u^2(c^2 + d^2) - d^2(u^2 + v^2)$$
$$\equiv 0 \pmod{p^k}.$$

It will be readily seen that just one of the numbers $uc - vd$ and $uc + vd$ can be, and is, divisible by p^k. Assuming that it is $uc + vd$ that is divisible by p^k, it follows from (11) that $(ud - vc)$ is also divisible by p^k. Thus, we get

$$t = \{(ud - vc)/p^k\}^2 + \{(uc + vd)/p^k\}^2.$$

If $ud - vc$ is positive, we take $A(a, b)$ as the point

$$(a, b) = \left(\frac{ud - vc}{p^k}, \frac{uc + vd}{p^k}\right).$$

If $ud - vc$ is negative, we take for A the point

$$(a, b) = \left(\frac{uc + vd}{p^k}, \frac{vc - ud}{p^k}\right).$$

In either case we take B as the point (c, d). In the former case, we have

$$ac - bd = -v, \quad ad + bc = u.$$

In the latter case, we have

$$ac - bd = u, \quad ad + bc = v.$$

We leave it to the reader to deal with the case when p^k divides $uc - vd$ and not $uc + vd$. Since the number of point-pairs A, B is $N(p^k) \,.\, N(t)$, the one–one association we have established proves that

$$N(m) = N(p^k) \,.\, N(t)$$

and we are through.

COROLLARY. *$N(m) = \prod_p (1 + (-1/p))$, where p runs over the odd prime divisors of m and $(-1/p)$ is Legendre's symbol.*

Exercises

1. If m is a product of r distinct primes of the form $4j + 1$, show that

$$\sum_{d|m} N(d) = 3^r.$$

2. For $m = 1105, p = 13, t = 85, k = 1$, find the point-pair associated with the point $(31, 12)$ on the circle $x^2 + y^2 = 1105$. Find also the point-pair associated with the point $(12, 31)$.
3. Show that theorem 27 implies that $r(n)$ is multiplicative.
4. If m is odd and has no prime divisors of the form $4j + 3$, then prove that

$$\sum_{d|m} N(d) = \text{number of divisors of } m^2.$$

13. A geometrical interpretation

There is a nice geometrical way of looking at the problem of multiplicativity of $r(m)$. Take the three circles

$$x^2 + y^2 = p^k, \qquad p \text{ a prime of the form } 4j + 1, \tag{1}$$

$$x^2 + y^2 = t, \qquad t \text{ having no prime divisor of the form } 4j + 3, t > 1, \tag{2}$$

$$x^2 + y^2 = m, \qquad \text{where } m = p^k t, \quad (p, t) = 1. \tag{3}$$

Of the two primitive lattice points on the arc of (1) in the first quadrant let $A(a, b)$ be the one with $a > b$. Let $Q(q, s)$ be any primitive lattice point on (2) not necessarily in the first quadrant.

We assert that if OQ is rotated round O through an angle $= XOA$ whether in the counter-clockwise or the clockwise direction then it cuts (3) in primitive lattice points.

(In the proof we write plp for primitive lattice point.)

Proof. Let

$$XOA = \phi_1; \quad XOQ = \phi_2; \quad 0 < \phi_1 < \pi/4, \quad 0 < \phi_2 < 2\pi.$$

Denote the radii of the three circles by r_1, r_2, r_3 respectively. (Evidently the r's are not necessarily integers.) Then we have

$$\begin{aligned} a &= r_1 \cos \phi_1, \quad b = r_1 \sin \phi_1, \\ q &= r_2 \cos \phi_2, \quad s = r_2 \sin \phi_2. \end{aligned} \tag{4}$$

Hence

$$\begin{aligned} aq - bs &= r_1 r_2 \cos(\phi_2 + \phi_1), \quad as + bq = r_1 r_2 \sin(\phi_2 + \phi_1), \\ aq + bs &= r_1 r_2 \cos(\phi_2 - \phi_1), \quad as - bq = r_1 r_2 \sin(\phi_2 - \phi_1). \end{aligned} \tag{5}$$

Since $r_1 r_2 = r_3$, it is evident that the points $P(aq - bs, as + bq)$ and $P'(aq + bs, as - bq)$ lie on (3) and are the points referred to in the assertion. That they are primitive follows from the relations

$$\begin{aligned} a(aq - bs) + b(as + bq) &= q(a^2 + b^2), \\ a(as + bq) - b(aq - bs) &= s(a^2 + b^2), \\ q(aq - bs) + s(as + bq) &= a(q^2 + s^2), \\ q(as + bq) - s(aq - bs) &= b(q^2 + s^2). \end{aligned}$$

We thus see that corresponding to each plp on (2) there are two plp's on (3) which are uniquely determined in the manner described above.

Conversely, let $P(u, v)$ be any plp on (3). Then we show that it can arise from a unique plp $Q(q, s)$ on (2) and has a unique companion $P'(u', v')$ on (3) which too is a plp.

We assert that only one of the relations

(i) $\quad aq - bs = u, \quad as + bq = v,$

and

(ii) $\quad aq + bs = u, \quad as - bq = v,$

can give integral values for q and s with $(q, s) = 1$. This follows from the following considerations. We have

$$\begin{aligned} (au + bv)(au - bv) &= a^2(u^2 + v^2) - v^2(a^2 + b^2) \\ &\equiv 0 \pmod{p^k}. \end{aligned}$$

Since $(a, b) = 1 = (u, v)$, only one of the relations

(iii) $\quad au + bv \equiv 0 \pmod{p^k},$

(iv) $au - bv \equiv 0 \pmod{p^k}$,

can hold good. Since (i) implies (iii), and (ii) implies (iv), it follows that if (iii) holds then q, s are determined from (i), and if (iv) holds then q, s are given by (ii). In either case q, s are uniquely determined. Since (i) and (ii) each imply that the g.c.d. of q and s divides both u and v, and since $(u, v) = 1$, we must have $(q, s) = 1$. If q and s are given by (i), $P'(u', v')$ is given by

$$u' = aq + bs, \quad v' = as - bq.$$

If q, s are given by (ii), we have

$$u' = aq - bs, \quad v' = as + bq.$$

That P' is a plp follows without any trouble.

Observe that there are $4N(t)$ plp's on (2), which means there are $8N(t)$ plp's on (3) obtained in the above manner. Of these exactly $2N(t)$ lie on the first quadrant of (3). We have thus established that

$$N(m) = 2N(t) = N(p^k)N(t).$$

Once Q is determined, the point P' is geometrically obtained by rotating OQ in the direction away from P through an angle $= \phi_1$.

Example. Let $m = 1105, p = 13, k = 1$ and $t = 85$.

Take $P = (24, 23)$ and $A = (3, 2)$. Since $au - bv = 3.24 - 2.23 = 26 \equiv 0 \pmod{13}$, the coordinates of $Q(q, s)$ are given by the relations

$$3q + 2s = 24, \quad 3s - 2q = 23.$$

Hence Q is the point $(2, 9)$. Now $P'(u', v')$ is given by

$$u' = aq - bs = 3.2 - 2.9 = -12,$$

$$v' = as + bq = 3.9 + 2.2 = 31.$$

13.1. Formula for $r(n)$

The relation (12.2), together with the corollary to theorem 27, shows that $r(n)$ is a multiplicative function. It would, therefore, be enough to evaluate $r(p^k)$ for primes $p \geqslant 2$. Thus, we have

$$r(p^k) = \begin{cases} 1 & p = 2, \quad k \geqslant 0, \\ k+1 & p \equiv 1 \pmod 4, \quad k \geqslant 0, \\ 1 & p \equiv 3 \pmod 4, \quad k \text{ even}, \\ 0 & p \equiv 3 \pmod 4, \quad k \text{ odd}. \end{cases}$$

There is, however, a nice formula for $r(n)$ in terms of the divisor function. This is stated in

THEOREM 28. *Let $d_1(n)$ denote the number of divisors of n which are* $\equiv 1 \pmod 4$ *and $d_3(n)$ the number of those which are* $\equiv 3 \pmod 4$, *then*

$$r(n) = d_1(n) - d_3(n).$$

Proof. We first show that the function

$$d_1(n) - d_3(n)$$

is multiplicative. Let $n = m_1 m_2$, where $(m_1, m_2) = 1$. Evidently, then

$$d_1(n) = d_1(m_1)d_1(m_2) + d_3(m_1)d_3(m_2),$$
$$d_3(n) = d_1(m_1)d_3(m_2) + d_3(m_1)d_1(m_2).$$

Subtracting, we get

$$d_1(n) - d_3(n) = \{d_1(m_1) - d_3(m_1)\}\{d_1(m_2) - d_3(m_2)\}$$

This proves our assertion.

To prove the theorem, all that is necessary now is to show that it holds for powers of primes.

(i) For $n = 2^k, k \geqslant 0$,

$$d_1(n) = 1 \quad \text{and} \quad d_3(n) = 0$$

because 1 is the only divisor of 2^k which $\equiv 1 \pmod 4$, and no divisor of 2^k is $\equiv 3 \pmod 4$. Thus, the theorem holds in this case.

(ii) For $n = p^k, k \geqslant 0$, when p is an odd prime, we observe that each of the divisors

$$1, p^2, p^{2.2}, \ldots, p^{2[k/2]}$$

is $\equiv 1 \pmod 4$, while each of the divisors

$$p, p^3, p^5, \ldots, p^{2[(k-1)/2]+1}$$

is $\equiv 1$ or $3 \pmod 4$, according as $p \equiv 1$ or $3 \pmod 4$. Hence

$$d_1(n) - d_3(n) = \left[\frac{k+2}{2}\right] + (-1)^{(p-1)/2}\left[\frac{k+1}{2}\right]. \tag{6}$$

This result is also in agreement with those already given. Hence the theorem follows.

Formula (6) is due to Gupta and Vaidya. [*American Math. Monthly*, **70** (1963), 1081–1082.]

13.2. The generating function of $r_2(n)$

Taking $r_2(0) = 1$ (and it is reasonable to do so), we can now state that $r_2(n)$ is the coefficient of x^n in the expansion of the series

$$1 + 4(U_1 - U_3 + U_5 - \cdots + (-1)^m U_{2m+1} + \cdots) \tag{7}$$

where

$$U_k = x^k/(1 - x^k).$$

Since x^n does not appear in U_j for any $j > n$, the convergence of (7) is of no importance to us.

Our statement is a direct consequence of the fact that the coefficient of x^n in U_k is 1 or 0 according as k is or is not a divisor of n.

14. The number of representations of n as a sum of k squares

Let $r_k(n)$ denote the number of solutions in integers, not necessarily positive, of the equation

$$x_1^2 + x_2^2 + x_3^2 + \cdots + x_k^2 = n, \tag{1}$$

or what is the same thing, the number of lattice points on the k-dimensional sphere (1). Then, it is clear that

$$\sum_{n=0}^{\infty} r_k(n)x^n = \left\{ \sum_{j=-\infty}^{\infty} x^{j^2} \right\}^k. \tag{2}$$

We considered the case $k = 2$, at some length, in the preceding sections. Jacobi, Liouville, Glaisher and others have considered the problem for even values of $k \leqslant 18$, but the study requires the use of elliptic functions and is beyond the scope of this book. But the case $k = 4$ can be treated in an elementary way as was done by Ramanujan. This we take up in the following section. In fact, we show that for $n \geqslant 1$,

$$r_4(n) = 8 \sum_{\substack{d \mid n \\ d \not\equiv 0 \pmod 4}} d.$$

15. Number of representations as sums of four squares

The proof of the main result will need the following lemmas.

LEMMA 1. *For any positive integer n,*

$$\cot(\tfrac{1}{2}\theta) \sin n\theta = 1 + 2\cos\theta + 2\cos 2\theta + \cdots + 2\cos(n-1)\theta + \cos n\theta$$

Proof. For any integer $m \geqslant 0$, we have

$$\begin{aligned}\cos m\theta + \cos(m+1)\theta &= 2\cos(\tfrac{1}{2}\theta)\cos(m+\tfrac{1}{2})\theta \\ &= \cot(\tfrac{1}{2}\theta) \,.\, 2\sin(\tfrac{1}{2}\theta)\cos(m+\tfrac{1}{2})\theta \\ &= \cot(\tfrac{1}{2}\theta)\{\sin(m+1)\theta - \sin m\theta\}.\end{aligned}$$

Replacing m by $0, 1, 2, \ldots, (n-1)$ in succession in this identity and adding the results so obtained, the lemma follows immediately.

Writing U_k for $x^k/(1-x^k)$, we have the following lemmas.

LEMMA 2. *For* $|x| < 1$,

$$\sum_{k=1}^{\infty} U_k(1 + U_k) = \sum_{k=1}^{\infty} kU_k.$$

Proof. The left side of the identity

$$= \sum_{k=1}^{\infty} x^k/(1 - x^k)^2$$

$$= \sum_{k=1}^{\infty} \sum_{r=1}^{\infty} rx^{rk}$$

$$= \sum_{r=1}^{\infty} \sum_{k=1}^{\infty} rx^{rk}$$

$$= \sum_{r=1}^{\infty} rx^r/(1 - x^r).$$

Hence the lemma follows.

LEMMA 3.

(i) $U_{r+n}(1 + U_r) = U_n(U_r - U_{r+n}), \quad r, n \geqslant 1;$

(ii) $U_k U_{n-k} = U_n(1 + U_k + U_{n-k}), \quad 1 \leqslant k < n.$

The verification is left to the reader.

LEMMA 4. *For* $|x| < 1$,

$$\sum_{n=1}^{\infty} (-1)^{n-1} U_{2n}(1 + U_{2n}) = \sum_{n=1}^{\infty} (2n - 1) U_{4n-2}.$$

Proof. The left side of the identity

$$= \sum_{n=1}^{\infty} (-1)^{n-1} \sum_{r=1}^{\infty} rx^{2nr}$$

$$= \sum_{r=1}^{\infty} r \sum_{n=1}^{\infty} (-1)^{n-1} x^{2nr}$$

$$= \sum_{r=1}^{\infty} rx^{2r}/(1 + x^{2r})$$

$$= \sum_{r=1}^{\infty} (rU_{2r} - 2rU_{4r})$$

$$= \sum_{m=1}^{\infty} (2m-1)U_{4m-2}.$$

This proves the lemma.

15.1. Ramanujan's identity

For $|x| < 1$ *and* $U_k = x^k/(1-x^k), k \geqslant 1$,

$$\left\{\cot \tfrac{1}{2}\theta + 4 \sum_{k=1}^{\infty} U_k \sin k\theta \right\}^2$$

$$= \cot^2 \tfrac{1}{2}\theta + 16 \sum_{k=1}^{\infty} U_k(1+U_k)\cos k\theta + 8 \sum_{k=1}^{\infty} kU_k(1-\cos k\theta).$$

Proof. The left side of the identity

$$= \cot^2 \tfrac{1}{2}\theta + 8 \sum_{k=1}^{\infty} U_k \cot \tfrac{1}{2}\theta \sin k\theta + 16 \sum_{k,r=1}^{\infty} U_k U_r \sin k\theta \sin r\theta$$

$$= \cot^2 \tfrac{1}{2}\theta + 8A + 8B.$$

From lemma 1, we have

$$A = U_1(1+\cos\theta) + U_2(1+2\cos\theta+\cos 2\theta) + \cdots +$$

$$+ U_k(1+2\cos\theta+2\cos 2\theta+2\cos 3\theta+\cdots+2\cos(k-1)\theta+\cos k\theta)+\cdots$$

$$= C_0 + C_1 \cos\theta + C_2\cos 2\theta + \cdots + C_k \cos k\theta + \cdots,$$

where

$$C_0 = \sum_{k=1}^{\infty} U_k,$$

$$C_n = U_n + 2\sum_{k=1}^{\infty} U_{n+k}, \quad n = 1, 2, 3, \ldots$$

Again

$$B = \sum_{k,r=1}^{\infty} U_k U_r\{\cos(k-r)\theta - \cos(k+r)\theta\}$$

$$= \sum_{k=1}^{\infty} U_k^2 + 2\sum_{\substack{k=1\\ r<k}}^{\infty} U_k U_r \cos(k-r)\theta - \sum_{k,r=1}^{\infty} U_k U_r \cos(k+r)\theta$$

$$= D_0 + D_1\cos\theta + D_2\cos 2\theta + \cdots + D_k\cos k\theta + \cdots,$$

where

$$D_0 = \sum_{k=1}^{\infty} U_k^2,$$

$$\begin{aligned}
D_n &= 2 \sum_{k-r=n}^{\infty} U_k U_r - \sum_{k+r=n}^{\infty} U_k U_r, \quad n = 1, 2, 3, \ldots, \\
&= 2 \sum_{r=1}^{\infty} U_{r+n} U_r - \sum_{r=1}^{n-1} U_r U_{n-r} \\
&= 2 \sum_{r=1}^{\infty} U_{r+n}(1 + U_r) - 2 \sum_{r=1}^{\infty} U_{r+n} - \sum_{r=1}^{n-1} U_r U_{n-r} \\
&= 2U_n \sum_{r=1}^{\infty} (U_r - U_{r+n}) - 2 \sum_{r=1}^{\infty} U_{r+n} - U_n \sum_{r=1}^{n-1} (1 + U_r + U_{n-r}) \\
&= 2U_n \sum_{r=1}^{n} U_r - 2 \sum_{r=1}^{\infty} U_{r+n} - (n-1)U_n - 2U_n \sum_{r=1}^{n-1} U_r \\
&= 2U_n^2 - 2 \sum_{r=1}^{\infty} U_{r+n} - nU_n + U_n.
\end{aligned}$$

Ramanujan's identity follows directly now.

COROLLARY. *For* $|x| < 1$,

$$\left\{ 1 + 4 \sum_{\substack{n=1 \\ n \text{ odd}}}^{\infty} (-1)^{\frac{1}{2}(n-1)} U_n \right\}^2 = 1 + 8 \sum_{\substack{n=1 \\ 4 \nmid n}}^{\infty} nU_n.$$

Proof. Replace θ by $\frac{1}{2}\pi$ in Ramanujan's identity and use lemma 4 proved earlier in this section.

15.2. Formula for $r_4(n)$

From what has been proved in sections 13.2 and 14, it follows that

$$\begin{aligned}
\left\{ \sum_{j=-\infty}^{\infty} x^{j^2} \right\}^2 &= 1 + \sum_{n=1}^{\infty} r_2(n) x^n \\
&= 1 + 4 \sum_{\substack{k=1 \\ k \text{ odd}}}^{\infty} (-1)^{\frac{1}{2}(k-1)} U_k.
\end{aligned}$$

In view of the corollary proved in 14.1, therefore, $r_4(n)$ is the coefficient of x^n in

$$1 + 8 \sum_{\substack{k=1 \\ k \not\equiv 0 \pmod 4}}^{\infty} kU_k.$$

Hence for $n \geqslant 1$,

$$r_4(n) = 8 \sum_{\substack{d \mid n \\ 4 \nmid d}} d.$$

Since the expression on the right is definitely positive, we have the

THEOREM 29. *Every natural number can be expressed as the sum of at most four squares of integers.*

This result can also be proved by the 'method of descent'. [See for example, *Introduction to the Theory of Numbers* by G. H. Hardy and E. M. Wright, fourth edition, p. 302.]

Problems

1. Show that no number of the form $8t + 7$ can be represented as a sum of three squares.
2. Investigate the solvability of the following congruences:

   (i) $x^2 \equiv 35 \pmod{71}$; (ii) $x^2 \equiv 365 \pmod{1951}$.

3. If p is a prime of the form 3 (mod 4), then show that integers a, b, c prime to p exist such that

   $$a^2 + b^2 + c^2 \equiv 0 \pmod p.$$

4. Use Lehmer's property to evaluate the Jacobi symbol (3/55).
5. Show that all the divisors of $a^2 - a - 1$ other than 5 are of the form $10t + 1$ or $10t - 1$.
6. If p is a prime of the form 3 (mod 4), then a necessary and sufficient condition for $2p + 1$ to be a prime is that

   $$2^p - 1 \equiv 0 \pmod{2p + 1}.$$

7. Show that the triangular number $n(n + 1)/2$ is a sum of two non-zero triangular numbers if and only if $n^2 + (n + 1)^2$ is not a prime.
8. If n and $n + 1$ are two consecutive numbers in the series $1, 2, \ldots, p - 1$ find how often they have prescribed quadratic characters.
9. Show that the sum of the quadratic residues of an odd prime $p > 3$ is a multiple of p and that their product is $\equiv 1$ or $-1 \pmod p$ according as $p \equiv 3$ or $1 \pmod 4$. Show also that for $p \equiv 1 \pmod 4$ the sum of the quadratic residues is $p(p - 1)/4$.
10. Show that if each of two natural numbers can be represented in the form $x^2 + 2y^2$, then their product can also be so represented. Show that a number which can be represented in this form in two essentially different ways, is composite.
11. Develop on the lines of section 10, the theory of representations of numbers in the form $x^2 + 2y^2$.
12. If p is an odd prime and D_p is the determinant of order $(p-1)$ which has the

Legendre symbol $\left(\frac{r-c}{p}\right)$ as the element in the rth row and cth column, then $D_p = p^{(p-3)/2}$.

13. If $p = 4n + 1$ is a prime, then

$$p = a^2 + b^2,$$

where a and b are the least absolute residues of

$$(2n)!/2(n!)^2 \quad \text{and} \quad (2n)!\, a \text{ modulo } p \qquad \text{[Gauss]}$$

respectively. Use this result to express 17 and 29 as sums of two squares.

14. Use Jacobsthal's formula to represent 17 and 29 as sums of two squares.

CHAPTER 4

CLASS ALGEBRA OF *k*-ic RESIDUES

1. Introduction

In this chapter, we study an algebra arising from the proper k-ic residues of a prime p. We then apply this algebra to the determination of the quadratic, cubic and biquadratic characters of small primes.

1.1. Notation

Throughout this chapter, we use the word set in the sense of a multiset, that is a set in which the elements are not necessarily distinct. Our sets will be sets of numbers and the definition of any set S will make it clear whether any given number does or does not belong to the set and, if it does, the number of times it occurs therein.

Since we deal with only proper k-ic residues of primes p, our p will invariably be of the form $kq + 1$ with $k > 1$ and q some positive integer. Since p shall be odd, one at least of the numbers k and q will have to be even.

Again since the k-ic residue character of two numbers congruent modulo p is the same, in what follows we freely replace any member of a set by an integer congruent to it modulo p and adopt the convention that in doing so the identity of the set is not affected.

If A and B are two sets, we write

$$A \oplus B$$

for the set of numbers obtained by adding each member of A to each of the members of B. This means that

$$A \oplus B = \{a + b : a \in A, b \in B\}.$$

Thus, for example, if

$$A = \{1, 2, 2\}, \quad B = \{1, 2, 3, 4\}$$

and $p = 5$, then

$$\begin{aligned} A \oplus B &= \{2, 3, 4, 5, \ \ 3, 4, 5, 6, \ \ 3, 4, 5, 6\} \\ &= \{0, 0, 0, \ \ 1, 1, \ \ 2, \ \ 3, 3, 3, \ \ 4, 4, 4\}. \end{aligned}$$

Also, we write tA for the set of numbers obtained on multiplying each member of

A by t, that is

$$tA = \{ta : a \in A\}.$$

Evidently the binary operation $\oplus$ is both commutative and associative. Moreover

$$t(A \oplus B) = tA \oplus tB.$$

If each of the elements of A appears exactly t times in S, we say t is the frequency of A in S, and write

$$A(t) \subset S.$$

On the other hand, the relation

$$A(t_1) + B(t_2) = S$$

indicates that the set S consists of the elements of A repeated t_1 times each and of those of B each repeated t_2 times. If A and B are disjoint sets only then can we say that t_1 and t_2 are the frequencies of A and B in S. For brevity, we usually write A for $A(1)$. We further write

$$A - B = C \quad \text{when } A = B + C.$$

For example, if

$$A = \{1, 1, 2, 3, 3, 4, 5, 5, 5\} \quad \text{and} \quad B = \{1, 2, 5, 5\}$$

then

$$C = \{1, 3, 3, 4, 5\}.$$

Finally, if A is a set of numbers and x is a real variable, then

$$F(A) = F(A, x) = \sum_{a \in A} x^a$$

is said to be the generating function of A. Evidently,

$$F(A) \,.\, F(B) = F(A \oplus B)$$

This provided the motivation for our definition of $A \oplus B$.

Throughout this chapter, g is used to denote a primitive root of p.

2. The classes

The set

$$S = \{1, 2, 3, \ldots, p - 1\}$$

can be divided into k classes C_j, $j = 0, 1, 2, \ldots, k - 1$ given by

$$\begin{aligned} C_j &= \{r_j : \mathrm{Ind}_g\, r_j \equiv j \pmod k\} \\ &= \{g^j, g^{j+k}, g^{j+2k}, \ldots, g^{j+(q-1)k}\} \end{aligned}$$

with

$$qk = p - 1.$$

Equivalently, we can write

$$C_j = \{c_{j,0}, c_{j,1}, c_{j,2}, \ldots, c_{j,q-1}\},$$

where

$$c_{j,t} = g^j \cdot g^{tk}, \quad t = 0, 1, 2, \ldots, q-1.$$

Note that C_j is completely determined by any one of its elements. Thus, for example if r_j is any one of its elements, then

$$C_j = \{r_j, r_j g^k, r_j g^{2k}, \ldots, r_j g^{(q-1)k}\}.$$

Note further that C_0 consists of all the k-ic residues of p.

Into this family of k mutually disjoint classes or subsets of S, we introduce a set C (without a subscript) consisting of exactly q zeros. Why we need C will be clear from the following.

Example. Take $p = 13, k = 3, g = 2$. Then, we have

$$C_0 = \{1, \ 8, \ 12, \ 5\}$$
$$C_1 = \{2, \ 3, \ 11, \ 10\}$$
$$C_2 = \{4, \ 6, \ 9, \ 7\}$$

so that

$$C_0 \oplus C_0 = \{2, 9, 0, 6, \ \ 9, 3, 7, 0, \ \ 0, 7, 11, 4, \ \ 6, 0, 4, 10\}$$
$$= C(1) + C_1(1) + C_2(2).$$

Without the introduction of C, which in this case is a set with four zeros as elements, we could not have expressed $C_0 \oplus C_0$ as a linear combination of C's.

Exercise

Examine if $C_0 \oplus C_1$, $C_0 \oplus C_2$, $C_1 \oplus C_1$, $C_1 \oplus C_2$, $C_2 \oplus C_2$ can be expressed as linear combinations of C's.

The following relations are easy to prove:

(i) $C_r = C_s$ if and only if $r \equiv s \pmod{k}$;

(Thus, one can write $C_{-j} = C_{k-j}$.)

(ii) $g^h C_j = C_{j+h}$;

(iii) $g^h(C_0 \oplus C_{j-h}) = C_h \oplus C_j$;

(iv) For any r_0 in C_0,

$$r_0(C_h \oplus C_j) = C_h \oplus C_j;$$

(v) $C \oplus C_j = C_j(q)$.

In view of these relations, it will be enough to study if $C_0 \oplus C_j$ can be expressed as a linear combination of C's. In the next section we prove that the answer to this is in the affirmative.

3. The study of $C_0 \oplus C_j$

DEFINITION. *Two classes C_j and C_h are said to be conjugate when the members of one are the same as those of the other but with opposite signs.*

Denoting the conjugate of C_j by C_j^*, we have

$$C_j^* = \{-r_j : r_j \in C_j\}.$$

Since

$$g^{(p-1)/2} \equiv -1 \pmod p$$

the conjugate of every C_j exists, is unique, and in fact is given by

$$C_j^* = C_{j+(p-1)/2},$$

where the subscript can be reduced modulo k. Now

$$(p-1)/2 = kq/2 \equiv 0 \quad \text{or} \quad k/2 \pmod k$$

according as q is even or odd. Hence

$$C_j^* = \begin{cases} C_j & \text{if } q \text{ is even,} \\ C_{j+k/2} & \text{if } q \text{ is odd.} \end{cases}$$

THEOREM 1. *The set*

$$C_0 \oplus C_j$$

can be expressed as a linear combination of the C's in the form:

$$C_0 \oplus C_j = C(e_{j,0}) + \sum_{t=0}^{k-1} C_t(a_{j,t})$$

where the a's are integers $\geqslant 0$, and $e_{j,0}$ is 1 *or* 0 *according as C_j is or is not the conjugate of C_0.*

Proof. Let

$$C_j = \{c_{j,0}, c_{j,1}, c_{j,2}, \ldots, c_{j,q-1}\}, \quad 0 \leqslant j \leqslant k-1,$$

where the c's have the same meaning as in the preceding section.

For any r, $0 < r \leqslant q$, consider the family of sets given by

$$\begin{aligned} T_r &= \{c_{0,r-1} + c_{j,t+r-2} : t = 1, 2, 3, \dots, q\} \\ &= \{g^{(r-1)k}(c_{0,0} + c_{j,t-1}) : t = 1, 2, 3, \dots, q\}. \end{aligned}$$

If for a certain t, $c_{0,0} + c_{j,t-1}$ belongs to some C (with or without subscript), then so does

$$g^{(r-1)k}(c_{0,0} + c_{j,t-1})$$

belong to the same C for each $r \leqslant q$. This means that the family of sets T_r contains all the elements of any C to which

$$c_{0,0} + c_{j,t-1}$$

belongs, as t runs over the values $1, 2, 3, \dots, q$. To know the frequency of C or any of the C's with subscripts, all that is necessary is to find how many members of any such C belong to the set

$$U = \{c_{0,0} + c_{j,t-1} : t = 1, 2, 3, \dots, q\}.$$

So far as C itself is concerned

$$c_{0,0} + c_{j,t-1}$$

can be zero only if C_0 and C_j are conjugates and then too only for $t = 1$. If $c_{0,0} + c_{j,t-1}$ belongs to C_m for $a_{j,m}$ ($\geqslant 0$) values of t, then the frequency of C_m in $C_0 \oplus C_j$ is $a_{j,m}$. This proves the theorem completely.

It will thus be seen that $a_{j,m}$ is the number of values of t for which

$$c_{0,0} + c_{j,t-1} = c_{m,s-1} \quad \text{for some } s, \quad 1 \leqslant s \leqslant q.$$

COROLLARY. *We have*

$$\sum_{m=0}^{k-1} a_{j,m} = q - 1 \quad \text{or} \quad q$$

according as C_j is or is not the conjugate of C_0.

The proof is simple and is left to the reader.

3.1. Examples

The following two examples will clarify the ideas and lead to new ones.

Example 1. Let $p = 31$, $k = 5$ and $g = 3$. Then the classes are given by

$$C_0 = \{\ 1, 26, 25, 30, \ 5, \ 6\}$$

$$C_1 = \{\ 3, 16, 13, 28, 15, 18\}$$

$$C_2 = \{\ 9, 17, \ 8, 22, 14, 23\}$$

$$C_3 = \{27, 20, 24, \ \ 4, 11, \ \ 7\}$$

$$C_4 = \{19, 29, 10, 12, \ \ 2, 21\}.$$

Evidently it will be a good idea to write the entries on the right of the equations in the form of a matrix which we could call the class matrix. This we propose to do in our next example.

To find the frequency of these classes and C in the expressions for $C_0 \oplus C_j$, $j = 0, 1, 2, 3, 4$, the theorem requires us to add $c_{0,0}$ i.e. 1 to each of the elements of C_j and see how often these sums belong to any given class or to C. Adding 1 to each of the elements of the class matrix, we get the matrix:

$$\begin{bmatrix} 2 & 27 & 26 & 0 & 6 & 7 \\ 4 & 17 & 14 & 29 & 16 & 19 \\ 10 & 18 & 9 & 23 & 15 & 24 \\ 28 & 21 & 25 & 5 & 12 & 8 \\ 20 & 30 & 11 & 13 & 3 & 22 \end{bmatrix}$$

We have now to note the class to which each of these elements belongs. This we do by writing r to indicate that the element is in C_r. This gives us the matrix:

$$\begin{bmatrix} 4 & 3 & 0 & . & 0 & 3 \\ 3 & 2 & 2 & 4 & 1 & 4 \\ 4 & 1 & 2 & 2 & 1 & 3 \\ 1 & 4 & 0 & 0 & 4 & 2 \\ 3 & 0 & 3 & 1 & 1 & 2 \end{bmatrix}$$

Hence, we obtain

$$C_0 \oplus C_0 = C(1) + C_0(2) + C_1(0) + C_2(0) + C_3(2) + C_4(1)$$

$$C_0 \oplus C_1 = C(0) + C_0(0) + C_1(1) + C_2(2) + C_3(1) + C_4(2)$$

$$C_0 \oplus C_2 = C(0) + C_0(0) + C_1(2) + C_2(2) + C_3(1) + C_4(1)$$

$$C_0 \oplus C_3 = C(0) + C_0(2) + C_1(1) + C_2(1) + C_3(0) + C_4(2)$$

$$C_0 \oplus C_4 = C(0) + C_0(1) + C_1(2) + C_2(1) + C_3(2) + C_4(0).$$

Leaving out the frequency of C, the frequencies of $C_0, C_1, \ldots, C_4$ can again be recorded in the form of a matrix to be called the frequency matrix. Denoting it by A_0 here, we have

$$A_0 = \begin{bmatrix} 2 & 0 & 0 & 2 & 1 \\ 0 & 1 & 2 & 1 & 2 \\ 0 & 2 & 2 & 1 & 1 \\ 2 & 1 & 1 & 0 & 2 \\ 1 & 2 & 1 & 2 & 0 \end{bmatrix}$$

Thus the frequency matrix can be obtained in four steps. In this example, q was even. We now consider an example where q is odd.

Example 2. Take $p = 31, k = 6$, and $g = 3$. Without repeating the details, we will simply record the four matrices in order:

$$\text{(i)}\quad \begin{bmatrix} 1 & 16 & 8 & 4 & 2 \\ 3 & 17 & 24 & 12 & 6 \\ 9 & 20 & 10 & 5 & 18 \\ 27 & 29 & 30 & 15 & 23 \\ 19 & 25 & 28 & 14 & 7 \\ 26 & 13 & 22 & 11 & 21 \end{bmatrix}$$

$$\text{(ii)}\quad \begin{bmatrix} 2 & 17 & 9 & 5 & 3 \\ 4 & 18 & 25 & 13 & 7 \\ 10 & 21 & 11 & 6 & 19 \\ 28 & 30 & 0 & 16 & 24 \\ 20 & 26 & 29 & 15 & 8 \\ 27 & 14 & 23 & 12 & 22 \end{bmatrix}$$

$$\text{(iii)}\quad \begin{bmatrix} 0 & 1 & 2 & 2 & 1 \\ 0 & 2 & 4 & 5 & 4 \\ 2 & 5 & 5 & 1 & 4 \\ 4 & 3 & . & 0 & 1 \\ 2 & 5 & 3 & 3 & 0 \\ 3 & 4 & 3 & 1 & 5 \end{bmatrix}$$

$$\text{(iv)}\quad \begin{bmatrix} 1 & 2 & 2 & 0 & 0 & 0 \\ 1 & 0 & 1 & 0 & 2 & 1 \\ 0 & 1 & 1 & 0 & 1 & 2 \\ 1 & 1 & 0 & 1 & 1 & 0 \\ 1 & 0 & 1 & 2 & 0 & 1 \\ 0 & 1 & 0 & 2 & 1 & 1 \end{bmatrix}$$

As the reader could not have failed to observe, for q even, the frequency matrix was symmetric. That welcome feature is missing for q odd. But that can be restored if we place the lower half of the frequency matrix at the top. We thus have the matrix:

$$\text{(v)}\quad \begin{bmatrix} 1 & 1 & 0 & 1 & 1 & 0 \\ 1 & 0 & 1 & 2 & 0 & 1 \\ 0 & 1 & 0 & 2 & 1 & 1 \\ 1 & 2 & 2 & 0 & 0 & 0 \\ 1 & 0 & 1 & 0 & 2 & 1 \\ 0 & 1 & 1 & 0 & 1 & 2 \end{bmatrix}$$

3.2. Relationships between frequencies

THEOREM 2. *The frequency of C_h in $C_0 \oplus C_j$ is the same as the frequency of C_j in $C_0^* \oplus C_h$.*

Proof. In the proof of theorem 1, we have seen that the frequency $a_{j,h}$ of C_h in $C_0 \oplus C_j$ is the same as the number of values of t for which $c_{0,0} + c_{j,t-1}$ i.e. $1 + c_{j,t-1}$ belongs to C_h. In other words, $a_{j,h}$ is the number of solutions of the equation

$$1 + c_{j,t-1} = c_{h,s-1}, \tag{1}$$

i.e. the number of number-pairs (t, s) for which (1) holds with

$$1 \leqslant t \leqslant q, \quad 1 \leqslant s \leqslant q.$$

Writing (1) in the form

$$-1 + c_{h,s-1} = c_{j,t-1} \tag{2}$$

the theorem follows immediately, as -1 is a typical member of C_0^*. If q is even, C_0 is self-conjugate, and -1 in equation (2) can be replaced by 1. The number of solutions of (2), therefore, is $a_{h,j}$. Hence

$$a_{j,h} = a_{h,j}$$

when q is even. For q even, therefore, the symmetry of (iv) is established. If q is odd, we have

$$C_0^* = C_{k/2}.$$

Hence

$$\begin{aligned} C_0^* \oplus C_h = C_{k/2} \oplus C_h &= g^{k/2}(C_0 \oplus C_{h-k/2}) \\ &= g^{k/2}\left(C(e_{h-k/2}) + \sum_{r=0}^{k-1} C_r(a_{h-k/2,r})\right) \\ &= C(e_{h-k/2}) + \sum_{r=0}^{k-1} C_{r+k/2}(a_{h-k/2,r}) \end{aligned}$$

$$= C(e_{h-k/2}) + \sum_{r=0}^{k-1} C_r(a_{h-k/2,\, r-k/2});$$

and we get

$$a_{j,h} = a_{h-k/2,\, j-k/2}.$$

This establishes the symmetry of (v) when q is odd.

THEOREM 3. *We have*

$$a_{j,h} = a_{k-j,\, h-j}.$$

Proof. Since

$$g^j(C_0 \oplus C_{k-j}) = C_j \oplus C_k = C_j \oplus C_0 = C_0 \oplus C_j,$$

and

$$g^j(C_0 \oplus C_{k-j}) = g^j\left\{C(?) + \sum_{r=0}^{k-1} C_r(a_{k-j,r})\right\}$$

$$= C(?) + \sum_{r=0}^{k-1} C_{r+j}(a_{k-j,r})$$

$$= C(?) + \sum_{r=0}^{k-1} C_r(a_{k-j,\, r-j})$$

the theorem follows readily.

To visualize the situation which theorem 3 creates, proceed as follows. Parallel to the principal diagonal to its immediate left, draw in the frequency matrix, a line as shown in the following diagram:

$$\begin{bmatrix} a_{0,0} & a_{0,1} & a_{0,2} & \cdot & \cdot & a_{0,k-1} \\ a_{1,0} & a_{1,1} & a_{1,2} & \cdot & \cdot & a_{1,k-1} \\ a_{2,0} & a_{2,1} & a_{2,2} & \cdot & \cdot & a_{2,k-1} \\ \cdot & \cdot & \cdot & \cdot & \cdot & \cdot \\ \cdot & \cdot & \cdot & \cdot & \cdot & \cdot \\ a_{k-1,0} & a_{k-1,1} & a_{k-1,2} & \cdot & \cdot & a_{k-1,k-1} \end{bmatrix}$$

Now leave out the first row. Then what theorem 3 implies is that the two halves into which the line divides the rest of the matrix are identical if one half is looked at from the bottom up and the other from the top down. This holds irrespective of whether q is odd or even. Thus in the examples in the preceding subsection, we had

$$\begin{bmatrix} 2 & 0 & 0 & 2 & 1 \\ 0 & 1 & 2 & 1 & 2 \\ 0 & 2 & 2 & 1 & 1 \\ 2 & 1 & 1 & 0 & 2 \\ 1 & 2 & 1 & 2 & 0 \end{bmatrix}, \quad \begin{bmatrix} 1 & 2 & 2 & 0 & 0 & 0 \\ 1 & 0 & 1 & 0 & 2 & 1 \\ 0 & 1 & 1 & 0 & 1 & 2 \\ 1 & 1 & 0 & 1 & 1 & 0 \\ 1 & 0 & 1 & 2 & 0 & 1 \\ 0 & 1 & 0 & 2 & 1 & 1 \end{bmatrix}$$

Theorems 2 and 3 greatly reduce the number of possible frequencies that can appear in a $k \times k$ frequency matrix. This we consider in the next subsection.

3.3. The fomula for the maximum number of frequencies

For q even, let us consider the form of a 9×9 frequency matrix. For our present purpose, it will be most convenient to replace the symbols $a_{0,0}, a_{0,1}, \ldots, a_{k-1,k-1}$ by numbers the same number being allotted to those a's that are necessarily equal by theorems 2 and 3.

The said matrix is then of the form:

$$\begin{bmatrix} 1 & 2 & 3 & 4 & 5 & 6 & 7 & 8 & 9 \\ 2 & 9 & 10 & 11 & 12 & 13 & 14 & 15 & 10 \\ 3 & 10 & 8 & 15 & 16 & 17 & 18 & 16 & 11 \\ 4 & 11 & 15 & 7 & 14 & 18 & 19 & 17 & 12 \\ 5 & 12 & 16 & 14 & 6 & 13 & 17 & 18 & 13 \\ 6 & 13 & 17 & 18 & 13 & 5 & 12 & 16 & 14 \\ 7 & 14 & 18 & 19 & 17 & 12 & 4 & 11 & 15 \\ 8 & 15 & 16 & 17 & 18 & 16 & 11 & 3 & 10 \\ 9 & 10 & 11 & 12 & 13 & 14 & 15 & 10 & 2 \end{bmatrix}$$

Notice that the entries satisfy the requirements of theorems 2 and 3, and we have had to use only 19 of the 81 frequency symbols. We could not have used less. We now prove

THEOREM 4. *Let $N(k)$ denote the maximum necessary number of distinct symbols for the $k \times k$ frequency matrix. Then*

$$N(k) = (k^2 + 3k + 6)/6 \quad \text{or} \quad (k^2 + 3k + 2)/6$$

according as k is or is not a multiple of 3.

Proof. For q even, we have in general

$$a_{j,h} = a_{h,j} = a_{k-j,h-j} = a_{h-j,k-j} = a_{k-h,j-h} = a_{j-h,k-h}$$

it being understood that the subscripts are reducible modulo k. This means that any frequency appears in general in six places in the matrix. But there are certain frequencies which appear in one, two or three places only.

These are easy to specify:

$a_{0,0}$ appears only once;

$$a_{0,j} = a_{j,0} = a_{k-j,k-j}, \quad 1 \leqslant j \leqslant k-1.$$

For all the remaining cases, $a_{j,h}$ appears in six places, except when $k = 3u$. In the exceptional case, we have

$$a_{2t,t} = a_{t,2t}.$$

Since the total number of entries in the matrix is k^2, we get

$$N(k) = 1 + (k-1) + 1 + (k^2 - 3k + 0)/6 \quad \text{when} \quad 3 \mid k;$$

$$\text{or} \quad 1 + (k-1) + (k^2 - 3k + 2)/6 \quad \text{when} \quad 3 \nmid k.$$

This proves the theorem when q is even.

The proof for q odd is exactly similar and is left to the reader. It might however be noted that in this case k is necessarily even. If one cuts out the right half of the frequency matrix and places it on the left, one obtains from the matrix for q even, the matrix for q odd.

Example. The frequency matrix for $k = 8$, q even, is of the form

$$\begin{bmatrix} 1 & 2 & 3 & 4 & 5 & 6 & 7 & 8 \\ 2 & 8 & 9 & 10 & 11 & 12 & 13 & 9 \\ 3 & 9 & 7 & 13 & 14 & 15 & 14 & 10 \\ 4 & 10 & 13 & 6 & 12 & 15 & 15 & 11 \\ 5 & 11 & 14 & 12 & 5 & 11 & 14 & 12 \\ 6 & 12 & 15 & 15 & 11 & 4 & 10 & 13 \\ 7 & 13 & 14 & 15 & 14 & 10 & 3 & 9 \\ 8 & 9 & 10 & 11 & 12 & 13 & 9 & 2 \end{bmatrix}$$

For $k = 8$, q odd, this provides the form

$$\begin{bmatrix} 5 & 6 & 7 & 8 & 1 & 2 & 3 & 4 \\ 11 & 12 & 13 & 9 & 2 & 8 & 9 & 10 \\ 14 & 15 & 14 & 10 & 3 & 9 & 7 & 13 \\ 12 & 15 & 15 & 11 & 4 & 10 & 13 & 6 \\ 5 & 11 & 14 & 12 & 5 & 11 & 14 & 12 \\ 11 & 4 & 10 & 13 & 6 & 12 & 15 & 15 \\ 14 & 10 & 3 & 9 & 7 & 13 & 14 & 15 \\ 12 & 13 & 9 & 2 & 8 & 9 & 10 & 11 \end{bmatrix}$$

Exercises

1. For $p = 37$ a primitive root is 2. Obtain the frequency matrices when $k = 2, 3, 4$ or 6 and verify the theorems of section 3.
2. Show that

$$g^h(C_0 \oplus C_{j-h}) = g^j(C_0 \oplus C_{h-j}).$$

Hence obtain a relation between the a's.

3. Express $C_0 \oplus C_j \oplus C_h$ as a linear combination of C's both when C_j and C_h are conjugate and when they are not.
4. Verify theorem 4 for $k = 6$, 7 and 11. Give the form of the matrix when q is even and also when q is odd.

4. Use of generating functions

For brevity, we write

$$F_j = F(C_j) = \sum_{t \in C_j} x^t.$$

While dealing with generating functions, we adopt the following conventions:

(i) $x^u = x^v$ whenever $u \equiv v \pmod p$.

(ii) $x^u x^v = x^s$ when $u + v \equiv s \pmod p$.

(iii) $F_j = F_h$ when $j \equiv h \pmod k$.

(iv) $a_{u,t} = a_{v,s}$ whenever $u \equiv v$ and $t \equiv s \pmod k$.

In view of these conventions and the results of the preceding section we have

$$F_j F_h = q e_{j,h} + \sum_{t=0}^{k-1} a_{j-h,k-h+t} F_t,$$

where $e_{j,h} = 1$ or 0 according as C_j and C_h are or are not conjugate.

4.1. The extended frequency matrix

For our study of the k-ic residue character of small primes and for small values of k, we shall have to deal with products and powers of several F's. The following formula will then be very useful.

$$\begin{bmatrix} q \\ F_0 \\ \cdot \\ F_h \\ \cdot \\ F_{k-1} \end{bmatrix} F_j = \begin{bmatrix} 0 & 0 & \cdots & q & \cdots & 0 \\ e_{j,0} & a_{j,0} & \cdots & a_{j,j} & \cdots & a_{j,-1} \\ \cdot & \cdot & \cdots & \cdot & \cdots & \cdot \\ e_{j,h} & a_{j-h,-h} & \cdots & a_{j-h,j-h} & \cdots & a_{j-h,-h-1} \\ \cdot & \cdot & \cdots & \cdot & \cdots & \cdot \\ e_{j,k-1} & a_{j+1,1} & \cdots & a_{j+1,j+1} & \cdots & a_{j+1,0} \end{bmatrix} \begin{bmatrix} q \\ F_0 \\ \cdot \\ F_h \\ \cdot \\ F_{k-1} \end{bmatrix}$$

The square matrix on the right is a $(k + 1) \times (k + 1)$ matrix. This is nothing but the familiar frequency matrix A_j to which have been added a row at the top and a column on the left. The row at the top has a q in the $(j + 2)$th column and a zero everywhere else. The column on the left has a 1 in the $(r + 2)$th row where C_j and C_r are conjugate and zero at all other places. This is our extended frequency matrix which we denote by M_j.

Denoting the column matrix on the right by Q, the above equation takes the form:

$$QF_j = M_j Q. \tag{1}$$

This result admits of extension and we readily obtain

$$QF_{j_1}F_{j_2} \ldots F_{j_m} = M_{j_1}M_{j_2} \ldots M_{j_m}Q. \tag{2}$$

Of particular interest to us will be result

$$QF_0^u = M_0^u Q. \tag{3}$$

The best way of writing out the expression for M_j is to write the last row first and then move upwards increasing each subscript of a's by one. Thus, for $k = 4$, q even, we have

$$M_0 = \begin{bmatrix} 0 & q & 0 & 0 & 0 \\ 1 & a_{0,0} & a_{0,1} & a_{0,2} & a_{0,3} \\ 0 & a_{3,3} & a_{3,0} & a_{3,1} & a_{3,2} \\ 0 & a_{2,2} & a_{2,3} & a_{2,0} & a_{2,1} \\ 0 & a_{1,1} & a_{1,2} & a_{1,3} & a_{1,0} \end{bmatrix}$$

while

$$M_2 = \begin{bmatrix} 0 & 0 & 0 & q & 0 \\ 0 & a_{2,0} & a_{2,1} & a_{2,2} & a_{2,3} \\ 0 & a_{1,3} & a_{1,0} & a_{1,1} & a_{1,2} \\ 1 & a_{0,2} & a_{0,3} & a_{0,0} & a_{0,1} \\ 0 & a_{3,1} & a_{3,2} & a_{3,3} & a_{3,0} \end{bmatrix}$$

5. The k-ic residue character of small primes

We can now utilize the class algebra developed in the foregoing sections to answer the question:

'For what primes p is a small prime p_1 (different from p) a k-ic residue of p?'

We will confine ourselves to quadratic, cubic and biquadratic residues only. The answer is provided by

THEOREM 5. *If p_1 is a prime different from p, then $p_1 \in C_j$, if and only if*

$$F_0^{p_1} \equiv F_j \pmod{p_1}.$$

Proof. We have

$$F_0^{p_1} = \left\{ \sum_{r \in C_0} x^r \right\}^{p_1}$$

$$\equiv \sum_{r \in C_0} x^{rp_1} \pmod{p_1}.$$

Now $rp_1 \in C_j$ if and only if $p_1 \in C_j$. Hence the theorem follows.

Remarks

The theorem implies that $p_1 \in C_j$, if and only if the coefficient of F_j in the expression for

$$F_0^{p_1} \tag{1}$$

as a linear function of $F_0, F_1, F_2, \ldots, F_{k-1}$ is congruent to 1 modulo p_1 while the coefficients of all other F's are divisible by p_1. In particular, p_1 is a k-ic residue of p if and only if the coefficient of F_0 in (1) is congruent to 1 (mod p_1).

5.1. Application to quadratic residues

Let us consider the question:

'For what primes p of the form $4m + 1$ is 7 a quadratic residue?'

We notice first that here our q is even. Therefore M_0 is of the form

$$M_0 = \begin{bmatrix} 0 & 2m & 0 \\ 1 & a & b \\ 0 & b & b \end{bmatrix}$$

By the corollary to theorem 1, we have

$$1 + a + b = 2m = b + b.$$

Hence

$$QF_0 = \begin{bmatrix} 0 & 2m & 0 \\ 1 & m-1 & m \\ 0 & m & m \end{bmatrix} Q$$

This means that

$$F_0^2 = 2m + (m-1)F_0 + mF_1,$$

$$F_0F_1 = \qquad mF_0 + mF_1.$$

These two relations can be used to find in succession the expressions for $F_0^3, F_0^4, \ldots, F_0^7$ and then the coefficient of F_0 in the last expression. This coefficient is congruent to 1 modulo 7. Alternatively, one can find the expression for F_0^7 from the relation

$$QF_0^6 = M_0^6 Q.$$

The second row of M_0^6 is all that we need to find the coefficient of F_0 in the expression for F_0^7. We leave these two courses for the reader to try.

Here we adopt a third course which leads to a lot of simplification in the calculations. Write

$$I = 1 + F_0 + F_1$$
$$= 1 + x + x^2 + \cdots + x^{p-1}.$$

In view of our conventions,

$$x^t . I = I, \quad \text{for all integers } t.$$

Hence

$$F_0 I = 2mI.$$

Moreover if

$$F_0^t = u_t + v_t F_0 + w_t I,$$

then

$$F_0^{t+1} = u_t F_0 + v_t F_0^2 + w_t \,.\, 2mI,$$
$$= mv_t + (u_t - v_t)F_0 + m(v_t + 2w_t)I,$$

because

$$F_0^2 = m - F_0 + mI.$$

Hence, we have

$$u_{t+1} = mv_t, \quad v_{t+1} = u_t - v_t, \quad w_{t+1} = m(v_t + 2w_t).$$

Finally, we will have

$$F_0^7 = u_7 + v_7 F_0 + w_7(1 + F_0 + F_1).$$

In the expression on the right, the coefficient of F_0 must be $\equiv 1 \pmod 7$, and that of F_1 to $0 \pmod 7$. Hence,

$$v_7 \equiv 1 \pmod 7.$$

All that we need compute, therefore, are the values of u_7 and v_7. The following table gives the values of u_t and v_t for $2 \leqslant t \leqslant 7$.

t	u_t	v_t
2	m	-1
3	$-m$	$m + 1$
4	$m^2 + m$	$-2m - 1$
5	$-2m^2 - m$	$m^2 + 3m + 1$
6	$m^3 + 3m^2 + m$	$-3m^2 - 4m - 1$
7	$-3m^3 - 4m^2 - m$	$m^3 + 6m^2 + 5m + 1$

Solving the congruence

$$m^3 + 6m^2 + 5m + 1 \equiv 1 \pmod{7},$$

we get

$$m \equiv 0, 2 \text{ or } 6 \pmod{7}.$$

Hence primes of which 7 is a quadratic residue are of the form

$$28j + 1, \quad 28j + 9, \quad 28j - 3.$$

Note that for m of the form 0, 2 or 6 modulo 7, u_t is zero modulo 7.

Exercises

1. Use the above table to find the form of primes for which 2, 3 or 5 are quadratic residues.
2. Consider on the same lines the case of primes of the form $4m - 1$.

5.2. Application to biquadratic residues

Let us now consider for what primes p of the form $8m + 5$, 3 is a biquadratic residue. In this case, $q = 2m + 1$. Therefore, we must have

$$M_0 = \begin{bmatrix} 0 & q & 0 & 0 & 0 \\ 0 & a_3 & a_4 & a_1 & a_2 \\ 0 & a_5 & a_5 & a_2 & a_4 \\ 1 & a_3 & a_5 & a_3 & a_5 \\ 0 & a_5 & a_2 & a_4 & a_5 \end{bmatrix}$$

$$M_1 = \begin{bmatrix} 0 & 0 & q & 0 & 0 \\ 0 & a_5 & a_5 & a_2 & a_4 \\ 0 & a_2 & a_3 & a_4 & a_1 \\ 0 & a_4 & a_5 & a_5 & a_2 \\ 1 & a_5 & a_3 & a_5 & a_3 \end{bmatrix}$$

From the corollary to theorem 1, we have

$$\begin{aligned} a_3 + a_4 + a_1 + a_2 &= 2m + 1 \\ a_5 + a_5 + a_2 + a_4 &= 2m + 1 \\ a_3 + a_5 &= m \end{aligned}$$

so that

$$\begin{aligned} a_5 &= m - a_3 \\ a_2 + a_4 &= 1 + 2a_3 \\ a_1 &= 2m - 3a_3. \end{aligned} \tag{1}$$

Equating the elements in the second row and third column of M_0M_1 and M_1M_0 we obtain

$$a_3a_5 + a_4a_3 + a_1a_5 + a_2a_3 = a_5a_5 + a_5a_2 + a_2a_4 + a_4a_5.$$

Relations (1), then give

$$5a_3^2 - 2a_3a_4 + a_4^2 - (4m - 2)a_3 - a_4 + m^2 - m = 0,$$

which is the same thing as

$$(8a_3 - 4m + 1)^2 + 4(2a_3 - 2a_4 + 1)^2 = p.$$

Now

$$QF_0^2 = M_0^2Q.$$

Hence

$$F_0^3 = a_1q + u_0F_0 + u_1F_1 + u_2F_2 + u_3F_3,$$

where

$$\begin{aligned} u_0 &= a_3^2 + a_4a_5 + a_1a_3 + a_2a_5 \\ &\equiv m + (m - 1)a_3 - a_3^2 \pmod 3, \\ u_1 &= a_3a_4 + a_4a_5 + a_1a_5 + a_2a_2 \\ &\equiv (a_3 + a_4 - m - 1)^2 + m^2 - 2m \pmod 3, \\ u_2 &= a_3a_1 + a_4a_2 + a_1a_3 + a_2a_4 \\ &\equiv a_4^2 + a_3a_4 - a_4 + ma_3 \pmod 3, \\ u_3 &= a_3a_2 + a_4a_4 + a_1a_5 + a_2a_5 \\ &\equiv (a_4 + m)^2 + m^2 + m \pmod 3. \end{aligned}$$

For 3 to be a biquadratic residue of $p = 8m + 5$, it is necessary that it should be a quadratic residue of p. Hence, we must have

$$m \equiv 1 \pmod 3.$$

We must also have

$$u_0 - 1 \equiv u_1 \equiv u_2 \equiv u_3 \equiv 0 \pmod 3.$$

These conditions imply that

$$a_3 \equiv 0 \pmod 3, \quad a_4 \equiv 0 \text{ or } 1 \pmod 3.$$

This means that p is of the forms:

$$(12t + 9)^2 + (6s + 1)^2 \quad \text{or} \quad (12t + 9)^2 + (6s - 1)^2.$$

Exercises

1. In the expression for F_0^3 is a_1q divisible by 3?
2. For what primes of the form $8m + 1$ is 3 a biquadratic residue?

5.3. Application to cubic residues

In this case q is necessarily even. We ask

'For what primes p of the form $6m + 1$ is 2 a cubic residue?'

We have

$$M_0 = \begin{bmatrix} 0 & q & 0 & 0 \\ 1 & a & b & c \\ 0 & b & c & d \\ 0 & c & d & b \end{bmatrix}, \qquad M_1 = \begin{bmatrix} 0 & 0 & q & 0 \\ 0 & b & c & d \\ 1 & c & a & b \\ 0 & d & b & c \end{bmatrix}$$

Now

$$a + b + c = 2m - 1,$$

$$b + c + d = 2m.$$

Moreover equating elements in the products M_0M_1 and M_1M_0, we have

$$b^2 + c^2 + d^2 = 2m + ac + ba + cb.$$

This is the same thing as

$$3a^2 + 3ab + 3b^2 - (6m - 5)a - (6m - 3)b + 4m^2 - 6m + 2 = 0,$$

which on multiplication by 36 can be written as

$$(9a - 6m + 7)^2 + 27(a + 2b - 2m + 1)^2 = 4p.$$

For 2 to be a cubic residue of p, it is necessary and sufficient that a is odd. Hence

$$p = u^2 + 27v^2.$$

Exercise

1. For what primes of the form $6m + 1$ is 3 a cubic residue?

CHAPTER 5

STIRLING NUMBERS AND BERNOULLI NUMBERS

1. Introduction

Many arithmetical functions are defined by means of recurrence relations which they satisfy. Thus, for example, the Fibonacci and Lucas numbers both satisfy the recurrence

$$u(n) + u(n+1) = u(n+2) \tag{1}$$

for all integral values of n. For the Fibonacci numbers, we have $u(0) = 0, u(1) = 1$; for the Lucas numbers, $u(0) = 1, u(1) = 3$. Defining $u(0)$ and $u(1)$ differently, any number of sequences satisfying (1) can be obtained. The recurrence relation (1) involves only one variable viz. n. The combinatorial functions satisfy the recurrence relation

$$C(n, r) + C(n, r-1) = C(n+1, r). \tag{2}$$

This involves two variables. For the combinatorial functions, we have

$$C(n, 0) = 1 \quad \text{for all integers } n,$$

$$C(1, 1) = 1 \quad \text{and} \quad C(1, r) = 0 \quad \text{for } r > 1.$$

It must be clearly understood that the combinatorial functions are not the only ones that satisfy (2).

In this chapter, we study another recurrence relation involving two integral variables. The motivation is provided by Stirling numbers of the first kind, defined by the relation

$$(x+1)(x+2)\ldots(x+n) = \sum_{r=0}^{n} G(n, r)x^{n-r} \tag{3}$$

with $G(n, 0) = 1, n \geqslant 1$. We take $G(n, r) = 0$ for $0 < n < r$. Evidently, $G(n, r)$ denotes the sum of the products of the first n natural numbers taken r at a time. For example,

$$G(3, 2) = 1.2 + 1.3 + 2.3 = 11.$$

Since the product on the left of (3) can be written

$$(x+n).(x+1)(x+2)\ldots(x+n-1) = (x+n).\sum_{r=0}^{n-1} G(n-1, r)x^{n-r-1}, \tag{4}$$

on comparing the coefficients of x^{n-r} on the right of (3) and (4), we immediately get the recurrence relation

$$G(n, r) = G(n-1, r) + nG(n-1, r-1), \tag{5}$$

which the Stirling numbers of the first kind satisfy.

Given $G(j, 0) = a_j$ for all integral values of j, and $G(0, k) = b_k$ for all integral values of k other than 0, the recurrence relation

$$G(x, y) = G(x-1, y) + xG(x-1, y-1) \tag{6}$$

defines $G(j, k)$ completely for all integral values of j, k. For Stirling numbers, we have

$$a_j = 1, \quad b_k = 0 \quad \text{for} \quad k \neq 0. \tag{7}$$

For $n > 0, r > 0$, $G(n, r)$ are Stirling numbers of the first kind; $G(-n, r)$ are Stirling numbers of the second kind; and $G(-n, -r)$ are what we shall call Stirling numbers of the third kind.

In what follows, we first obtain expressions for $G(n, r)$ in terms of combinatory functions. If n is replaced by $-n$ in these expressions we will get $G(-n, r)$. Independent closed expressions are then found for $G(-n, r)$. These are utilized for defining $G(-n, -r)$.

A compact formula for $G(n, r)$ is obtained and a very general addition formula for $G(i+j, r)$ is given. $S(n, r)$, which is defined as the sum of the rth powers of the first n natural numbers, is expressed in terms of Stirling numbers of the second kind and an expression for Bernoulli numbers derived.

The symbol $x^{(k)}$ is invariably used to denote the product

$$x(x-1)(x-2)\ldots(x-k+1),$$

x being any rational number other than zero and k a positive integer. We define

$$x^{(0)} = 1$$

for all rational values of x including zero. We may now write

$$\binom{x}{r} = \frac{x^{(r)}}{r!},$$

which generalizes the combinatorial function $C(n, r)$.

We leave it to the reader to show that (6) and (7) imply that

$$G(n, r) = 0 \quad \text{for} \quad 0 \leqslant n < r. \tag{8}$$

2. The difference operator Δ

Several of our results in this chapter take a compact form if we employ the difference operator Δ.

If $F(x)$ is a polynomial in x, then we define

$$\Delta F(x) = F(x+1) - F(x). \tag{1}$$

Also we define inductively

$$\Delta^r F(x) = \Delta\Delta^{r-1} F(x), \quad r > 1. \tag{2}$$

Thus

$$\Delta^2 F(x) = \Delta\{F(x+1) - F(x)\} = F(x+2) - 2F(x+1) + F(x),$$

$$\Delta^3 F(x) = \Delta\{F(x+2) - 2F(x+1) + F(x)\}$$
$$= F(x+3) - 3F(x+2) + 3F(x+1) - F(x).$$

In general, it is easy to show by induction that

$$\Delta^r F(x) = \sum_{j=0}^{r} (-1)^j \binom{r}{j} F(x + r - j). \tag{3}$$

Exercises

Show that

(i) $\Delta x^{(j)} = j x^{(j-1)}$;

(ii) $\Delta^r x^{(j)} = j^{(r)} x^{(j-r)}$;

(iii) $\Delta \binom{x}{j} = \binom{x}{j-1}$;

(iv) $\Delta^r \binom{x}{j} = \binom{x}{j-r}, \quad r \leqslant j$;

(v) $(1 + \Delta) x^s = (x+1)^s$;

(vi) $(1 + \Delta)^r x^s = (x + r)^s$;

(vii) $(1 + \Delta)^r F(x) = F(x + r)$.

THEOREM 1. *If* $\Delta F(x) = F_1(x)$, *then*

$$\sum_{x=1}^{n} F_1(x) = F(n+1) - F(1).$$

Proof. By definition, we have

$$F_1(x) = F(x+1) - F(x).$$

Substituting in succession $1, 2, 3, \ldots, n$ for x in this relation and adding the results, we get

$$F_1(1) + F_1(2) + F_1(3) + \cdots + F_1(n) = F(n+1) - F(1).$$

Example

$$\Delta \binom{x}{j} = \binom{x}{j-1}, \quad j \geqslant 1.$$

Hence

$$\sum_{x=1}^{n} \binom{x}{j-1} = \binom{n+1}{j} - \binom{1}{j}.$$

For $j > 1$, $\binom{1}{j} = 0$, and we have

$$\sum_{x=1}^{n} \binom{x}{j-1} = \binom{n+1}{j}, \quad \text{for } j > 1.$$

We shall have occasion to use the operator Δ again in section 7.

3. The two expressions for $G(n, r)$

From section 1, we have

$$G(n, r) = G(n-1, r) + nG(n-1, r-1) \tag{1}$$

and

$$G(n, r) = 0 \quad \text{for} \quad 0 \leqslant n < r.$$

Hence

$$G(n, r) = \sum_{m=1}^{n} mG(m-1, r-1). \tag{2}$$

We use (2) to express $G(n, r)$ in terms of combinatory functions. Since $G(m-1, 0) = 1$ for each $m \geqslant 1$, we have

$$G(n, 1) = \sum_{m=1}^{n} m = \sum_{m=1}^{n} \binom{m}{1} = \binom{n+1}{2}.$$

For $G(n, 2)$, $G(n, 3)$ . . . we use the results of theorem 1 and the relation

$$m \,.\, \binom{m}{j} = j \,.\, \binom{m}{j} + (j+1) \,.\, \binom{m}{j+1}. \tag{3}$$

Thus, we get

$$\begin{aligned} G(n, 2) &= \sum_{m=1}^{n} mG(m-1, 1) = \sum_{m=1}^{n} m\binom{m}{2} \\ &= \sum_{m=1}^{n} \left\{2\binom{m}{2} + 3 \,.\, \binom{m}{3}\right\} \\ &= 2 \,.\, \binom{n+1}{3} + 3 \,.\, \binom{n+1}{4}. \end{aligned}$$

$$G(n, 3) = \sum_{m=1}^{n} m\left\{2 \, . \binom{m}{3} + 3 \, . \binom{m}{4}\right\},$$

$$= \sum_{m=1}^{n} \left\{6 \, . \binom{m}{3} + 20 \, . \binom{m}{4} + 15 \, . \binom{m}{5}\right\}$$

$$= 6 \, . \binom{n+1}{4} + 20 \, . \binom{n+1}{5} + 15 \, . \binom{n+1}{6}.$$

Similarly

$$G(n, 4) = 24\binom{n+1}{5} + 130\binom{n+1}{6} + 210\binom{n+1}{7} + 105\binom{n+1}{8}.$$

In general, let

$$G(n, r-1) = \sum_{j=1}^{r-1} F(r-1, j)\binom{n+1}{r-1+j},$$

where F's are positive integers independent of n. Then

$$G(n, r) = \sum_{m=1}^{n} mG(m-1, r-1)$$

$$= \sum_{m=1}^{n} \sum_{j=1}^{r-1} F(r-1, j) \, . \, m\binom{m}{r-1+j}$$

$$= \sum_{m=1}^{n} \sum_{j=1}^{r-1} F(r-1, j)\left\{(r-1+j)\binom{m}{r-1+j} + (r+j)\binom{m}{r+j}\right\}$$

$$= \sum_{m=1}^{n} \sum_{j=1}^{r} (r-1+j)\{F(r-1, j-1) + F(r-1, j)\}\binom{m}{r-1+j},$$

where we take $F(r-1, 0) = 0 = F(r-1, r)$. Thus

$$G(n, r) = \sum_{j=1}^{r} (r-1+j)\{F(r-1, j-1) + F(r-1, j)\}\binom{n+1}{r+j}. \tag{4}$$

Hence the F's are given recursively by the formula

$$F(r, j) = (r-1+j)\{F(r-1, j-1) + F(r-1, j)\}. \tag{5}$$

In particular, since $F(1, 1) = 1$, we have for $r > 1$,

$$F(r, 1) = r \, . \, F(r-1, 1) = r!$$

and

$$F(r,r) = (2r-1)F(r-1,r-1) = (2r-1)(2r-3)F(r-2,r-2) = \cdots$$
$$= (2r-1)(2r-3)(2r-5)\ldots 5 \,.\, 3 \,.\, 1.$$

Result (4) provides one expression for $G(n, r)$ in terms of combinatory functions. For the other, we use in place of (3) the relation

$$m \,.\, \binom{m+k-1}{j-1} = j \,.\, \binom{m+k}{j} - k \,.\, \binom{m+k-1}{j-1}. \tag{6}$$

We have, as before

$$G(n, 1) = \binom{n+1}{2}.$$

Hence

$$G(n, 2) = \sum_{m=1}^{n} m\binom{m}{2} = \sum_{m=1}^{n} \left\{3 \,.\, \binom{m+1}{3} - \binom{m}{2}\right\}$$
$$= 3 \,.\, \binom{n+2}{4} - \binom{n+1}{3}.$$

Similarly

$$G(n, 3) = 15 \,.\, \binom{n+3}{6} - 10 \,.\, \binom{n+2}{5} + \binom{n+1}{4}.$$

Proceeding in this manner, we can show that

$$G(n, r) = \sum_{j=0}^{r-1} (-1)^j H(r, j)\binom{n+r-j}{2r-j}, \tag{7}$$

where the H's are positive integers independent of n, given recursively by the formula

$$H(r, j) = (2r-j-1) \,.\, H(r-1, j) + (r-j) \,.\, H(r-1, j-1) \tag{8}$$

with

$$H(r-1, -1) = 0 = H(r-1, r-1), \quad r > 1,$$

and

$$H(0, 0) = 1, \quad H(0, -1) = 0.$$

In particular, since $H(1, 0) = 1$, we have

$$H(r, 0) = (2r-1)H(r-1, 0) = (2r-1)(2r-3)\ldots 5 \,.\, 3 \,.\, 1;$$

and

$$H(r, r-1) = H(r-1, r-2) = \cdots = 1.$$

From the results obtained above, it follows that $G(n, r)$ is a polynomial in n of degree $2r$, with fractional coefficients in general. The result (7) is substantially due to Morgan Ward.*

4. The polynomial $G(x, r)$

We now extend the definition of $G(n, r)$ by defining $G(x, r)$ for all integers x and all nonnegative integral values of r as

$$G(x,r) = \begin{cases} 1 & r = 0, \\ \sum_{j=0}^{r} F(r,j)\binom{x+1}{r+j} & r > 0, \end{cases} \tag{1}$$

where $F(r, j)$ has the same meaning as in the preceding section. Obviously for $r > 0$, $G(x, r)$ is of the form

$$G(x,r) = (x+1)^{(r+1)}(a_1x^{r-1} + a_2x^{r-2} + \cdots + a_r)/(2r)! \tag{2}$$

where the a's are integers. In particular,

$$G(x, 1) = (x+1)x/2!$$

$$G(x, 2) = (x+1)x(x-1)(3x+2)/4!$$

$$G(x, 3) = (x+1)x(x-1)(x-2)(15x^2+15x)/6! \tag{3}$$

and

$$G(x, 4) = (x+1)x(x-1)(x-2)(x-3)(105x^3+105x^2-70x-56)/8!.$$

Notice that $G(x, r)$ as defined above, still satisfies the fundamental relation

$$G(x+1, r) = G(x, r) + (x+1)G(x, r-1). \tag{4}$$

5. $G(x, r)$ for negative x

We proceed to obtain an expression for $G(x, r)$ when x is a negative integer < -1.

Consider the series

$$g(n+1) = y^{-n-1} + G(-n-1, 1)y^{-n-2} + G(-n-1, 2)y^{-n-3} + \cdots$$

$$= \sum_{r=0}^{\infty} G(-n-1, r)y^{-n-r-1}, \quad y > n \geqslant 0, \tag{1}$$

where $G(-n-1, 0) = 1$, as already stated. Since $G(-1, r) = 0$ for $r > 0$, we have

$$g(1) = 1/y. \tag{2}$$

*Stirling's numbers and polynomials. *Amer. J. Math.* **56** (1934), 87–95.

Multiplying both sides of (1) by $(y - n)$, we get

$$(y-n)g(n+1) = \sum_{r=0}^{\infty} \{G(-n-1, r) - nG(-n-1, r-1)\} y^{-n-r}$$

$$= \sum_{r=0}^{\infty} G(-n, r) y^{-n-r} = g(n). \tag{3}$$

Here we have used the fundamental relation (4.4) with $(-n - 1)$ for x. From (2) and (3), we obtain

$$\begin{aligned} g(n+1) &= g(n)/(y-n) = g(n-1)/(y-n)(y-n+1) = \cdots = \\ &= g(1)/(y-1)^{(n)} = 1/y^{(n+1)}. \end{aligned} \tag{4}$$

Breaking up into partial fractions, we get

$$\begin{aligned} g(n+1) &= \frac{(-1)^n}{n!} \left\{ \frac{\binom{n}{0}}{y} - \frac{\binom{n}{1}}{y-1} + \frac{\binom{n}{2}}{y-2} - \cdots + (-1)^n \frac{\binom{n}{n}}{y-n} \right\} \\ &= \frac{(-1)^n}{n!y} \left\{ \binom{n}{0} - \binom{n}{1}(1-y^{-1})^{-1} + \binom{n}{2}(1-2y^{-1})^{-1} - \cdots \right. \\ &\quad \left. + (-1)^n \binom{n}{n}(1-ny^{-1})^{-1} \right\}. \end{aligned} \tag{5}$$

Comparing the coefficients of y^{-n-r-1} in (1) and (5), we get

$$n!G(-n-1, r) = \sum_{k=0}^{n-1} (-1)^k \binom{n}{k} (n-k)^{n+r}, \quad n \geqslant 1, r \geqslant 0, \tag{6}$$

which is the same thing as

$$(n-1)!G(-n-1, r) = \sum_{k=0}^{n-1} (-1)^k \binom{n-1}{k} (n-k)^{n+r-1}, \quad n \geqslant 1, r \geqslant 0. \tag{7}$$

Thus

$$\begin{aligned} G(-2, r) &= 1, \\ 1!G(-3, r) &= 2^{r+1} - 1, \\ 2!G(-4, r) &= 3^{r+2} - 2 \,.\, 2^{r+2} + 1. \end{aligned} \tag{8}$$

Taking $r = 0$ in (7), we further have

$$\begin{aligned} (n-1)! &= \binom{n-1}{0} n^{n-1} - \binom{n-1}{1}(n-1)^{n-1} + \binom{n-1}{2}(n-2)^{n-1} - \cdots \\ &\quad + (-1)^{n-1}\binom{n-1}{n-1} 1^{n-1}. \end{aligned} \tag{9}$$

Exercises

1. In the expressions for $G(x, 1)$, $G(x, 2)$, $G(x, 3)$ and $G(x, 4)$ given in (4.3), replace x by $-2, -3, -4$ and verify the results in (5.8).
2. When x is a negative integer $-n$, $n \geqslant 1$, the fundamental relation (4.4) takes the form:

$$G(-n, r) = G(-n - 1, r) - nG(-n - 1, r - 1).$$

Use this relation to prove that

$$G(-n - 1, r) = G(-n, r) + nG(-n, r - 1) + n^2G(-n, r - 2) + \cdots + n^rG(-n, 0).$$

Hence obtain expressions for $G(-2, r)$, $G(-3, r)$ and $G(-4, r)$.

6. $G(n, r)$ for negative n and r

We now define $G(-n - 2, r)$ for values of $n \geqslant 1$ and all integral values of r by the relation

$$n!G(-n - 2, r) = \sum_{k=0}^{n} (-1)^k \binom{n}{k} (n - k + 1)^{n+r}, \tag{1}$$

which can also be written as

$$(n + 1)!G(-n - 2, r) = \sum_{k=0}^{n} (-1)^k \binom{n+1}{k} (n - k + 1)^{n+r+1}. \tag{2}$$

Observe that for negative r and $n \geqslant 1$, $G(-n - 2, r)$ may not be an integer. For example, one can show that

$$G(-n - 2, -n - 1) = (-1)^n/(n + 1)!, \quad n \geqslant 1. \tag{3}$$

From the definition, it is clear that the fundamental relation (4.4) for $G(x, r)$ holds even when r is a negative integer.

We have seen that

$$2!G(-4, r) = 3^{r+2} - 2 . 2^{r+2} + 1.$$

Therefore

$$G(-4, -1) = 0 = G(-4, -2) \quad \text{while} \quad G(-4, -3) = 1/6.$$

This is a particular case of the interesting

THEOREM 2. *$G(-n - 2, -r)$ vanishes for $1 \leqslant r \leqslant n$.*

Proof. We write the fundamental relation (4.4) in the form

$$(n + 1)G(-n - 2, -r) = G(-n - 2, -r + 1) - G(-n - 1, -r + 1), \quad n \geqslant 1. \tag{4}$$

Replacing r by 1, 2, 3, . . . in succession in (4), we get

$$(n + 1)G(-n - 2, -1) = G(-n - 2, 0) - G(-n - 1, 0) = 1 - 1 = 0,$$

so that for $n \geqslant 1$,

$$G(-n-2, -1) = 0.$$

Again

$$(n+1)G(-n-2, -2) = G(-n-2, -1) - G(-n-1, -1) = 0, \quad n \geqslant 2.$$

Hence

$$G(-n-2, -2) = 0 \quad \text{for} \quad n \geqslant 2.$$

We now use induction on r and suppose that

$$G(-n-2, -r+1) = 0 \quad \text{when} \quad n \geqslant (r-1).$$

Then the first term on the right of (4) vanishes for $n \geqslant (r-1)$ and the second for $n \geqslant r$, so that

$$G(-n-2, -r) = 0 \quad \text{for} \quad n \geqslant r \geqslant 1.$$

7. Expressions for $G(-n, r)$ in terms of Δ

The expressions for $(n-1)!\,G(-n-1, r)$ and $n!\,G(-n-1, r)$ can be made to look very much more elegant and simple than they do in (5.7) and (5.6) if we make use of the operator Δ.

For $j \geqslant 0$, we have

$$\Delta^j(n-j)^s = n^s - \binom{j}{1}(n-1)^s + \binom{j}{2}(n-2)^s - \cdots + (-1)^j\binom{j}{j}(n-j)^s,$$

where Δ operates on n. Also, from (5.7) and (5.6),

$$j!\,G(-j-1, r) = \sum_{k=0}^{j} (-1)^k \binom{j}{k}(j-k)^{j+r}, \quad j \geqslant 1, r \geqslant 0,$$

and

$$j!\,G(-j-2, r) = \sum_{k=0}^{j} (-1)^k \binom{j}{k}(j-k+1)^{j+r}, \quad j \geqslant 0, r \geqslant 0.$$

Hence, we can write

$$j!\,G(-j-1, r) = \{\Delta^j n^{j+r}\}_{n=0}, \quad j \geqslant 1, r \geqslant 0, \tag{1}$$

and

$$j!\,G(-j-2, r) = \{\Delta^j n^{j+r}\}_{n=1}, \quad j \geqslant 0, r \geqslant 0, \tag{2}$$

where $\{\ \}_{n=t}$ indicates that n is to be replaced by t at the end of the operation by Δ.

As an application of these results, we prove

THEOREM 3. *For all values of $r > 0$,*

$$G(-2, r) - 1!G(-3, r-1) + 2!G(-4, r-2) - \cdots + (-1)^r r!G(-r-2, 0) = 0.$$

Proof. We have

$$j!G(-j-2, r) = \{\Delta^j n^{j+r}\}_{n=1}, \quad j \geqslant 0, r \geqslant 0.$$

Hence

$$\begin{aligned} \sum_{j=0}^{r} (-1)^j j!G(-j-2, r-j) &= \sum_{j=0}^{r} \{(-1)^j \Delta^j n^r\}_{n=1} \\ &= \{(1 - \Delta + \Delta^2 - \cdots + (-1)^r \Delta^r) n^r\}_{n=1}. \end{aligned} \tag{3}$$

Since $\Delta^s n^r = 0$ for $s > r$, (3) can be written as

$$\begin{aligned} \sum_{j=0}^{r} (-1)^j j!G(-j-2, r-j) &= \{(1 - \Delta + \Delta^2 - \cdots) n^r\}_{n=1} \\ &= \left\{\frac{1}{1+\Delta} n^r\right\}_{n=1} \\ &= \{(n-1)^r\}_{n=1} = 0. \end{aligned}$$

8. The function $S(n, r)$

We define $S(n, r)$ as the sum of the rth powers of the first n natural numbers, $r \geqslant 1$. To obtain an expression for $S(n, r)$, we need

THEOREM 4. *For all positive integral values of s,*

$$n^s = n^{(s)} + G(-s, 1)n^{(s-1)} + G(-s+1, 2)n^{(s-2)} + \cdots + G(-2, s-1)n^{(1)}.$$

Proof. Let

$$n^s = \sum_{k=1}^{s} c_k n^{(k)}, \tag{1}$$

where the c's are independent of n. Then, since

$$\Delta^j n^{(k)} = k^{(j)} n^{(k-j)}, \tag{2}$$

we get

$$\begin{aligned} \Delta^j n^s &= \Delta^j \sum_{k=1}^{s} c_k n^{(k)} \\ &= \sum_{k=1}^{s} c_k k^{(j)} n^{(k-j)}. \end{aligned} \tag{3}$$

Hence

$$j!G(-j-1, s-j) = \{\Delta^j n^s\}_{n=0} = \left\{\sum_{k=1}^{s} c_k k^{(j)} n^{(k-j)}\right\}_{n=0}. \tag{4}$$

Every term of the sum in the bracket on the right side of (4), except the one with $k = j$, vanishes for $n = 0$. Hence

$$j!G(-j-1, s-j) = c_j j^{(j)} = j!c_j,$$

so that

$$c_j = G(-j-1, s-j). \tag{5}$$

We can now evaluate $S(n, r)$ for $r \geqslant 1$, as follows. We have

$$\begin{aligned} S(n, r) &= \sum_{m=1}^{n} m^r \\ &= \sum_{m=1}^{n} \left\{\sum_{k=1}^{r} G(-k-1, r-k) m^{(k)}\right\} \\ &= \sum_{k=1}^{r} G(-k-1, r-k) \frac{(n+1)^{(k+1)}}{k+1}. \end{aligned} \tag{6}$$

Using the relations

$$(n+1)^{(k+1)} = n^{(k+1)} + (k+1) \,.\, n^{(k)}, \tag{7}$$

and

$$G(-k, r-k+1) + kG(-k-1, r-k) = G(-k-1, r-k+1), \tag{8}$$

we can write (6) in the form

$$S(n, r) = \sum_{k=1}^{r+1} G(-k-1, r-k+1) \frac{n^{(k)}}{k}. \tag{9}$$

This shows that $S(n, r)$ is a polynomial in n of degree $(r + 1)$ with fractional coefficients in general.

Example. We have

$$\begin{aligned} S(n, 2) &= \sum_{k=1}^{3} G(-k-1, 3-k) \,.\, n^{(k)}/k \\ &= G(-2, 2)n + G(-3, 1)n(n-1)/2 + G(-4, 0)n(n-1)(n-2)/3 \\ &= n + 3n(n-1)/2 + n(n-1)(n-2)/3 \\ &= n(n+1)(2n+1)/6. \end{aligned}$$

Exercise

Find the sums of the cubes and the fourth powers of the first n natural numbers.

9. Bernoulli numbers

As already noted in section 8, $S(n, r)$ is a polynomial in n of degree $(r + 1)$. We can, therefore, write

$$S(n,r) = \left\{ B_0 + \frac{B_1}{1!} D + \frac{B_2}{2!} D^2 + \cdots + \frac{B_r}{r!} D^r \right\} \frac{n^{r+1}}{r+1}, \tag{1}$$

where D stands for the differential operator $\dfrac{d}{dn}$ and the B's (called the Bernoulli numbers) are certain constants independent of n and not necessarily integers, not necessarily positive either. This result can be written symbolically in the form

$$S(n, r) \triangleq \left\{ e^{BD} \frac{m^{r+1}}{r+1} \right\}_0^n,$$

where $D = \dfrac{d}{dm}$, and it is understood that the powers of B in the expansion of e^{BD} are to be written as subscripts of B. The symbol $\{f(m)\}_0^n$ stands, as usual, for $f(n) - f(0)$.

Comparing the coefficients of n in the two expressions for $S(n, r)$ obtained in this section and the preceding one, we get

$$B_r = G(-2, r) - \frac{1!}{2} G(-3, r-1) + \frac{2!}{3} G(-4, r-2) - \cdots$$
$$+ (-1)^r \frac{r!}{r+1} G(-r-2, 0). \tag{2}$$

This formula will be of use to us when we discuss properties of Bernoulli numbers in the next chapter. The first few Bernoulli numbers are

$$B_0 = 1, \quad B_1 = \tfrac{1}{2}, \quad B_2 = \tfrac{1}{6}, \quad B_3 = 0.$$

Exercises

1. Calculate the values of B_4, B_5 and B_6.
2. Instead of comparing the coefficients of n in (8.9) and (9.2) compare the coefficients of $n^2, n^3, \ldots$ in these expressions and calculate the first few Bernoulli numbers.
3. Show that for $r \geqslant 1$,

$$S(n, r) = B_r n + r \,.\int_0^1 S(m, r-1)\, dm$$

where $S(n, 0) = n$. Using the fact that $S(1, r) = 1$ for each $r \geqslant 0$, calculate the value of B_r for $r \leqslant 6$.

10. The expressions for $H(r, j)$ and $G(n, r)$

The H's in (3.7) are not easy to determine except for the first few values of r. We therefore proceed to obtain a compact formula for $H(r, j)$. In its turn, this will lead to a compact expression for $G(n, r)$.

For $r \geqslant 1$, we have

$$G(n, r) = \sum_{j=0}^{r-1} (-1)^j H(r, j) \binom{n + r - j}{2r - j}. \tag{1}$$

Replacing n by $(-n - 1)$ and j by $(r - j)$ in (1), we get

$$\begin{aligned} G(-n-1, r) &= \sum_{j=1}^{r-1} (-1)^{r-j} H(r, r-j) \binom{-n+j-1}{r+j}, \quad n \geqslant 1, \\ &= \sum_{j=1}^{r} (-1)^{r-j} H(r, r-j)(-1)^{r+j} \binom{n+r}{r+j} \\ &= \sum_{j=1}^{r} H(r, r-j) \binom{n+r}{n-j}. \end{aligned} \tag{2}$$

Replacing n by $1, 2, 3, \ldots, j$ in succession in (2), we get

$$\begin{aligned} G(-2, r) &= \binom{r+1}{0} H(r, r-1) \\ G(-3, r) &= \binom{r+2}{1} H(r, r-1) + \binom{r+2}{0} H(r, r-2) \\ G(-4, r) &= \binom{r+3}{2} H(r, r-1) + \binom{r+1}{1} H(r, r-2) + \binom{r+3}{0} H(r, r-3) \\ &\vdots \\ G(-j-1, r) &= \binom{r+j}{j-1} H(r, r-1) + \binom{r+j}{j-2} H(r, r-2) + \cdots + \binom{r+j}{0} H(r, r-j). \end{aligned}$$

Hence

$$\begin{aligned} H(r, r-1) &= G(-2, r) \\ H(r, r-2) &= G(-3, r) - \binom{r+2}{1} G(-2, r) \\ H(r, r-3) &= G(-4, r) - \binom{r+3}{1} G(-3, r) + \binom{r+3}{2} G(-2, r), \end{aligned}$$

and generally, by induction,

$$H(r, r-j) = \sum_{k=1}^{j} (-1)^{j-k} \binom{r+j}{j-k} G(-k-1, r), \quad 1 \leqslant j \leqslant r. \tag{3}$$

THEOREM 5. *For* $1 \leqslant r \leqslant n$,

$$G(n, r) = (2r+1)\binom{n+r+1}{2r+1}\sum_{k=1}^{r}(-1)^{r-k}\binom{2r}{r-k}G(-k-1, r)/(n+k+1).$$

Proof. We have

$$\begin{aligned} G(n, r) &= \sum_{j=1}^{r}(-1)^{r-j}H(r, r-j)\binom{n+j}{r+j} \\ &= \sum_{j=1}^{r}(-1)^{r-j}\binom{n+j}{r+j}\sum_{k=1}^{j}(-1)^{j-k}\binom{r+j}{j-k}G(-k-1, r) \\ &= \sum_{j=1}^{r}\sum_{k=1}^{j}(-1)^{r-k}\binom{n+j}{r+j}\binom{r+j}{r+k}G(-k-1, r). \end{aligned} \tag{4}$$

Now since

$$\binom{n+j}{r+j}\binom{r+j}{r+k} = \binom{n+j}{n+k}\binom{n+k}{r+k}$$

and

$$\binom{n+j}{n+k} = 0 \quad \text{for} \quad 1 \leqslant j < k,$$

we deduce, from (4), that

$$\begin{aligned} G(n, r) &= \sum_{j=1}^{r}\sum_{k=1}^{r}\left\{(-1)^{r-k}\binom{n+k}{r+k}\binom{n+j}{n+k}G(-k-1, r)\right\} \\ &= \sum_{k=1}^{r}\left\{(-1)^{r-k}\binom{n+k}{r+k}G(-k-1, r)\sum_{j=1}^{r}\binom{n+j}{n+k}\right\} \\ &= \sum_{k=1}^{r}\left\{(-1)^{r-k}\binom{n+k}{r+k}\binom{n+r+1}{n+k+1}G(-k-1, r)\right\} \\ &= (2r+1)\binom{n+r+1}{2r+1}\sum_{k=1}^{r}\left\{(-1)^{r-k}\binom{2r}{r-k}G(-k-1, r)/(n+k+1)\right\}. \end{aligned}$$

This answers a question of M. Ward, who had been looking for such an expression for $G(n, r)$.

Exercises

1. Use theorem 5 to obtain expressions for $G(n, r)$, $r \leqslant 4$.
2. Find expressions for $H(r, r-j)$, $j \leqslant 5$.
3. Show that for $1 \leqslant j \leqslant r$,

$$H(r, r-j) = (r+j-1)H(r-1, r-j) + jH(r-1, r-j-1).$$

11. Addition formula for $G(n, r)$

We proceed to prove the important

THEOREM 6. *For all integral values of j and k,*

$$G(j+k, r) = \sum_{t=0}^{r} \theta(t) \binom{j+k-t}{r-t} j^{r-t}, \quad r \geqslant 0$$

where

$$\theta(t) = \sum_{s=0}^{t} (-1)^{t-s} G(k, s) G(j-1, t-s),$$

and

$$\theta(0) = 1.$$

Proof. When both j and k are positive, we consider the product

$$P = (x+1)(x+2)(x+3)\ldots(x+j)(x+j+1)\ldots(x+j+k)$$
$$= \sum_{r=0}^{j+k} G(j+k, r) x^{j+k-r}. \tag{1}$$

Let $y = x + j$, then we have

$$P = (y-j+1)(y-j+2)\ldots(y-1)y\,.\,(y+1)(y+2)\ldots(y+k)$$
$$= \sum_{u=0}^{j-1} (-1)^u G(j-1, u) y^{j-u} \,.\, \sum_{v=0}^{k} G(k, v) y^{k-v}$$
$$= \sum_{t=0}^{j+k-1} \theta(t) y^{j+k-t} = \sum_{t=0}^{j+k-1} \theta(t)(x+j)^{j+k-t}, \tag{2}$$

where $\theta(t)$ has the same meaning as stated in the theorem. Comparing the coefficients of x^{j+k-r} in (1) and (2), we have

$$G(j+k, r) = \sum_{t=0}^{r} \theta(t) \binom{j+k-t}{r-t} j^{r-t}. \tag{3}$$

When j is positive and k negative, we have

$$\sum_{v=0}^{\infty} G(k, v) y^{k-v} = 1/y(y-1)(y-2)\ldots(y+k+1)$$

while

$$\sum_{u=0}^{j-1} (-1)^u G(j-1, u) y^{j-u} = (y-j+1)(y-j+2)\ldots(y-1)y,$$

so that

$$\sum_{u=0}^{j-1} (-1)^u G(j-1, u) y^{j-u} \cdot \sum_{v=0}^{\infty} G(k, v) y^{k-v}$$

$$= \begin{cases} (y-j+1)(y-j+2)\ldots(y+k) & \text{if} \quad -k<j, \\ 1 & \text{if} \quad -k=j, \\ 1/(y+k+1)(y+k+2)\ldots(y-j) & \text{if} \quad -k>j, \end{cases}$$

$$= \begin{cases} (x+1)(x+2)\ldots(x+j+k) & \text{if} \quad -k<j, \\ 1 & \text{if} \quad -k=j, \\ 1/(x+j+k+1)(x+j+k+2)\ldots x & \text{if} \quad -k>j; \end{cases}$$

and the theorem follows as before. Other cases can be dealt with in a similar way.

12. Further extension of the definition of G-functions

We could have defined the functions $G(n, r)$ and $G(-n-1, r)$ by the relations

$$(1+x)(1+2x)(1+3x)\ldots(1+nx) = \sum_{r=0}^{n} G(n, r)x^r \tag{1}$$

and

$$1/(1+x)(1+2x)(1+3x)\ldots(1+nx) = \sum_{r=0}^{\infty} G(-n-1, r)(-x)^r, \tag{2}$$

which hold for $|x| < 1/n$.

In this form, a direct relationship between the two functions becomes apparent. In fact, we have

$$\sum_{r=0}^{n} G(n, r)x^r \cdot \sum_{s=0}^{\infty} G(-n-1, s)(-x)^s = 1; \tag{3}$$

which means that the coefficient of x^t in the product on the left is zero for each $t > 0$. For example,

$$G(4, 0)G(-5, 4) - G(4, 1)G(-5, 3) + G(4, 2)G(-5, 2) - \cdots$$
$$+ G(4, 4)G(-5, 0)$$
$$= 1701 - 3500 + 2275 - 500 + 24$$
$$= 0.$$

On the same lines, one could take a set of nonnegative integers

$$A_k = \{a_1, a_2, a_3, \ldots, a_k\},$$

where

$$a_j \leqslant a_{j+1}, \quad j = 1, 2, \ldots, k-1,$$

and define the functions $G(A_k, r)$ and $G^*(A_k, r)$ by the relations

$$(1 + a_1 x)(1 + a_2 x) \ldots (1 + a_k x) = \sum_{r=0}^{k} G(A_k, r) x^r, \tag{4}$$

$$1/(1 + a_1 x)(1 + a_2 x) \ldots (1 + a_k x) = \sum_{r=0}^{\infty} G^*(A_k, r)(-x)^r. \tag{5}$$

These relations hold for $|x| < 1/a_k$. The two functions G, G^* are not only generalizations of the Stirling numbers of the first and second kind but they also include as special cases the binomial coefficients.

Exercise

Find $U(m, a)$, the number of ways in which m persons can be accommodated in a like compartments, no compartment being left vacant. Identify these numbers.

Appendix

The following is a short table of values of

$$G(n, r), -5 \leqslant n \leqslant 5, -4 \leqslant r \leqslant 4.$$

						r						
	1701	301	31	1	0	4	0	0	0	0	24	274
	350	90	15	1	0	3	0	0	0	6	50	225
	65	25	7	1	0	2	0	0	2	11	35	85
	10	6	3	1	0	1	0	1	3	6	10	15
	1	1	1	1	1	0	1	1	1	1	1	1
n	−5	−4	−3	−2	−1		0	1	2	3	4	5
	0	0	0	1	0	−1						
	0	0	−1/2	1	0	−2						
	0	1/6	−3/4	1	0	−3						
	−1/24	11/36	−7/8	1	0	−4						

CHAPTER 6

CONGRUENCE PROPERTIES OF SYMMETRIC FUNCTIONS

1. Notations

In this chapter, the letter n will denote an integer >2. We will continue to denote by p a prime $\geqslant 2$. Other small letters will, in general, denote nonnegative integers.

2. Generalizations of some known theorems

In the preceding chapter we have proved that, for *all* integral values of j and k,

$$G(j+k, r) = \sum_{t=0}^{r} \theta(t) \binom{j+k-t}{r-t} j^{r-t}, \quad r \geqslant 0,$$

where

$$\theta(t) = \sum_{s=0}^{t} (-1)^{t-s} G(k, s) G(j-1, t-s), \quad t \geqslant 0.$$

We utilize this result to obtain generalizations of several theorems in number theory.

THEOREM 1. *For* $r = 1, 2, 3, \ldots, p-2$, where $p \geqslant 3$,

$$G(p^u - 1, r) \equiv 0 \pmod{p^u}.$$

Proof. Substituting 1 for j in the addition formula above, we get

$$G(k+1, r) = \sum_{t=0}^{r} \theta(t) \binom{k-t+1}{r-t}, \tag{1}$$

with

$$\theta(t) = G(k, t). \tag{2}$$

Since

$$G(k+1, r) = G(k, r) + (k+1) G(k, r-1),$$

(1) gives after a little simplification

$$(r-1)G(k, r-1) = \binom{k-r+3}{2} G(k, r-2) + \binom{k-r+4}{3} G(k, r-3) + \cdots + \binom{k+1}{r}. \tag{3}$$

In (3), replace k by $p^u - 1$ and r by $2, 3, 4, \ldots, p-1$ in succession. Then, we have

$$G(p^u - 1, 1) = \binom{p^u}{2} \equiv 0 \pmod{p^u}, \quad p > 2,$$

$$2G(p^u - 1, 2) = \binom{p^u - 1}{2} G(p^u - 1, 1) + \binom{p^u}{3}, \quad p > 3,$$

so that

$$G(p^u - 1, 2) \equiv 0 \pmod{p^u}, \quad p > 3,$$

and finally

$$(p-2)G(p^u - 1, p-2) = \binom{p^u - p + 3}{2} G(p^u - 1, p-3) + \cdots + \binom{p^u}{p-1}.$$

Thus, for $p \geqslant 3$,

$$G(p^u - 1, p-2) \equiv 0 \pmod{p^u}.$$

This is a generalization of Lagrange's Theorem: *For* $1 \leqslant r \leqslant p-2$,

$$G(p-1, r) \equiv 0 \pmod{p}. \tag{4}$$

THEOREM 2.

$$G(p^u - 2, p-2) \equiv 1 \pmod{p^u}.$$

Proof. In the addition formula, take $j = p^u$ and $k = -2$. Then for $r = p-2$, utilizing theorem 1, we get

$$\begin{aligned} G(p^u - 2, p-2) &\equiv \theta(p-2) \pmod{p^u} \\ &\equiv G(-2, p-2) \pmod{p^u} \\ &\equiv 1 \pmod{p^u}. \end{aligned}$$

This is a generalization of
Wilson's Theorem:

$$(p-1)! \equiv -1 \pmod{p}.$$

Proof. Take $u = 1$ in the theorem, then

$$G(p-2, p-2) \equiv 1 \pmod{p}$$

or

$$(p-2)! \equiv 1 \pmod p$$

so that

$$(p-1)! \equiv (p-1) \equiv -1 \pmod p.$$

THEOREM 3. *For* $1 \leqslant r \leqslant \dfrac{p-3}{2}$, $p \geqslant 5$,

$$G(p^u-1, 2r+1) \equiv 0 \pmod{p^{2u}}.$$

Proof. In the addition theorem, let $j = p^u$, $k = -1$, then

$$G(p^u-1, 2r+1) \equiv \theta(2r+1) + \theta(2r)\binom{p^u-2r-1}{1} p^u \pmod{p^{2u}}$$

$$\equiv -G(p^u-1, 2r+1) + G(p^u-1, 2r)\binom{p^u-2r-1}{1} p^u \pmod{p^{2u}}.$$

Hence from theorem 1,

$$2G(p^u-1, 2r+1) \equiv 0 \pmod{p^{2u}}$$

or

$$G(p^u-1, 2r+1) \equiv 0 \pmod{p^{2u}}.$$

This is a generalization of
Wolstenholme's Theorem: *For* $p \geqslant 5$,

$$G(p-1, p-2) \equiv 0 \pmod{p^2}.$$

THEOREM 4. *For* $p \geqslant 3$,

$$G(p+k, r) \equiv \begin{cases} G(k, r) \pmod p, & r = 1, 2, 3, \ldots, p-2, \\ G(k, r) - G(k, r-p+1) \pmod p, & r \geqslant p-1. \end{cases}$$

Proof. In the addition formula, put p for j, then since

$$G(p-1, r) \equiv \begin{cases} 0 \pmod p, & r = 1, 2, 3, \ldots, p-2, \quad \text{by theorem 1}, \\ -1 \pmod p, & r = p-1, \quad \text{by Wilson's Theorem}, \end{cases}$$

we have

$$G(p+k, r) \equiv \theta(r) \pmod p,$$

where

$$\theta(r) \equiv G(k, r) \pmod p \quad \text{when} \quad 1 \leqslant r \leqslant p-2,$$

and

$$\equiv G(k,r) + G(k,r-p+1)G(p-1,p-1) \pmod p, \quad r \geqslant p-1,$$
$$\equiv G(k,r) - G(k,r-p+1) \pmod p \quad r \geqslant p-1.$$

In particular, replacing k by $-k-1$ and r by $p-k$ in theorem 4, we obtain

$$G(-k-1,p-k) \equiv G(p-k-1,p-k) \pmod p, \quad 2 \leqslant k \leqslant p-1,$$
$$\equiv 0 \pmod p,$$

or

$$(k-1)!G(-k-1,p-k) \equiv 0 \pmod p. \tag{5}$$

We leave it to the reader, to derive from (5)
Fermat's Theorem: *If* $(a,p) = 1$, *then*

$$a^{p-1} \equiv 1 \pmod p, \quad p \geqslant 2. \tag{6}$$

The following theorems are a direct consequence of theorem 4.

THEOREM 5. *For* $p \geqslant 2$,

$$G(-p,r) \equiv 0 \quad \text{or} \quad 1 \pmod p$$

according as $(p-1)$ *does not or does divide* r.

Proof. In theorem 4, replace k by $-p$, then

$$G(-p,r) \equiv \begin{cases} G(0,r) \pmod p, & 1 \leqslant r \leqslant p-2, \\ G(0,r) + G(-p,r-p+1) \pmod p, & r \geqslant p-1, \end{cases}$$

that is

$$G(-p,r) \equiv 0 \pmod p, \quad 1 \leqslant r \leqslant p-2,$$

and

$$G(-p,r) \equiv G(-p,r-p+1) \pmod p, \quad r \geqslant p-1. \tag{7}$$

If $r \equiv s \pmod{(p-1)}$, $0 \leqslant s < p-1$, we get

$$G(-p,r) \equiv G(-p,s) \pmod p$$
$$\equiv 0 \pmod p, \quad s \neq 0,$$

and

$$\equiv 1 \pmod p, \quad s = 0.$$

THEOREM 6. *If n and m are any positive integers, then for* $p \geqslant 2$,

$$G(np-1, m(p-1)) \equiv (-1)^m \binom{n}{m} \pmod p.$$

Proof. Applying the reduction

$$G(p+k,r) \equiv G(k,r) - G(k,r-p+1) \pmod p$$

h times in succession to $G(np-1, m(p-1))$, we get

$$G(np-1, m(p-1)) \equiv \sum_{t=0}^{h} (-1)^t \binom{h}{t} G((n-h)p-1, (m-t)(p-1)) \pmod p.$$

For $h = n$, we obtain

$$G(np-1, m(p-1)) \equiv \sum_{t=0}^{n} (-1)^t \binom{n}{t} G(-1, (m-t)(p-1)) \pmod p,$$

$$\equiv 0 \pmod p \quad \text{if} \quad n < m; \quad [G(-1,r) = 0, r \geqslant 1],$$

and

$$\equiv (-1)^m \binom{n}{m} \pmod p \quad \text{if} \quad n \geqslant m; \quad [G(-1,0) = 1].$$

Since

$$\binom{n}{m} = 0 \quad \text{for} \quad m > n > 0;$$

we can write

$$G(np-1, m(p-1)) \equiv (-1)^m \binom{n}{m} \pmod p.$$

This is another generalization of Wilson's Theorem.

As corollaries of theorem 6, we have

$$G(p^u - 1, \phi(p^u)) \equiv -1 \pmod p, \quad p \geqslant 2, \quad u \geqslant 1,$$
$$G(p^u - 1, m\phi(p)) \equiv 0 \pmod p, \quad 0 < m < p^{u-1}, \quad u > 1. \tag{8}$$

Exercises

1. Apply the reduction of theorem 4 to show that

$$G(np-1, r) \equiv 0 \pmod p \quad \text{if} \quad (p-1) \nmid r.$$

2. Show that, for $u \geqslant 1$,

$$\operatorname{pot}_p G(p^u - 1, \phi(p^u)) = 0.$$

3. A theorem of Gauss

In this section, we shall prove a theorem of Gauss. We shall denote by $P(t)$ the product of numbers less than t and prime to it, when $t \geqslant 3$. We shall take

$$P(1) = 1 = P(2).$$

For our proof, we shall need the

LEMMA. *If* $(m, n) = 1$, *then*

$$P(mn) \equiv \{P(n)\}^{\phi(m)} (\text{mod } n).$$

Proof. If $b <' n$, then in the series of m terms given by

$$b, b + n, b + 2n, \ldots, b + (m - 1)n$$

there are exactly $\phi(m)$ numbers less than and prime to mn. Since each of these numbers is congruent to b modulo n, their product is congruent to $b^{\phi(m)}$ modulo n. Letting b run through all numbers less than and prime to n, the lemma follows immediately.

THEOREM 7 *(Gauss).*

$$P(n) \equiv \begin{cases} -1 \,(\text{mod } n) & \text{if} \quad n = 2, 2^2, p^u, 2p^u; \quad p \geqslant 3, u \geqslant 1; \\ 1 \,(\text{mod } n) & \text{otherwise.} \end{cases}$$

Proof. First we take the case when n is a power of 2. The theorem evidently holds for $n = 2, 2^2$. For $n = 2^u$, $u > 2$, we use induction. We have

$$P(2^3) = 1 \,.\, 3 \,.\, 5 \,.\, 7 \equiv 1 \,(\text{mod } 2^3).$$

Suppose

$$P(2^q) \equiv 1 \,(\text{mod } 2^q), \quad q = 3, 4, 5, \ldots, u - 1.$$

Then

$$\begin{aligned} P(2^u) &\equiv \{1 \,.\, 3 \,.\, 5 \,.\, \cdots \,.\, (2^{u-1} - 1)\} \{(2^u - 1)(2^u - 3)(2^u - 5) \cdots \\ &\qquad \cdots (2^u - 2^{u-1} + 1)\} \\ &\equiv \{1 \,.\, 3 \,.\, 5 \,.\, \cdots \,.\, (2^{u-1} - 1)\} \{(-1)(-3)(-5) \cdots (-2^{u-1} + 1)\} (\text{mod } 2^u) \\ &\equiv (-1)^{2^{u-2}} \{P(2^{u-1})\}^2 (\text{mod } 2^u), \quad u > 3, \\ &\equiv (2^{u-1} h + 1)^2 (\text{mod } 2^u) \text{ for some integer } h \\ &\equiv 1 \,(\text{mod } 2^u). \end{aligned}$$

This proves the theorem for powers of 2.

Secondly, let

$$n = p^u, \quad p \geqslant 3, \quad u \geqslant 1.$$

For $u = 1$,

$$P(p) = (p - 1)! \equiv -1 \,(\text{mod } p).$$

Suppose

$$P(p^q) \equiv -1 \,(\text{mod } p^q) \quad \text{for} \quad q = 1, 2, 3, \ldots, u - 1.$$

Let a be any integer less than and prime to p^{u-1}. Then

$$\begin{aligned} &a(a+p^{u-1})(a+2p^{u-1})\ldots(a+(p-1)p^{u-1}),\\ &\quad=\sum_{r=0}^{p-1} G(p-1,r)a^{p-r}p^{(u-1)r}\\ &\quad\equiv a^p \pmod{p^u}, \end{aligned} \tag{1}$$

by theorem 1.

Letting a run through all numbers less than and prime to p^{u-1}, and multiplying the resulting congruences, we get, for some integer h,

$$\begin{aligned} P(p^u) &\equiv \{P(p^{u-1})\}^p \pmod{p^u}\\ &\equiv (p^{u-1}h-1)^p \pmod{p^u}\\ &\equiv -1 \pmod{p^u}. \end{aligned}$$

The theorem follows by induction for all powers of p.

Next let

$$n = 2p^u, \quad p \geqslant 3, \quad u \geqslant 1.$$

Then, by the lemma, we have

$$\begin{aligned} P(2p^u) &\equiv \{P(p^u)\}^{\phi(2)} \pmod{p^u}\\ &\equiv -1 \pmod{p^u}. \end{aligned} \tag{2}$$

Also

$$\begin{aligned} P(2p^u) &\equiv \{P(2)\}^{\phi(p^u)} \pmod{2}\\ &\equiv 1 \equiv -1 \pmod{2}. \end{aligned} \tag{3}$$

Hence

$$P(2p^u) \equiv -1 \pmod{2p^u}.$$

Lastly, let $n = p^u m$, where $(m, p) = 1, m > 2, p \geqslant 2$. Then, since $\phi(m)$ is even for $m > 2$, we have

$$\begin{aligned} P(p^u m) &\equiv \{P(p^u)\}^{\phi(m)} \pmod{p^u}\\ &\equiv 1 \pmod{p^u}. \end{aligned}$$

Thus, if

$$n = p_1^{u_1} p_2^{u_2} p_3^{u_3} \ldots p_q^{u_q},$$

where p's are distinct primes and $p_2^{u_2} \ldots p_q^{u_q} > 2$, we shall have

$$\begin{aligned} P(n) &\equiv 1 \pmod{p_s^{u_s}}, \quad s = 1, 2, 3, \ldots, q,\\ &\equiv 1 \pmod{n}. \end{aligned}$$

4. Coefficients in $G(n,r)$

In the preceding chapter, we had expressed $G(n,r)$ in terms of combinatory functions and then as a polynomial of degree $2r$ in n. Here, we prove some interesting properties of the coefficients involved.

THEOREM 8. *If $p \geqslant 2$ and $0 < \alpha \leqslant u$, then*
$\text{pot}_p\, G(p^u - 1, \phi(p^\alpha)) = u - \alpha.$

Proof. From section 3 of the last chapter, we have for $r \geqslant 1$,

$$G(n,r) = \sum_{j=1}^{r} F(r,j)\binom{n+1}{r+j} \tag{1}$$

where F's are positive integers independent of n. Hence

$$G(p^u - 1, \phi(p^\alpha)) = \sum_{j=1}^{\phi(p^\alpha)} F(\phi(p^\alpha), j)\binom{p^u}{\phi(p^\alpha)+j}. \tag{2}$$

The p-potency of $G(p^u - 1, \phi(p^\alpha))$ will be definitely known if from the terms on the right side of (2), we can single out one with a potency less than that of any other of the sum.

We assert that such a term is the one with

$$j = p^{\alpha-1}. \tag{3}$$

In view of (2.8), we have for $\alpha > 0$,

$$\text{pot}_p\, G(p^\alpha - 1, \phi(p^\alpha)) = 0, \tag{4}$$

that is

$$\text{pot}_p \sum_{j=1}^{\phi(p^\alpha)} F(\phi(p^\alpha), j)\binom{p^\alpha}{\phi(p^\alpha)+j} = 0.$$

Since every term in the bracket excepting the one with

$$\phi(p^\alpha) + j = p^\alpha$$

is necessarily divisible by p, we must have

$$\text{pot}_p\, F(\phi(p^\alpha), j) = \text{pot}_p\, F(\phi(p^\alpha), p^{\alpha-1}) = 0. \tag{5}$$

Hence

$$\begin{aligned}\text{pot}_p\, G(p^u - 1, \phi(p^\alpha)) &= \text{pot}_p \left\{F(\phi(p^\alpha), p^{\alpha-1})\binom{p^u}{p^\alpha}\right\}\\ &= u - \alpha.\end{aligned}$$

Remark

For values of j in the range 1 to $\phi(p^\alpha)$, $\binom{p^u}{p^\alpha}$ has less p-potency than $\binom{p^u}{\phi(p^\alpha)+j}$

for any j other than $p^{\alpha-1}$.

THEOREM 9. *In the expression*

$$\frac{(n+1)^{(r+1)}}{(2r)!}\{a_1 n^{r-1} + a_2 n^{r-2} + a_3 n^{r-3} + \cdots + a_r\}$$

for $G(n, r)$, $r > 0$, the a's are divisible by each of the primes p greater than $(r+1)$ but not exceeding $2r$.

Proof. Since $r + 1 < p < 2r$, we must have

$$0 < r \leqslant p - 2.$$

Hence, by theorem 4,

$$G(p + j, r) \equiv G(j, r) (\text{mod } p).$$

Since

$$G(j, r) = 0, \quad -1 \leqslant j \leqslant r - 1,$$

we have

$$G(p + j, r) \equiv 0 \,(\text{mod } p), \quad -1 \leqslant j \leqslant r - 1.$$

Again since $2r > p$,

$$(2r)! \equiv 0 \,(\text{mod } p).$$

Hence

$$(2r)! G(p + j, r) \equiv 0 \,(\text{mod } p^2) \quad \text{for} \quad -1 \leqslant j \leqslant r - 1,$$

so that

$$(p + j + 1)^{(r+1)}\{a_1 (p + j)^{r-1} + a_2 (p + j)^{r-2} + \cdots + a_r\}$$
$$\equiv 0 \,(\text{mod } p^2) \quad \text{for} \quad -1 \leqslant j \leqslant r - 1.$$

Now, for $-1 \leqslant j \leqslant r - 1 \leqslant p - 3$

$$(p + j + 1)^{(r+1)} \equiv 0 \,(\text{mod } p), \quad \text{but} \not\equiv 0 \,(\text{mod } p^2)$$

we must therefore have, for $j = -1, 0, 1, 2, 3, \ldots, r - 1$,

$$a_1 j^{r-1} + a_2 j^{r-2} + a_3 j^{r-3} + \cdots + a_r \equiv 0 \,(\text{mod } p).$$

Since $p > r + 1$, the numbers $-1, 0, 1, 2, \ldots, r - 1$ are incongruent modulo p. This means that the congruence

$$a_1 x^{r-1} + a_2 x^{r-2} + a_3 x^{r-3} + \cdots + a_r \equiv 0 \,(\text{mod } p) \tag{6}$$

has more than $(r - 1)$ incongruent roots. This is possible only if

$$a_k \equiv 0 \,(\text{mod } p), \quad k = 1, 2, 3, \ldots, r.$$

Exercise
Where does the argument fail when $r + 1 = p$?

THEOREM 10. *For odd values of* $r \geqslant 3$,

$$(2r)!G(n, r) \equiv 0 \pmod{n^2(n + 1)^2}.$$

Proof. Let p be an odd prime $\geqslant q$, where

$$q = \text{Max}\{2r, |a_r|, |a_1 - a_2 + a_3 - \cdots + a_r|\}.$$

Then, since r is odd and $\geqslant 3$,

$$G(p - 1, r) \equiv 0 \pmod{p^2}. \tag{7}$$

Also

$$(2r)! \not\equiv 0 \pmod{p}. \tag{8}$$

Therefore, we must have

$$p^{(r+1)}\{a_1(p-1)^{r-1} + a_2(p-1)^{r-2} + a_3(p-1)^{r-3} + \cdots + a_r\} \equiv 0 \pmod{p^2}. \tag{9}$$

However

$$p^{(r+1)} \equiv 0 \pmod{p}, \quad \text{but} \not\equiv 0 \pmod{p^2};$$

therefore from (9), we have

$$a_1 - a_2 + a_3 - \cdots + a_r \equiv 0 \pmod{p} \tag{10}$$

for every prime $p > q$.

Since there are infinitely many primes $>q$, (10) can hold only if

$$a_1 - a_2 + a_3 - \cdots + a_r = 0. \tag{11}$$

Hence $(n + 1)$ must be a factor of the expression

$$a_1 n^{r-1} + a_2 n^{r-2} + a_3 n^{r-3} + \cdots + a_r.$$

This proves the theorem so far as the factor $(n + 1)^2$ in the modulus is concerned.
Again, let p be any prime $>q$ as before. Then, since

$$G(p - 1, r - 1) \equiv 0 \pmod{p} \quad \text{for} \quad r \geqslant 3,$$

and

$$G(p, r) = G(p - 1, r) + pG(p - 1, r - 1),$$

we have

$$G(p, r) \equiv 0 \pmod{p^2} \quad \text{for} \quad r \text{ odd and } \geqslant 3.$$

Thus, for r odd and $\geqslant 3$,

$$(2r)!G(p, r) \equiv 0 \pmod{p^2}$$

so that

$$(p+1)^{(r+1)}\{a_1p^{r-1}+a_2p^{r-2}+a_3p^{r-3}+\cdots+a_r\}\equiv 0\ (\mathrm{mod}\ p^2). \tag{12}$$

Now

$$(p+1)^{(r+1)}\equiv 0\ (\mathrm{mod}\ p),\quad \text{but} \not\equiv 0\ (\mathrm{mod}\ p^2),$$

therefore from (12), we get

$$a_r\equiv 0\ (\mathrm{mod}\ p). \tag{13}$$

As before we may conclude that $a_r = 0$. This completes the proof of the theorem.

5. A congruence for $S(n-1, 2j)$

In what follows, we write

$$N_{2j}=\prod_{(p-1)|2j} p,\quad p\geqslant 2.$$

For example, $N_6 = 2\,.\,3\,.\,7 = 42$.

THEOREM 11. *For $j > 0$,*

$$(n, N_{2j})S(n-1, 2j)\equiv 0\ (\mathrm{mod}\ n).$$

Proof. We first state two lemmas.

LEMMA. *Unless k is 4 or is a prime $p \geqslant 2$,*

$$k \mid (k-1)!.$$

The proof is easy and is left to the reader.

LEMMA. *For any n and k, we have*

$$\frac{n^{(k)}}{k}\equiv\begin{cases}0\left(\mathrm{mod}\ \dfrac{n}{2}\right) & \text{if}\quad k=2,4 \text{ and } k\mid n,\\[2ex] 0\left(\mathrm{mod}\ \dfrac{n}{p}\right) & \text{if}\quad k \text{ is a prime } p\geqslant 3 \text{ and } p\mid n,\\[2ex] 0\ (\mathrm{mod}\ n) & \text{otherwise.}\end{cases}$$

Proof. First, if $k\neq 2, 4, p; p\geqslant 3$, then

$$(k-1)! \mid (n-1)^{(k-1)}$$

and, by the previous lemma,

$$k \mid (k-1)!$$

Therefore

$$k \mid (n-1)^{(k-1)},$$

so that

$$\frac{n^{(k)}}{k} = n \,.\, \frac{(n-1)^{(k-1)}}{k} \equiv 0 \pmod{n}.$$

Next, if $k = 4$ and $4 \nmid n$, then

$$(n-1)^{(3)} = (n-1)(n-2)(n-3) \equiv 0 \pmod{4};$$

for if n is odd then $(n-1)$ and $(n-3)$ are even, while if n is even but not a multiple of 4, then $4 \mid (n-2)$. Hence

$$\frac{n^{(4)}}{k} = n\,\frac{(n-1)^{(3)}}{4} \equiv 0 \pmod{n}.$$

Again, if k is a prime $p \geqslant 2$ and $p \nmid n$, then

$$(n-1)^{(p-1)} = (n-1)(n-2)(n-3)\ldots(n-p+1) \equiv 0 \pmod{p},$$

because one of the $(p-1)$ factors on the right must be divisible by p. Hence

$$\frac{n^{(p)}}{p} = n\,\frac{(n-1)^{(p-1)}}{p} \equiv 0 \pmod{n}.$$

Also, if $k = 4$ and $4 \mid n$, then

$$\frac{n^{(4)}}{4} = \frac{n}{2} \cdot \frac{(n-1)(n-2)(n-3)}{2} \equiv 0 \left(\bmod \frac{n}{2}\right)$$

but $\qquad\qquad \not\equiv 0 \pmod{n}.$

Finally if $k = p$ and $p \mid n, p \geqslant 2$, then none of the numbers $(n-1), (n-2), (n-3), \ldots, (n-p+1)$ is divisible by p. Hence

$$\frac{n^{(p)}}{p} \equiv 0 \left(\bmod \frac{n}{p}\right) \quad \text{but} \quad \not\equiv 0 \pmod{n}.$$

Proof of theorem 11. We have from (6) of section 8 of the preceding chapter

$$S(n-1, 2j) = \sum_{k=2}^{2j+1} G(-k, 2j-k+1)\,\frac{n^{(k)}}{k}.$$

Now

$$\frac{n^{(k)}}{k} \equiv 0 \pmod{n}$$

except when $k \mid n$ and $k = 4$ or is a prime $p \geqslant 2$. When $k = p$,

$$G(-k, 2j-k+1) \equiv 0 \pmod{p},$$

unless $(p-1) \mid 2j$. When $(p-1) \mid 2j, k \mid n$ and $k = 2, 4$ or a prime $p \geqslant 3$,

$$G(-k, 2j-k+1)\frac{n^{(k)}}{k} \equiv 0\left(\operatorname{mod}\frac{n}{p}\right).$$

Hence the theorem.

6. Some definitions

If

$$A_k = \{a_1, a_2, a_3, \ldots, a_k\},$$

then we write

$$tA_k = \{ta_1, ta_2, ta_3, \ldots, ta_k\}.$$

Suppose there exists a set $T = \{t : t \not\equiv 1 \pmod{n}\}$ such that for each $t \in T$, the members of tA_k are congruent modulo n to the members of set A_k, though not necessarily in the same order. Then the set A_k is said to be self-contained modulo n, with respect to the set T. As examples of sets self-contained modulo n, we may mention the complete residue set modulo n, the reduced residue set modulo n, the set of quadratic residues of n and many others. We leave it to the reader to give more examples of such sets.

If the members of a set A_k satisfy the relations

$$a_1 + a_k = a_2 + a_{k-1} = a_3 + a_{k-2} = \cdots = a_r + a_{k-r+1} = \cdots = n,$$

then the set A_k is said to be a balanced n-set. As examples of such sets, we can mention the sets

$$0, 1, 2, 3, \ldots, n;$$

the set of integers $<' n$ arranged in ascending or descending order of magnitude; and many others.

Thus, the set

$$1, 3, 4, 9, 10, 12$$

of quadratic residues of 13 is a balanced 13-set.

In the preceding section, we have defined

$$N_j = \prod_{(p-1)|j} p, \quad p \geqslant 2.$$

We shall continue to use N_j in this sense in the following sections.

If $\text{piq}_p\, j = \alpha$, then we define

$$M_j = \prod_{\alpha \geqslant 1} p^{\alpha+\lambda},$$

where

$$\lambda = \begin{cases} 1 & \text{if } j \text{ is even and } p = 2; \\ 0 & \text{otherwise.} \end{cases}$$

Thus

$$M_6 = 2^{2+1} \,.\, 3^2 \,.\, 7 = 504,$$

while $M_3 = 2$.

Exercises

1. Is the set of quadratic non-residues of n a self-contained set modulo n?
2. Give examples of sets self-contained modulo 13.
3. Is the set of cubic residues of 19 a self-contained set modulo 19? Can the six cubic residues of 19 be written so as to form a balanced 19-set?
4. Evaluate N_8, M_5 and M_{12}.

7. Congruences for self-contained and balanced sets

THEOREM 12. *If A_k is a set self-contained modulo n, with respect to T, then for each $t \in T$,*

$$(n, t^{2j} - 1)G(A_k, 2j) \equiv 0 \pmod{n}, \quad 1 \leqslant j \leqslant \left[\frac{k}{2}\right].$$

Proof. Since A_k is self-contained modulo n with respect to T, we have for each $t \in T$,

$$tA_k \equiv A_k \pmod{n}.$$

Hence

$$t^{2j}G(A_k, 2j) = G(tA_k, 2j) \equiv G(A_k, 2j) \pmod{n},$$

so that

$$(t^{2j} - 1)G(A_k, 2j) \equiv 0 \pmod{n}.$$

THEOREM 13. *If A_k is self-contained modulo n, with respect to a set T, then for each $t \in T$,*

$$(n, t^{2j} - 1)S(A_k, 2j) \equiv 0 \pmod{n}.$$

The proof is left to the reader as an exercise.

THEOREM 14. *If A_k is a balanced n-set, then for* $1 \leqslant j \leqslant \left[\dfrac{k-1}{2}\right]$,

$$2G(A_k, 2j + 1) \equiv (k - 2j)nG(A_k, 2j) \left(\bmod \frac{n^3}{c}\right),$$

where

$$c = \begin{cases} 2 & \text{if } n \text{ is even and } k \text{ odd;} \\ 1 & \text{otherwise.} \end{cases}$$

Proof. We have

$$\prod_{r=1}^{k} (x + a_r) = \prod_{r=1}^{k} (x + n - a_r),$$

where

$$a_1, a_2, a_3, \ldots, a_k$$

are the members of set A_k. Therefore,

$$\sum_{r=0}^{k} G(A_k, r)x^{k-r} = \sum_{r=0}^{k} \{(-1)^r G(A_k, r)(x+n)^{k-r}\}. \tag{1}$$

Equating the coefficients of x^{k-2j-1} in (1), we get

$$2G(A_k, 2j+1) = \binom{k-2j}{1} nG(A_k, 2j) - \binom{k-2j+1}{2} n^2 G(A_k, 2j-1) + \cdots$$
$$\cdots + \binom{k}{2j+1} n^{2j+1}. \tag{2}$$

Hence

$$G(A_k, 2j+1) \equiv 0 \left(\bmod \frac{n}{c}\right), \quad 0 < 2j+1 \leqslant k;$$

that is,

$$G(A_k, 2j-1) \equiv 0 \left(\bmod \frac{n}{c}\right), \quad 1 \leqslant 2j-1 \leqslant k. \tag{3}$$

Thus from (2) and (3), we have

$$2G(A_k, 2j+1) \equiv (k-2j)nG(A_k, 2j) \left(\bmod \frac{n^3}{c}\right).$$

Analogous to theorem 14, we have the

THEOREM 15. *If A_k is a balanced n-set, then for $j \geqslant 0$*

$$2S(A_k, 2j+1) \equiv (2j+1)nS(A_k, 2j) \left(\bmod \frac{n^3}{(n, 2)}\right).$$

Proof. Evidently

$$2S(A_k, 2j+1) = \sum_{r=1}^{k} (a_r^{2j+1} + (n - a_r)^{2j+1})$$

$$= \binom{2j+1}{1} nS(A_k, 2j) - \binom{2j+1}{2} n^2 S(A_k, 2j-1)$$
$$+ \cdots + \binom{2j+1}{2j+1} n^{2j+1}.$$

Hence

$$S(A_k, 2j+1) \equiv 0 \left(\bmod \frac{n}{(n,2)}\right), \quad j \geqslant 0,$$

or what is the same thing,

$$S(A_k, 2j-1) \equiv 0 \left(\bmod \frac{n}{(n,2)}\right), \quad j \geqslant 1.$$

Thus, for $j \geqslant 0$,

$$2S(A_k, 2j+1) \equiv (2j+1)nS(A_k, 2j) \left(\bmod \frac{n^3}{(n,2)}\right).$$

Exercises

1. If A_k is self-contained modulo n with respect to a set T then, for $j > 0$,

$$hG(A_k, 2j) \equiv 0 \pmod n$$

where h is the g.c.d. of the numbers

$$t^{2j} - 1, \quad t \in T.$$

2. If A_k is a balanced n-set, self-contained modulo n with respect to a set T and h is as in exercise 1, then for $j > 0$,

$$2hG(A_k, 2j+1) \equiv 0 \pmod{n^2}.$$

3. Do the results of exercise 1 and 2 hold when G is replaced by S?

We can now offer an improvement on theorem 10 proved earlier. We observe that the set of natural numbers

$$1, 2, 3, \ldots, n-1$$

is a balanced n-set, self-contained modulo n with respect to the set of reduced residues of n. If $(t, n) = 1$, then the g.c.d. of the numbers

$$t^{2j} - 1$$

is a divisor of M_{2j}. This is because, if

$$\text{piq}_p(2j) = \alpha \geqslant 1, \quad p \geqslant 3,$$

and g is a primitive root of p^2, then

$$g^{2j} \equiv 1 \pmod{p^\alpha} \quad \text{but} \quad \not\equiv 1 \pmod{p^{\alpha+1}}.$$

When $p = 2$ and $t \equiv 3 \pmod 4$ we have

$$t^{2j} \equiv 1 \pmod{p^{\alpha+1}} \quad \text{but} \quad \not\equiv 1 \pmod{p^{\alpha+2}},$$

where

$$\alpha = \text{piq}_2(2j).$$

Theorem 12 now gives

THEOREM 16. *For* $j > 0$

$$(n, M_{2j})G(n-1, 2j) \equiv 0 \pmod{n}.$$

Hence we have from theorem 14:

THEOREM 17.

$$2(n, M_{2j})G(n-1, 2j+1) \equiv 0 \pmod{n^2}, \quad j > 0.$$

Making use of the fact that

$$0, 1, 2, 3, \ldots, n-1$$

is a balanced $(n-1)$-set, self-contained modulo $(n-1)$ with respect to every reduced residue set modulo $(n-1)$, theorem 17 gives

THEOREM 18.

$$2(n-1, M_{2j})G(n-1, 2j+1) \equiv 0 \pmod{(n-1)^2}, \quad j > 0.$$

Combining the two theorems 17 and 18, we finally have

THEOREM 19. *For* $j > 0$,

$$2(n^2-n, M_{2j})G(n-1, 2j+1) \equiv 0 \pmod{n^2(n-1)^2}.$$

We leave it to the reader to prove an analogue of this theorem for S.

Exercises

1. Prove that for $j > 0$,

$$2(n, N_{2j})S(n-1, 2j+1) \equiv 0 \pmod{n^2}.$$

2. Show that for $p \geqslant 3$,

$$2\left(n, \frac{N_{p-1}}{p}\right)S(n-1, p) \equiv 0 \pmod{n^2}.$$

In particular, deduce that

$$12S(n-1, 5) \equiv 0 \pmod{n^2}.$$

3. Prove that for $j > 0, p \geqslant 2, u \geqslant 1$,

$$S(p^u-1, 2j) \equiv 0 \pmod{p^{u-1} \text{ or } p^u},$$

according as $(p-1)$ does or does not divide $2j$.

4. If $p > 3$, $(p-1) \nmid 2j$, $j > 0$ and $u \geqslant 1$, then show that

$$G(p^u-1, 2i+1) \equiv 0 \pmod{p^{2u}}.$$

Hence deduce Wolstenholme's Theorem

$$G(p-1, p-2) \equiv 0 \pmod{p^2}, \quad p \geqslant 5.$$

5. In the identity

$$(x+1)(x+2)(x+3)\ldots(x+p-1) = \sum_{r=0}^{p-1} G(p-1, r)x^{p-r-1}, \quad p \geqslant 2;$$

put 1 for x and deduce Wilson's Theorem

$$(p-1)! + 1 \equiv 0 \pmod{p}.$$

6. If $p \geqslant 2$, $0 < 2j < p^u - 1$, $\text{piq}_p(2j) = \alpha$, $u \geqslant 1$, then

$$G(p^u - 1, 2j) \equiv 0 \pmod{p^\lambda},$$

where

$$\lambda = u - \alpha - \left[\frac{2}{p}\right].$$

8. $G'(n, r)$ and $S'(n, r)$

The set of integers $<'n, n > 2$, is a balanced n-set self-contained modulo n with respect to any reduced residue set modulo n. It has interesting congruence properties, some of which are very much more elegant than those for the set of natural numbers $1, 2, 3, \ldots, n$.

Let us denote the sum of the products taken r at a time of the numbers $<'n$, $n > 2$, by $G'(n, r)$, and the sum of their rth powers by $S'(n, r)$.

In this section, we prove some congruence properties of these functions.

THEOREM 20. *If $n > 2, j > 0$ and $h = (\phi(n), 2j)$, then*

$$2(n, 2N_h)G'(n, 2j+1) \equiv 0 \pmod{n^2}.$$

Proof. As a consequence of theorem 12, we have

$$(n, M_{2j})G'(n, 2j) \equiv 0 \pmod{n}. \tag{1}$$

Theorem 14 gives

$$2G'(n, 2j+1) \equiv (\phi(n) - 2j)nG'(n, 2j) \pmod{n^3}.$$

Let p be a prime $\geqslant 2$, such that

$$\text{pot}_p(n, M_{2j}) = \alpha > 1.$$

Then

$$\text{pot}_p(\phi(n), 2j) \geqslant \alpha - 1 - \left[\frac{2}{p}\right], \tag{2}$$

because

$$\operatorname{pot}_p \phi(n) \geqslant \operatorname{pot}_p(n) - 1,$$

and

$$\operatorname{pot}_p(2j) = \operatorname{pot}_p(M_{2j}) - 1 - \left[\frac{2}{p}\right].$$

Hence

$$\operatorname{pot}_p(\phi(n) - 2j) \geqslant \alpha - 1 - \left[\frac{2}{p}\right],$$

and we have

$$2(n, 2N_h)G'(n, 2j + 1) \equiv 0 \pmod{n^2}.$$

Exercises

1. If $(n, 6) = 1$ and $n > 2$, then
$$G'(n, k - 1) \equiv 0 \pmod{n^2},$$
where $k = \phi(n)$. (Leudesdorf's Theorem)
2. If $(n, 30) = 1$, $n > 2$ and $k = \phi(n)$, then
$$G'(n, k - 3) \equiv 0 \pmod{n^2}.$$
3. Show that Wolstenholme's Theorem is a particular case of Leudesdorf's Theorem.
4. Can G' be replaced by S' in theorem 20?

THEOREM 21. *If $n = p^u m$, $(m, p) = 1$, $p \geqslant 2$ and $u \geqslant 1$, then for $j > 0$,*

$$\operatorname{pot}_p S'(n, 2j) \geqslant \begin{cases} u + \operatorname{pot}_p \phi(m) - 1, & \text{if} \quad (p-1) \mid 2j, \\ u + \operatorname{pot}_p \phi(m), & \text{if} \quad (p-1) \nmid 2j. \end{cases}$$

Proof. If a is an integer $<' p^u$, then among the numbers

$$a, a + p^u, a + 2p^u, \ldots, a + (m-1)p^u$$

there are exactly $\phi(m)$ which are $<' n$. Hence if the numbers $<' n$ are divided by p^u, then the remainders will be all the numbers $<' p^u$, each repeated $\phi(m)$ times. Hence for $j > 0$,

$$\operatorname{pot}_p S'(n, 2j) \geqslant \operatorname{pot}_p \phi(m) + \operatorname{pot}_p S'(p^u, 2j). \tag{3}$$

Since

$$a^{\phi(p^u)} \equiv 1 \pmod{p^u} \quad \text{for} \quad (a, p) = 1,$$

we have, for $t \geqslant 1$,

$$S'(p^u, 2j) \equiv S'(p^u, 2j + t\phi(p^u)) \pmod{p^u}. \tag{4}$$

Moreover,

$$S(p^u - 1, r) = S'(p^u, r) + p^r S(p^{u-1} - 1, r), \quad r \geqslant 0, \tag{5}$$

so that for $r \geqslant u$,

$$S'(p^u, r) \equiv S(p^u - 1, r) (\bmod p^u). \tag{6}$$

Choose t so that

$$r = 2j + t\phi(p^u) \geqslant u.$$

Then we get

$$S'(p^u, 2j) \equiv S(p^u - 1, r) (\bmod p^u).$$

Now

$$S(p^u - 1, r) \equiv \begin{cases} 0 \,(\bmod p^{u-1}), & \text{if} \quad (p-1) \mid r, \\ 0 \,(\bmod p^u), & \text{if} \quad (p-1) \nmid r. \end{cases}$$

Hence the theorem.

Exercises

1. Show that for $j > 0$,

$$(n, N_{2j}) S'(n, 2j) \equiv 0 \,(\bmod n),$$

Deduce that

$$2(n, N_{2j}) S'(n, 2j+1) \equiv 0 \,(\bmod n^2), \quad \text{for} \quad j > 0.$$

2. If $n = p^u m$, $(p, m) = 1$, $p \geqslant 2$ and $u \geqslant 1$, then show that for $j > 0$

$$\operatorname{pot}_p S'(n, 2j+1) \geqslant \begin{cases} 2u - 2 & \text{when} \quad p = 2, \quad m = 1, \\ 2u - 1 & \text{when} \quad p > 2, \quad (p-1) \mid 2j, \\ 2u & \text{otherwise.} \end{cases}$$

What happens when $p = 2j + 1$?

3. If $(n, 6) = 1$, prove that

$$S'(n, -1) \equiv 0 \,(\bmod n^2). \quad \text{(Leudesdorf's Theorem)}$$

4. Find g for which

$$gS'(546, 5) \equiv 0 \,(\bmod 546^2).$$

5. Find a value of n not prime to 6, such that

$$S'(n, -2) \equiv 0 \,(\bmod n).$$

6. There always exists a number g independent of n, such that

$$gS(n-1, 2j+1) \equiv 0 \,(\bmod n^2), \quad j > 0.$$

Hence show that the Bernoulli numbers

$$B_3, B_5, B_7, \ldots, B_{2j+1}, \ldots$$

are zero.

9. A theorem on Bernoulli numbers

It will be recalled that the Bernoulli numbers are not necessarily integers. In fact the Bernoulli numbers with positive even subscripts are all fractional and so also is B_1. On the other hand $B_3, B_5, B_7, \ldots, B_{2j+1}, \ldots$ are each zero, and $B_0 = 1$.

Here we prove

THEOREM 22. *For values of* $j \geqslant 1$,

$N_{2j}B_{2j}$ *is an integer.*

Proof. From (9.2) of the preceding chapter, we have

$$B_{2j} = \sum_{k=0}^{2j} (-1)^k \frac{k!}{k+1} G(-k-2, 2j-k).$$

Now $\dfrac{k!}{k+1}$ is an integer except when $k = 3$ or $k = p - 1, p \geqslant 2$. When $k = 3$,

$$\begin{aligned} k!G(-k-2, 2j-k) &= 3!G(-5, 2j-3) \\ &= 4^{2j} - 3.3^{2j} + 3.2^{2j} - 1 \\ &\equiv 0 \pmod 4. \end{aligned}$$

When $k = p - 1$,

$$k!G(-k-2, 2j-k) = (p-1)!G(-p-1, 2j-p+1). \tag{1}$$

Since by theorem 4,

$$G(-1, r) \equiv \begin{cases} G(-p-1, r) \pmod p & \text{for} \quad 0 \leqslant r \leqslant p-2, \\ G(-p-1, r) - G(-p-1, r-p+1) \pmod p & \text{for} \quad r \geqslant p-1, \end{cases}$$

$$\text{the right side of (1)} \equiv \begin{cases} 0 \pmod p, & \text{if} \quad (p-1) \nmid 2j, \\ -1 \pmod p, & \text{if} \quad (p-1) \mid 2j. \end{cases}$$

Thus only those primes p for which $(p - 1) \mid 2j$ contribute to the denominator of B_{2j} and the theorem follows.

COROLLARY. *For* $j > 0$,

$$B_{2j} + \sum_{(p-1)\mid 2j} \frac{1}{p}$$

is an integer. (Von Staudt's Theorem).

Exercises

1. Prove that

$$2(2^{2j} - 1)B_{2j}, \quad j > 0,$$

is an odd integer. (Ramanujan)

2. Show that for $j > 0$

$$p(p^{2j} - 1)B_{2j}$$

is an integer.

3. Show that $N_1 B_1$ is an integer.
4. Prove that N_{2j} in theorem 22 is the best possible.

10. Indentical congruences

All polynomials in this and the following sections will be deemed to have integral coefficients.

If $f(x) = a_0 x^n + a_1 x^{n-1} + a_2 x^{n-2} + \cdots + a_n$, we write

$$f(x) \equiv 0 \pmod{m}$$

when $a_r \equiv 0 \pmod{m}$ for $r = 0, 1, 2, 3, \ldots, n$. In general, if $f(x)$ and $g(x)$ are two polynomials in x of the same or of different degrees, then we write

$$f(x) \equiv g(x) \pmod{m}$$

if $f(x) - g(x) \equiv 0 \pmod{m}$. Thus for $p \geqslant 2$ and $t \geqslant 1$, we have

$$(x + p^t)^p \equiv x^p \pmod{p^{t+1}}. \tag{1}$$

Again, if $f(x)$ is any polynomial in x, then

$$f(x^2) \equiv f(y^2) \pmod{2^{t+1}}, \quad \text{for } y = x + 2^t. \tag{2}$$

The results (1) and (2) will be useful to us in proving our theorems in this and the following sections.

THEOREM 23. *If $p \geqslant 3$, then*

$$\prod_{a=1}^{p^u} (x + a) \equiv (x^p - x)^{p^{u-1}} \pmod{p^u}, \quad u \geqslant 1.$$

Proof. When $u = 1$, we have

$$\prod_{a=1}^{p} (x + a) = \sum_{r=0}^{p} G(p, r) x^{p-r}$$
$$\equiv x^p - x \pmod{p},$$

because

$$G(p, r) = 1, \quad \text{for } r = 0$$

and

$$G(p, r) \equiv \begin{cases} 0 \pmod{p}, & \text{when} \quad r = 1, 2, 3, \ldots, p-2, \\ -1 \pmod{p}, & \text{when} \quad r = p - 1, \\ 0 \pmod{p}, & \text{when} \quad r = p. \end{cases}$$

Thus the theorem holds when $u = 1$. Suppose the theorem holds when

$u = 1, 2, 3, \ldots, t-1$, so that

$$\prod_{a=1}^{p^u} (x+a) = (x^p - x)^{p^{u-1}} + p^u f(x), \quad 1 \leqslant u \leqslant t-1,$$

where $f(x)$ is some polynomial in x. Then

$$\prod_{a=1}^{p^t} (x+a) = \prod_{a=1}^{p^{t-1}} \prod_{k=0}^{p-1} (x + a + kp^{t-1})$$

$$= \prod_{a=1}^{p^{t-1}} \left\{ \sum_{r=0}^{p-1} G(p-1, r) p^{(t-1)r} (x+a)^{p-r} \right\}$$

$$\equiv \prod_{a=1}^{p^{t-1}} (x+a)^p \pmod{p^t},$$

because

$$G(p-1, r) \equiv \begin{cases} 0 \pmod{p}, & \text{for} \quad r = 1, 2, 3, \ldots, p-2, \\ 1 \pmod{p}, & \text{for} \quad r = 0, \\ -1 \pmod{p}, & \text{for} \quad r = p-1. \end{cases}$$

Hence

$$\prod_{a=1}^{p^t} (x+a) \equiv \left\{ \prod_{a=1}^{p^{t-1}} (x+a) \right\}^p \pmod{p^t}$$

$$\equiv \{(x^p - x)^{p^{t-2}} + p^{t-1} f(x)\}^p \pmod{p^t}$$

$$\equiv (x^p - x)^{p^{t-1}} \pmod{p^t}.$$

The proof is completed by induction.

COROLLARY. *For $p \geqslant 3, u \geqslant 1$ and $r \leqslant p^{u-1}$,*

$$G(p^u, r(p-1)) \equiv (-1)^r \binom{p^{u-1}}{r} \pmod{p^u}.$$

COROLLARY. *For $p \geqslant 3, u \geqslant 1$,*

$$\operatorname{pot}_p G(p^u, r(p-1)) = u - 1 - \operatorname{pot}_p r, \quad r \leqslant p^{u-1}.$$

Remark

Theorem 23 holds trivially for $p = 2, u = 1$.

THEOREM 24. *For* $u \geqslant 2$,

$$\prod_{a=1}^{2^u} (x+a) \equiv \{(x+2)^{(4)}\}^{2^{u-2}} \pmod{2^u}.$$

Proof. For $u = 2$,

$$\prod_{a=1}^{2^2} (x+a) = (x+1)(x+2)(x+3)(x+4)$$

$$\equiv (x+2)^{(4)} \pmod 4.$$

Thus the theorem holds for $u = 2$. Suppose it holds for $u = 2, 3, 4, \ldots, t-1$, so that

$$\prod_{a=1}^{2^u} (x+a) = \{(x+2)^{(4)}\}^{2^{u-2}} + 2^u f(x), \quad 2 \leqslant u \leqslant t-1,$$

where $f(x)$ is a polynomial in x. Then

$$\prod_{a=1}^{2^t} (x+a) = \prod_{a=1}^{2^{t-1}} (x+a) \, . \prod_{a=1}^{2^{t-1}} (x + 2^{t-1} + a)$$

$$= \prod_{a=1}^{2^{t-1}} (x+a) \, . \{(x + 2^{t-1} + 2)^{(4)} + 2^{t-1} f(x + 2^{t-1})\}.$$

Now, for $t \geqslant 3$,

$$\begin{aligned}(x + 2^{t-1} + 2)^{(4)} &= (x + 2^{t-1} + 2)(x + 2^{t-1} + 1)(x + 2^{t-1})(x + 2^{t-1} - 1)\\ &= \{(x + 2^{t-1} + 2)(x + 2^{t-1})\}\{(x + 2^{t-1} + 1)(x + 2^{t-1} - 1)\}\\ &= \{x(x+2) + 2^t x + 2^t + 2^{2t-2}\}\{(x + 2^{t-1})^2 - 1\}\\ &\equiv \{x(x+2)\}\{(x^2 - 1)\} \pmod{2^t}\\ &\equiv (x+2)^{(4)} \pmod{2^t}.\end{aligned}$$

Also, for $t \geqslant 3, f(x + 2^{t-1}) \equiv f(x) \pmod 2$

so that $2^{t-1} f(x + 2^{t-1}) \equiv 2^{t-1} f(x) \pmod{2^t}$.

Hence

$$\prod_{a=1}^{2^t} (x+a) \equiv \{\{(x+2)^{(4)}\}^{2^{t-3}} + 2^{t-1} f(x)\}^2 \pmod{2^t}$$

$$\equiv \{(x+2)^{(4)}\}^{2^{t-2}} \pmod{2^t}.$$

The theorem readily follows by inductive reasoning.

COROLLARY. *For $u \geqslant 3$,*

$$G(2^u, 2r) \equiv (-1)^r \left\{ \binom{2^{u-2}}{r} + 2^{u-1} \binom{2^{u-2}}{r-1} \right\} \pmod{2^u}.$$

COROLLARY. *For $u \geqslant 2$ and $r \geqslant 0$,*

$$G(2^u, 2r+1) \equiv (-1)^r \binom{2^{u-2}}{r} 2^{u-1} \pmod{2^u}.$$

THEOREM 25. *If $m = p^u n$ where $(p, n) = 1, p \geqslant 2, u \geqslant 1$, then*

$$\prod_{a=1}^{m} (x+a) \equiv \left\{ \prod_{a=1}^{p^u} (x+a) \right\}^n \pmod{p^u}.$$

Proof. If the numbers $1, 2, 3, \ldots, m$ are divided by p^u, then we get as remainders the numbers $0, 1, 2, 3, \ldots, p^u - 1$ each repeated n times. The theorem follows directly from this.

11. Bauer's Theorems

As analogues of the theorems proved in the preceding section, we have the following due to Bauer.

THEOREM 26. *For $p \geqslant 3, u \geqslant 1$,*

$$\prod_{a <' p^u} (x+a) \equiv (x^{p-1} - 1)^{p^{u-1}} \pmod{p^u}.$$

THEOREM 27. *For $u \geqslant 2$,*

$$\prod_{a <' 2^u} (x+a) \equiv (x^2 - 1)^{2^{u-1}} \pmod{2^u}.$$

THEOREM 28. *If $m = p^u n$, where $(p, n) = 1, p \geqslant 2, u \geqslant 1$, then*

$$\prod_{a <' m} (x+a) \equiv \left\{ \prod_{a <' p^u} (x+a) \right\}^{\phi(n)} \pmod{p^u}.$$

The proofs of these theorems are similar to those in the preceding section and are left to the reader.

Remark

Theorem 26 holds trivially for $p = 2, u = 1$.

Exercises

1. Prove that for $m = 2^u j, j$ odd and $\geqslant 3$,

$$\prod_{a<'m} (x+a) \equiv (x-1)^{\phi(m)} (\text{mod } 2^u).$$

2. For $p \geqslant 3, u \geqslant 1$,

$$G'(p^u, r(p-1)) \equiv (-1)^r \binom{p^{u-1}}{r} (\text{mod } p^u).$$

3. Deduce the theorem of Gauss from Bauer's theorems.
4. Deduce theorem 24 from theorem 27.

12. A generalization of the combinatory function $\binom{n}{r}$

Let

$$(x+1)(x+a)(x+a^2)\ldots(x+a^{n-1}) = \sum_{r=0}^{n} \begin{bmatrix} n \\ r \end{bmatrix}_a x^{n-r}, \tag{1}$$

where $a > 0$. Then

$$\binom{n}{r} = \begin{bmatrix} n \\ r \end{bmatrix}_{a=1}.$$

In particular,

$$\begin{bmatrix} n \\ 0 \end{bmatrix}_a = 1, \quad \begin{bmatrix} n \\ 1 \end{bmatrix}_a = \frac{a^n - 1}{a - 1}, \quad \begin{bmatrix} n \\ n \end{bmatrix}_a = a^{n(n-1)/2}.$$

We take

$$\begin{bmatrix} n \\ r \end{bmatrix}_a = 0 \quad \text{for} \quad 0 < n < r.$$

Putting -1 for x in (1), we obtain

$$\sum_{r=0}^{n} (-1)^r \begin{bmatrix} n \\ r \end{bmatrix}_a = 0.$$

Henceforth, we write $\begin{bmatrix} n \\ r \end{bmatrix}$ *for* $\begin{bmatrix} n \\ r \end{bmatrix}_a$ for simplicity. Evidently, we have

$$\sum_{t=0}^{j+k-1} (x+a^t) = \sum_{t=0}^{j-1} (x+a^t) \,.\, \sum_{s=0}^{k-1} (x+a^j \,.\, a^s)$$

$$= \sum_{t=0}^{j} \begin{bmatrix} j \\ t \end{bmatrix} x^{j-t} \cdot \sum_{s=0}^{k} \begin{bmatrix} k \\ s \end{bmatrix} a^{js} x^{k-s}.$$

Hence

$$\begin{bmatrix} j+k \\ r \end{bmatrix} = \sum_{\substack{t+s=r \\ 0 \leqslant t \leqslant j}} \begin{bmatrix} j \\ t \end{bmatrix} \begin{bmatrix} k \\ s \end{bmatrix} a^{js}. \tag{2}$$

In particular, replacing j by n and k by 1 in (2), we have

$$\begin{bmatrix} n+1 \\ r \end{bmatrix} = \begin{bmatrix} n \\ r \end{bmatrix} + a^n \begin{bmatrix} n \\ r-1 \end{bmatrix}.$$

This readily gives

$$\begin{bmatrix} n+1 \\ r \end{bmatrix} = \sum_{m=r-1}^{n} a^m \begin{bmatrix} m \\ r-1 \end{bmatrix}. \tag{3}$$

Putting $r = 2, 3, 4, \ldots,$ in succession in (3), it can be shown that

$$\begin{aligned} \begin{bmatrix} n \\ r \end{bmatrix} &= \frac{(a^n-1)(a^n-a)(a^n-a^2)\ldots(a^n-a^{r-1})}{(a-1)(a^2-1)(a^3-1)\ldots(a^r-1)} \\ &= \frac{(a^n-1)(a^{n-1}-1)(a^{n-2}-1)\ldots(a^{n-r+1}-1)}{(a-1)(a^2-1)(a^3-1)\ldots(a^r-1)} a^{r(r-1)/2} \\ &= \frac{n]_a}{r]_a(n-r)]_a} \begin{bmatrix} r \\ r \end{bmatrix}_a, \quad r \geqslant 1, \end{aligned}$$

where

$$n]_a = \frac{(a-1)(a^2-1)(a^3-1)\ldots(a^n-1)}{(a-1)^n}, \quad 0]_a = 1.$$

When $a \to 1$, $n]_a = n!$. We now define $\begin{bmatrix} y \\ r \end{bmatrix}_a$ for all integral values of y by the relation

$$\begin{bmatrix} y \\ r \end{bmatrix}_a = \frac{(a^y-1)(a^{y-1}-1)(a^{y-2}-1)\ldots(a^{y-r+1}-1)}{(a-1)(a^2-1)(a^3-1)\ldots(a^r-1)} \begin{bmatrix} r \\ r \end{bmatrix}_a, \quad r \geqslant 1.$$

We take $\begin{bmatrix} y \\ 0 \end{bmatrix}_a = 1$, for all values of y including 0. It is easy to show that

$$\begin{bmatrix} -k \\ r \end{bmatrix}_a = (-1)^r \begin{bmatrix} k+r-1 \\ r \end{bmatrix}_a \cdot a^{-r(2k+r-1)/2}.$$

In particular

$$\begin{bmatrix} -1 \\ r \end{bmatrix}_a = (-a)^{-r}.$$

The results of this section will be useful to us in the section that follows.

13. Identical congruences for k-ic residues

THEOREM 29. *If a runs through the* k-ic *residues of* p^u, *where p is an odd prime and* $u \geqslant 1$, *then*

$$\prod_a (x-a) \equiv (x^h - 1)^{p^{u-c}} \pmod{p^u},$$

where

$$h = (p-1)/d, \quad d = (k, p-1),$$

and c is given by the relation

$$(k, p^{u-1}) = p^{c-1}, \quad 1 \leqslant c \leqslant u.$$

Proof. Let t be a primitive root of p^u, then the k-ic residues of p^u are congruent modulo p^u to the members of the set

$$\{t^{jg}, \quad j = 0, 1, 2, 3, \ldots, (m-1)\},$$

where

$$g = (k, \phi(p^u)) = p^{c-1}d,$$

and

$$m = \phi(p^u)/g = p^{u-c}h.$$

Hence

$$\prod_a (x-a) \equiv \prod_{j=0}^{m-1} (x - t^{jg}) \pmod{p^u},$$

$$\equiv \sum_{r=0}^{m} (-1)^r \begin{bmatrix} m \\ r \end{bmatrix}_{t^g} x^{m-r} \pmod{p^u}.$$

Now

$$\begin{bmatrix} m \\ r \end{bmatrix}_{t^g} = \frac{t^{mg}-1}{t^{rg}-1} \cdot \prod_{j=1}^{r-1} \frac{t^{mg}-t^{jg}}{t^{jg}-1}.$$

Since t is a primitive root of p^u,

$$t^{mg} - 1 \equiv t^{\phi(p^u)} - 1 \equiv 0 \pmod{p^u}.$$

Again

$$t^{jg} - 1 \equiv 0 \pmod{p},$$

if and only if $h \mid j$. Therefore, if $h \nmid j$, we have

$$\frac{t^{mg} - t^{jg}}{t^{jg} - 1} + 1 = \frac{t^{mg} - 1}{t^{jg} - 1} \equiv 0 \,(\text{mod}\, p^u).$$

From this, it readily follows that if $h \nmid r$, then

$$\begin{bmatrix} m \\ r \end{bmatrix}_{t^g} \equiv 0 \,(\text{mod}\, p^u). \tag{1}$$

When $h \mid j$, let $j = p^{b-c}hq$ where $(q, p) = 1$, $c \leqslant b \leqslant u$. Then, for some v prime to p,

$$\begin{aligned} t^{jg} = t^{q\phi(p^b)} &= (1 + p^b v)^q \\ &\equiv 1 + qp^b v \,(\text{mod}\, p^{2b}), \end{aligned}$$

so that

$$\text{pot}_p(t^{jg} - 1) = b.$$

Also

$$\begin{aligned} t^{mg} &= t^{p^{u-b}\phi(p^b)} = (1 + p^b v)^{p^{u-b}} \\ &\equiv (1 + p^u v)(\text{mod}\, p^{u+b}). \end{aligned}$$

Hence

$$q(t^{mg} - 1) \equiv p^{u-b}(t^{jg} - 1)(\text{mod}\, p^{u+b}).$$

Therefore, if $h \mid j$, we have

$$\begin{aligned} \frac{t^{mg} - 1}{t^{jg} - 1} &\equiv \frac{p^{u-b}}{q} \,(\text{mod}\, p^u) \\ &\equiv \frac{p^{u-c}}{(j/h)} \,(\text{mod}\, p^u), \end{aligned}$$

and

$$\frac{t^{mg} - t^{jg}}{t^{jg} - 1} \equiv \frac{p^{u-c} - (j/h)}{(j/h)} \,(\text{mod}\, p^u).$$

Hence, if $h \mid r$, we have

$$\begin{bmatrix} m \\ r \end{bmatrix}_{t^g} \equiv (-1)^{r-r_1} \binom{p^{u-c}}{r_1} \,(\text{mod}\, p^u), \tag{2}$$

where $r_1 = \dfrac{r}{h}$. The theorem now follows immediately from (1) and (2).

THEOREM 30. *If a runs through the k-ic residues of 2^u and k is odd, then, modulo 2^u,*

$$\sum_a (x - a) \equiv \begin{cases} (x^2 - 1)^{2^{u-2}} (\text{mod}\, 2^u), & u \geqslant 2, \\ (x - 1)(\text{mod}\, 2^u), & u = 1. \end{cases}$$

Proof. The second part of the theorem is obvious. As regards the first part, we notice that the k-ic residues of 2^u, for k odd and $u \geqslant 2$, are congruent modulo 2^u to the numbers

$$\pm(1, 3, 3^2, \ldots, 3^{m-1}),$$

where $m = 2^{u-2}$. Hence

$$\prod_a (x-a) \equiv \prod_{j=0}^{m-1} (x^2 - 3^{2j}) (\bmod 2^u)$$

$$\equiv \sum_{r=0}^{m} (-1)^r \begin{bmatrix} m \\ r \end{bmatrix}_{3^2} x^{2(m-r)} (\bmod 2^u).$$

In this case, we notice that for $j = 2^{c-2}q$, where $(q, 2) = 1$ and $c \geqslant 2$, and for some odd v,

$$3^{2j} = (1 + 2^{c+1}v)^q$$
$$\equiv (1 + 2^{c+1}qv) (\bmod 2^{2(c+1)});$$

while

$$3^{2m} = (1 + 2^{c+1}v)^{2^{u-c}}$$
$$\equiv (1 + 2^{u+1}v) (\bmod 2^{u+c+1}).$$

Hence we obtain, as in the proof of theorem 29,

$$\begin{bmatrix} m \\ r \end{bmatrix}_{3^2} \equiv \binom{m}{r} (\bmod 2^u),$$

and the theorem follows.

THEOREM 31. *If a runs through the k-ic residues of* 2^u *and k is even, then modulo* 2^u,

$$\prod_a (x-a) \equiv \begin{cases} (x-1)^{2^n} - 2^{u-1}(x-1)^{2^{n-1}}, & n \geqslant 1, \\ (x-1), \quad u = 1, 2, 3, \ldots, c+2, \end{cases}$$

where c is given by the relation

$$k = 2^c q, \quad q \text{ odd},$$

and

$$n = u - c - 2.$$

Proof. Since k is even, c is $\geqslant 2$. We further notice that the k-ic residues of 2^u are the same as the 2^c-ic residues of 2^u and are, modulo 2^u, given by the incongruent members of the set

$$\{3^{2^{c}j}, \quad j \geqslant 1\}.$$

If $c \geqslant u - 2$, the only k-ic residue of 2^u is 1. This proves the second part of the theorem. If $c < u - 2$, the k-ic residues of 2^u are the numbers

$$3^{2^{c-j}}, \quad j = 1, 2, 3, \ldots, 2^n.$$

Equivalently, the k-ic residues of 2^u are

$$1 + 2^{c+2}j, \quad j = 1, 2, 3, \ldots, 2^n.$$

(The proof of this is left to the reader as an exercise.) Hence, for $n \geqslant 1$,

$$\prod_a (x - a) \equiv \prod_{j=1}^{2^n} (x - 1 - 2^{c+2}j) (\text{mod } 2^u)$$

$$\equiv \sum_{r=0}^{2^n} (-1)^r G(2^n, r) 2^{r(c+2)} (x - 1)^{2^n - r} (\text{mod } 2^u).$$

The theorem is readily verified for $n = 1, 2$. For $n \geqslant 3$,

$$G(2^n, 1) = 2^{n-1}(2^n + 1),$$

while for $r \geqslant 2$,

$$\text{pot}_2\, G(2^n, r) \geqslant (n - 1) - \text{pot}_2\, r.$$

Hence for $r \geqslant 2$, each term of the sum is divisible by 2^u. The theorem follows directly as a consequence of this observation.

As an analogue of theorems 25 and 28 we have the following

THEOREM 32. *If $n = p^u t$, where $(t, p) = 1$ and $p \geqslant 2$, then*

$$\prod_{a_1} (x - a_1) \equiv \left\{ \prod_{a_2} (x - a_2) \right\}^{K(t)} (\text{mod } p^u)$$

where a_1 and a_2 run through the k-ic residues of n and p^u respectively, and $K(t)$ is the number of k-ic residues of t.

Proof. The k-ic residues of n when divided by p^u, leave as remainders the k-ic residues of p^u each repeated $K(t)$ times. Hence the theorem.

Exercises

1. Give an inductive proof of theorem 31.
2. Show that theorems 29 and 30 are generalizations of theorems 26 and 27.

14. Wilson's quotient

Wilson's Theorem states that for any prime p,

$$(p - 1)! + 1 \equiv 0 (\text{mod } p).$$

Here we prove

THEOREM 33. *For* $p \geqslant 3$,

$$(p-1)! \equiv (B_{p-1}-1)p \pmod{p^2},$$

where B_{p-1} *is a Bernoulli number.* (Glaisher and Lerch.)

Proof. We have

$$S(p-1,p-1) = \sum_{j=1}^{p-1} G(-j-1,p-j-1)\frac{p^{(j+1)}}{j+1}$$

$$= G(-p,0)\frac{p^{(p)}}{p} + \sum_{j=1}^{p-2} G(-j-1,p-j-1)\frac{p^{(j+1)}}{j+1}.$$

But

$$G(-p,0)\frac{p^{(p)}}{p} = (p-1)!$$

and for $1 \leqslant j \leqslant p-2$,

$$G(-j-1,p-j-1) \equiv G(p-j-1,p-j-1) \pmod{p}$$
$$\equiv (p-j-1)! \pmod{p}.$$

Hence

$$\frac{S(p-1,p-1)-(p-1)!}{p} \equiv \sum_{j=1}^{p-2} (p-j-1)!\frac{(p-1)^{(j)}}{j+1} \pmod{p}$$

$$\equiv \sum_{j=1}^{p-2} \frac{(p-1)!}{j+1} \pmod{p}$$

$$\equiv (p-1)!\left\{\frac{1}{2}+\frac{1}{3}+\frac{1}{4}+\cdots+\frac{1}{p-1}\right\} \pmod{p}$$

$$\equiv G(p-1,p-2)-(p-1)! \pmod{p}$$

$$\equiv 1 \pmod{p}.$$

So that

$$S(p-1,p-1) \equiv (p-1)! + p \pmod{p^2}. \tag{1}$$

Again, for $p \geqslant 3$,

$$S(p-1,p-1) = S(p,p-1) - p^{p-1}$$

$$\equiv S(p,p-1) \pmod{p^2}$$

$$\equiv \left\{\left(B_0 + \frac{B_1 D}{1!} + \frac{B_2 D^2}{2!} + \cdots + \frac{B_{p-1}D^{p-1}}{(p-1)!}\right)\frac{n^p}{p}\right\}_{n=p} \pmod{p^2},$$

where D stands for $\dfrac{d}{dn}$ as usual. Since p occurs in the denominator of B_{p-1} but not in the denominator of any of the earlier B's, we must have

$$S(p-1, p-1) \equiv B_{p-1}\, p (\text{mod } p^2), \tag{2}$$

because

$$\left\{\frac{B_j D^j}{j!}\,\frac{n^p}{p}\right\}_{n=p} = \frac{B_j}{j!} \cdot \frac{p^{(j)}}{p} \cdot p^{p-j}$$

$$\equiv \begin{cases} 0\,(\text{mod } p^2), & 0 \leqslant j \leqslant p-2; \\ \dfrac{B_{p-1}}{(p-1)!}\,\dfrac{p^{(p-1)}}{p} \cdot p = B_{p-1}\,p\,(\text{mod } p^2), & j = p-1. \end{cases}$$

The theorem follows from (1) and (2).

Exercise

1. Verify the theorem for $p = 5$ and $p = 7$.

Problems

1. If A denotes the set of k-ic residues of m and $a \in A$, show that for $r > 0$,

$$(a^r - 1) G(A, r) \equiv 0\,(\text{mod } m).$$

If $m = p^u$ where $p \geqslant 2$, $u \geqslant 1$, find the p-potency of $G(A, r)$.

2. If g is a primitive root of p^u, then the numbers

$$1, g, g^2, \ldots, g^{\phi(p^u)-1}$$

form a reduced residue system modulo p^u. Make use of this fact to get a proof of Gauss' Theorem.

CHAPTER 7

PARTITION FUNCTIONS OF VARIOUS TYPES, GENERATING FUNCTIONS, GRAPHS AND IDENTITIES

1. Introduction

A partition of a natural number n is a representation of n as a sum of one or more natural numbers, the order in which the summands appear in the representation being irrelevant.

When the order is relevant, the representation is called a decomposition of n. When there are no restrictions on the size or the number of summands in a partition, each $n > 1$ has more than one partition and this is true of the decompositions as well. For restricted partitions, this statement is not necessarily true.

Example. The unrestricted partitions of 6 are

$$1+1+1+1+1+1;\quad 1+1+1+1+2;\quad 1+1+1+3;\quad 1+1+2+2;$$

$$1+1+4;\quad 1+2+3;\quad 1+5;\quad 2+2+2;\quad 2+4;\quad 3+3;\quad 6.$$

This means 6 has 11 unrestricted partitions. But the partitions of 6 into odd parts are

$$1+1+1+1+1+1;\quad 1+1+1+3;\quad 1+5;\quad 3+3.$$

Again 6 has only one partition into distinct odd parts, viz. $1 + 5$.

The restrictions may sometimes be so stringent that some numbers have no partitions at all. Thus 10 cannot be partitioned into three odd parts.

Exercises

1. Write out all the partitions of 7 in which (i) the least part is 1; (ii) the largest part is 3; (iii) the largest part is not less than 3; (iv) the least part is not less than 2; (v) the parts are distinct; (vi) the parts are all even.
2. Find the number of partitions of 10 into (i) squares; (ii) cubes; (iii) triangular numbers; (iv) primes.
3. How many partitions do the numbers $1, 2, 3, \ldots, 15$ have when the summands belong to the set $\{2, 5, 8, 11\}$?
4. How many decompositions do the numbers $1, 2, \ldots, 5$ have? Can you make any guess about the number of decompositions of n? Can you prove your conjecture?

1.1. Counting partitions

If the reader is asked to write out all the unrestricted partitions of 10 (which is certainly not too big a number) and he starts putting them down on paper in a haphazard manner, it will then be a surprise if he does not repeat some partitions and miss some others. It will not, therefore, be without interest to have an infallible method of doing this sort of a job. This is described below.

We first note that 1 has only one partition, viz. 1. Suppose the partitions of all natural numbers $\leqslant m-1$, $m \geqslant 2$ have already been listed according to plan, then the partitions of m are obtained as follows.

Adjoin 1 (on the left) of each of the partitions of $(m-1)$. Now write 2 and adjoin to it *in turn* each partition of $(m-2)$ in which the least part is not less than 2. In the same manner, write 3 and tag on to it in order each of those partitions of $(m-3)$ in which the least part is not less than 3, and so on; write j and adjoin to it in turn all those partitions of $(m-j)$ in which the least part is not less than j. When you reach $[m/2]$, you will have just to adjoin to it the unique part $m-[m/2]$. The final partition to record will be m itself.

The following table gives the partitions of all numbers $m \leqslant 10$.

Serial No.	Partition									
1.	1	1	1	1	1	1	1	1	1	1
2.	1	1	1	1	1	1	1	1	2	
3.	1	1	1	1	1	1	1	3		
4.	1	1	1	1	1	1	2	2		
5.	1	1	1	1	1	1	4			
6.	1	1	1	1	1	2	3			
7.	1	1	1	1	1	5				
8.	1	1	1	1	2	2	2			
9.	1	1	1	1	2	4				
10.	1	1	1	1	3	3				
11.	1	1	1	1	6					
12.	1	1	1	2	2	3				
13.	1	1	1	2	5					
14.	1	1	1	3	4					
15.	1	1	1	7						
16.	1	1	2	2	2	2				
17.	1	1	2	2	4					
18.	1	1	2	3	3					
19.	1	1	2	6						
20.	1	1	3	5						
21.	1	1	4	4						
22.	1	1	8							
23.	1	2	2	2	3					
24.	1	2	2	5						
25.	1	2	3	4						
26.	1	2	7							
27.	1	3	3	3						
28.	1	3	6							
29.	1	4	5							
30.	1	9								

31	2	2	2	2	2
32.	2	2	2	4	
33.	2	2	3	3	
34.	2	2	6		
35.	2	3	5		
36.	2	4	4		
37.	2	8			
38.	3	3	4		
39.	3	7			
40.	4	6			
41.	5	5			
42.	10				

In listing the partitions the plus signs are omitted for convenience.

To find the unrestricted partitions of any $n \leqslant 10$, all that one has to do is to look down the nth column from the right till the entry n is reached. Now draw a vertical line on its immediate left directed upwards and a horizontal line immediately below it directed rightwards. The rows contained within the right angle so formed give all the partitions of n.

It is noteworthy that in this scheme of numbering the partitions the table can be extended without any loss of effort. Moreover, the parts in any partition are arranged in a non-descending order. Read from the right, the parts get arranged in a non-ascending order. Since, it is usual to arrange the parts in a partition in one of these two ways, a table such as the one presented here will be very useful.

Exercise

From the table, find all the partitions of 9 which have (i) distinct odd parts; (ii) parts none less than 2; (iii) the least part exactly 2; (iv) exactly three parts; (v) at most three parts.

1.2. Notation

Our main concern, in this chapter, will be not so much with the actual recording of partitions as with the number of partitions. For this purpose we will use the following notations:

$p(n)$ for the number of unrestricted partitions of n;

$q(n)$ for the number of partitions of n into distinct parts;

$p(n, m)$ for the number of partitions of n into exactly m parts;

$P(n, m)$ for the number of partitions of n in which the least part is m;

$q(n, m)$ for the number of partitions of n into exactly m distinct parts;

$p(n, A)$ for the number of partitions of n when the summands belong to a given set A and can be repeated;

$q(n, A)$ for the number of partitions of n into distinct elements of A.

The summands in a partition will usually be written in a non-descending order. For reasons that will be clear in the next section, we shall take

$$p(0) = p(0, 0) = q(0, 0) = q(0) = 1.$$

Other notations will be defined as need arises.

2. Generating functions

The problem of finding the number of partitions is most simply stated as follows.

Let

$$A = \{a_1, a_2, a_3, \ldots, a_k\}$$

be a multiset of numbers. Without loss of generality, we can assume that

$$a_1 \leqslant a_2 \leqslant a_3 \leqslant \cdots \leqslant a_k. \tag{1}$$

We will, however, regard even the equal a's as distinguishable. For example if

$$a_2 = a_3 = 5$$

while talking of a_2, we shall speak of 5 as of the first kind and while talking of a_3, we shall speak of 5 as of the second kind. It will be most convenient to write 5_1 for 5 of the first kind and 5_2 for 5 of the second kind. Thus, it will be more appropriate to consider A as a set rather than a multiset. The partitions

$$3 + 5_1 \quad \text{and} \quad 3 + 5_2$$

of 8 will be regarded as distinct and count as two and not one.

For $j = 1, 2, 3, \ldots, k$ let

$$A_j = \{a_j f_j : 0 \leqslant f_j \leqslant t_j\} \tag{2}$$

be a family of k sets, which will be regarded as distinct from one another as explained earlier. Then the problem is to find how often any natural number n occurs in the set (a multiset in general)

$$A_1 \oplus A_2 \oplus A_3 \oplus \cdots \oplus A_k \tag{3}$$

where $\oplus$ has the same meaning as in chapter 4. The number of times n occurs in (3) will be called its frequency therein and will be denoted by $f(n, a_j, t_j, k)$ or simply by $f(n)$ when there is no cause for confusion.

Evidently, $f(n)$ is the number of solutions of the Diophantine equation

$$a_1 f_1 + a_2 f_2 + a_3 f_3 + \cdots + a_k f_k = n \tag{4}$$

where f_j can take the values $0, 1, 2, \ldots, t_j$. Note that each solution of (4) provides a partition of n into parts belonging to A, the largest not exceeding a_k and any part a_j not occuring more than t_j times in the partition. Our $f(n)$ denotes the number of such partitions of n.

We can also say that $f(n)$ is the coefficient of x^n in the expansion of the product

$$\prod_{j=1}^{k} (1 + x^{a_j} + x^{2a_j} + \cdots + x^{t_j a_j}). \tag{5}$$

Writing

$$F_j = F(A_j) = \sum_{r \in A_j} x^r \tag{6}$$

as we did in the case of classes in chapter 4, we get

$$F_1 F_2 \ldots F_k = \sum_{n \geqslant 0} f(n) x^n. \tag{7}$$

It will be readily seen that for $n = 0$, equation (4) has only one solution. We, therefore, take $f(0) = 1$. If (4) has no solution for $n = n_1$, we take $f(n_1) = 0$. Note also that for

$$u = a_1 t_1 + a_2 t_2 + \cdots + a_k t_k,$$

(4) has no solution if $n > u$. The expression on the right of (7) is therefore a polynomial in x of degree exactly u. In fact for $n = u$, (4) has just one solution, so that $f(u) = 1$.

The product in (5) is called the generating function of $f(n)$. It will be in our best interest to impose on the arbitrary variable x the condition that $|x| < 1$, for this makes every factor in (5) absolutely convergent.

2.1. Examples

Selecting a's and t's suitably, several types of partitions can be handled in the manner described above. Letting t_j tend to infinity, the restriction on the number of times a_j can appear in a partition of n gets removed. Taking $t_j = 1$ for each $j \leqslant k$ ensures that no part in a partition is repeated. Letting

$$a_1 < a_2 < \cdots < a_k$$

will mean that no part in the partition is of more than one kind. Taking $a_j = j$ and allowing k to tend to infinity, the restriction on the size of parts will also disappear.

We mention here briefly some important cases.

(1) $A = \{1, 2, 3, \ldots, k\}$, t_j tending to infinity for each $j \leqslant k$, gives partitions of n into parts each $\leqslant k$. The related Diophantine equation is

$$1 \,.\, f_1 + 2 \,.\, f_2 + \cdots + k \,.\, f_k = n;$$

and the generating function is given by

$$1/(1-x)(1-x^2)(1-x^3)\ldots(1-x^k).$$

(2) $A = \{1, 2, 3, \ldots, j, \ldots\}$, k and t_j both tending to infinity, gives the unrestricted partitions of n. The related Diophantine equation is

$$1 . f_1 + 2 . f_2 + \cdots + n . f_n = n$$

with no restriction on f's; and the generating function of $p(n)$ is

$$1/(1-x)(1-x^2)(1-x^3) \ldots (1-x^j) \ldots$$

so that

$$\prod_{j \geqslant 1} (1-x^j)^{-1} = \sum_{n \geqslant 0} p(n) x^n.$$

(3) A as in (2), $t_j = 1$ for all $j \geqslant 1$. In this case, $f(n)$ counts the partitions of n into distinct parts, i.e. $f(n) = q(n)$. The related Diophantine equation now is

$$1 . f_1 + 2 . f_2 + \cdots + n . f_n = n, \quad f_j = 0 \text{ or } 1,$$

and we have

$$\prod_{j \geqslant 1} (1+x^j) = \sum_{n \geqslant 0} q(n) x^n.$$

2.2. Partitions into an odd or even number of parts

Let

$$A = \{a_1, a_2, \ldots, a_j, \ldots\}$$

be an infinite set of natural numbers such that

$$a_1 < a_2 < \cdots < a_j < \cdots$$

Denote by $q_o(n, A)$ the number of partitions of n into an odd number of distinct parts taken from A and by $q_e(n, A)$ the number of partitions of n into an even number of distinct parts taken from A. Then, it is not difficult to see that

$$\prod_{j \geqslant 1} (1-x^{a_j}) = \sum_{n \geqslant 0} \{q_e(n, A) - q_o(n, A)\} x^n,$$

where we take $q_e(0, A) = 1$ and $q_o(0, A) = 0$.

2.3. Curtailing the generating function

Sometimes we are not interested in the expansion of a generating function beyond a certain power of x, say the mth. This fact is indicated by writing (mod x^{m+1}) after the function. The appropriateness of this is fairly evident. This introduces several simplifications in our calculations. For example, we can delete from A all elements greater than m, from A_j all multiples of a_j exceeding m, and as the work of expansion proceeds all terms involving powers of x beyond the mth.

Exercises

1. Construct the sets A_j from the following data and obtain the related Diophantine equations and the generating functions:

(i) $k = 6, a_j = 2j - 1, t_j = 7 - j$.
(ii) $k = 4, a_j = 2, t_j = 3$.
(iii) $k = 3, a_1 = 5, a_2 = 8, a_3 = 11, t_j = j + 1$.
(iv) $k = 4, a_j = j^2, t_1 = 1, t_2 = 2, 1 \leqslant f_3 \leqslant 3, 2 \leqslant f_4 \leqslant 3$. Find in this case, the least value of n such that $f(m) = 0$ for each $m \geqslant n$. Obtain also the values of $f(0)$, $f(20)$ and $f(50)$.
(v) k tends to infinity, $t_j = 2$ and $a_j = 3^{j-1}$ for each j. Can you show that $f(n) = 1$ for all $n \geqslant 0$?
(vi) k tends to infinity, $a_j = j$, $f_4 = 2$ and for all j other than 4, t_j tends to infinity.

2. Find the generating functions for the following types of partitions:

(i) Partitions in which the largest part is m.
(ii) Partitions in which the least part is m.
(iii) Partitions into distinct powers of 2.
(iv) Partitions of n in which the first five odd numbers enter as parts at least once but not more than thrice, and all other numbers freely.

3. Comment on the usefulness of the generating functions in finding the number of partitions of specified numbers. Could not the related Diophantine equation serve the same purpose?

4. Expand the infinite product

$$\prod_{j \geqslant 1} (1 - x^j) \pmod{x^{18}}.$$

3. Partitions into exactly k summands

As before, let

$$A = \{a_1, a_2, \ldots, a_r, \ldots\} \quad \text{with} \quad a_r < a_{r+1}$$

be a set of natural numbers. Then, the number of partitions of n into exactly k summands chosen from the elements of A but not necessarily distinct, is the coefficient of $z^k x^n$ in the expansion of the product

$$\prod_{a_r \leqslant n} (1 - zx^{a_r})^{-1}. \tag{1}$$

On the other hand, the number of partitions of n into exactly k distinct summands taken from among the members of A, is the coefficient of $z^k x^n$ in the product

$$\prod_{a_r \leqslant n} (1 + zx^{a_r}). \tag{2}$$

In each of the two cases, z^k ensures that exactly k summands will be used while x^n ensures that the sum of the elements chosen will be n.

In particular, when A is the set of natural numbers, i.e. when $a_j = j$, we have

THEOREM 1. *The number $p(n, k)$ of partitions of n into exactly k summands is the coefficient of x^n in*

$$x^k/(1-x)(1-x^2)(1-x^3)\ldots(1-x^k).$$

Proof. Let

$$\prod_{r\geqslant 1}(1-zx^r)^{-1} = 1 + \sum_{t\geqslant 1} z^t\phi_t(x). \tag{3}$$

Replacing z by zx in (3), we get

$$\prod_{r\geqslant 2}(1-zx^r)^{-1} = 1 + \sum_{t\geqslant 1} z^t x^t\phi_t(x). \tag{4}$$

Hence

$$1 + \prod_{t\geqslant 1} z^t x^t \phi_t(x) = (1-zx)\left\{1 + \sum_{t\geqslant 1} z^t\phi_t(x)\right\}.$$

Equating the coefficients of z^t on the two sides, we obtain

$$x\phi_1(x) = \phi_1(x) - x,$$

and (5)

$$x^t\phi_t(x) = \phi_t(x) - x\phi_{t-1}(x), \quad t > 1.$$

Thus, we have the recurrence formula

$$\phi_t(x) = \frac{x}{(1-x^t)}\phi_{t-1}(x),$$

which readily gives

$$\phi_t(x) = x^t/(1-x)(1-x^2)(1-x^3)\ldots(1-x^t).$$

Since the coefficient of $z^k x^n$ in (3) is the same as the coefficient of x^n in $\phi_k(x)$, the theorem follows.

COROLLARY. *The number of partitions of n into exactly k summands is the same as the number of partitions of n in which the largest part is k, repetitions being allowed in both cases.*

Incidentally, the theorem gives the generating function for $p(n, k)$. The corollary shows that $p(n, k)$ denotes also the number of partitions of n in which the largest part is k.

THEOREM 2. *The number $q(n, k)$ of partitions of n into exactly k distinct summands is the coefficient of x^n in*

$$x^{k(k+1)/2}/(1-x)(1-x^2)(1-x^3)\ldots(1-x^k).$$

The proof proceeds on the same lines as that of theorem 1 and is left to the reader.

COROLLARY. *The number of partitions of n into k distinct parts is the same as the number of partitions of $n - k(k+1)/2$ into parts none of which exceeds k.*

COROLLARY. *The number of partitions of $n + k(k-1)/2$ into k distinct parts is the same as the number of partitions of n into parts the greatest of which is k. This means that we have*

$$q(n + k(k-1)/2, k) = p(n, k).$$

In the same manner, one could consider partitions into k distinct summands taken from a set of numbers in an arithmetical progression. Take for example the case

$$A = \{1, 3, 5, \ldots, 2r - 1, \ldots\}.$$

Then, we have

THEOREM 3. *The number of partitions of n into exactly k distinct odd numbers, is the coefficient of x^n in*

$$x^{k^2}/(1-x^2)(1-x^4)(1-x^6)\ldots(1-x^{2k}).$$

Proof. Let

$$\prod_{r\geqslant 1}(1 + zx^{2r-1}) = 1 + \sum_{t\geqslant 1} z^t g_t(x).$$

Then, replacing z by zx^2 one can proceed on the same lines as in proof of theorem 2 and complete the proof of this theorem.

Exercise

The number of partitions of n into exactly k odd parts all distinct is the same as the number of partitions of $(n + k)/2$ into k distinct parts.

4. Recurrence relations

Let $p(n, k)$ denote the number of partitions of n into exactly k parts, or what is the same thing into parts the largest of which is k.

If from any such partition of n, we take away one part equal to k, we are left with a partition of $(n - k)$ in which the largest part is k or some smaller number including 0 (as happens when $n = k$ itself). Hence for $n \geqslant k$, we have

$$p(n, k) = p(n-k, 0) + p(n-k, 1) + p(n-k, 2) + \cdots + p(n-k, k), \tag{1}$$

where we take

$$p(0, 0) = 1,$$

$$p(n, 0) = 0, \quad n \geqslant 1,$$

$$p(n, k) = 0, \quad 0 < n < k.$$

Changing n into $(n-1)$ and k into $(k-1)$ in (1), we get

$$p(n-1, k-1) = p(n-k, 0) + p(n-k, 1) + p(n-k, 2) + \cdots + p(n-k, k-1). \tag{2}$$

Hence

$$p(n, k) = p(n-1, k-1) + p(n-k, k). \tag{3}$$

Since $p(n, 1) = 1 = p(n, n)$ for each $n \geqslant 1$, the recurrence relation (3) can be used to compute the values of $p(n, k)$ as in the following table.

k \ n	1	2	3	4	5	6	7	8	9	10
1	1	1	1	1	1	1	1	1	1	1
2	0	1	1	2	2	3	3	4	4	5
3	0	0	1	1	2	3	4	5	7	8
4	0	0	0	1	1	2	3	5	6	9
5	0	0	0	0	1	1	2	3	5	7
6	0	0	0	0	0	1	1	2	3	5
7	0	0	0	0	0	0	1	1	2	3
8	0	0	0	0	0	0	0	1	1	2
9	0	0	0	0	0	0	0	0	1	1
10	0	0	0	0	0	0	0	0	0	1
$p(n)$	1	2	3	5	7	11	15	22	30	42

Extensive tables of values of $p(n, k)$ are given in the Royal Society 'Tables of Partitions' by H. Gupta, C. E. Gwyther and J. C. P. Miller, University Press, Cambridge, 1958.

The method used in the computation of these tables can be briefly described as follows. For integers $k, r > 1$, let $B_{k-1,r}$ denote the row matrix

$$[p(rk-1, k-1) \quad p(rk, k-1) \quad \ldots \quad p(rk+k-2, k-1)].$$

Then, we have

$$[p(rk, k) \quad p(rk+1, k) \quad p(rk+2, k) \quad \ldots \quad p(rk+k-1, k)] = B_{k-1,1} + B_{k-1,2} + \cdots + B_{k-1,r}.$$

The statement is a direct consequence of the recurrence relation

$$p(n, k) = p(n-1, k-1) + p(n-k, k).$$

Thus, for $k = 4$, the tables give

$$B_{3,1} = [\,1 \quad 1 \quad 2 \quad 3\,]$$

$$B_{3,2} = [\,4 \quad 5 \quad 7 \quad 8\,]$$

$$B_{3,3} = [10 \quad 12 \quad 14 \quad 16]$$

$$B_{3,4} = [19 \quad 21 \quad 24 \quad 27]$$

and we obtain

$$\begin{aligned}
[p(4,4) \quad p(5,4) \quad p(6,4) \quad p(7,4)\ ] &= [\ 1 \quad 1 \quad 2 \quad 3] \\
[p(8,4) \quad p(9,4) \quad p(10,4) \quad p(11,4)] &= [\ 5 \quad 6 \quad 9 \quad 11] \\
[p(12,4) \quad p(13,4) \quad p(14,4) \quad p(15,4)] &= [15 \quad 18 \quad 23 \quad 27] \\
[p(16,4) \quad p(17,4) \quad p(18,4) \quad p(19,4)] &= [34 \quad 39 \quad 47 \quad 54]
\end{aligned}$$

The process was called *block addition* for obvious reasons and proved useful even in the computation of the number of partitions of bipartite numbers [see Tables of Partitions of Gaussian Integers, prepared by M. S. Cheema under the guidance of H. Gupta; National Institute of Sciences of India (now Indian National Science Academy), 1956].

An alternative method of computing a table of values of $p(n, k)$ is the following. We have

$$\begin{aligned}
p(n+1, 1) &= p(n, 1) \\
p(n+2, 2) &= p(n, 1) + p(n, 2) \\
p(n+3, 3) &= p(n, 1) + p(n, 2) + p(n, 3) \\
&\vdots \\
p(2n, n) &= p(n, 1) + p(n, 2) + p(n, 3) + \cdots + p(n, n) = p(n).
\end{aligned}$$

Thus the successive sums of the entries in a row provide the 'diagonal entries' for the rows ahead.

Thus for $n = 5$, we have

$n\rightarrow$ / $k\downarrow$	5	6	7	8	9	10
1	1	1				
2	2		3			
3	2			5		
4	1				6	
5	1					7

4.1. The recurrence formula for $P(n, k)$

The reader has already obtained the generating function for $P(n, k)$ the number of partitions of n in which the least part is k. It is given by

$$x^k/(1-x^k)(1-x^{k+1})(1-x^{k+2})\ldots(1-x^{k+r})\ldots$$

If from any partition of n in which the least part is k, we delete one part equal to k, we are left with a partition of $(n-k)$ in which the least part is either k or some larger integer, provided of course that $n > 2k$. Hence

$$\begin{aligned} P(n, k) &= P(n-k, k) + P(n-k, k+1) + P(n-k, k+2) + \cdots \\ &\quad + P(n-k, n-k) \\ &= P(n-k, k) + P(n+1, k+1), \end{aligned} \tag{1}$$

where we take

$$P(m, k) = 0, \quad 0 \leqslant m < k; P(m, m) = 1.$$

Writing (1) in the form

$$P(n, k) - P(n+1, k+1) = P(n-k, k),$$

we can easily show that for any integer $h \geqslant 0$,

$$\begin{aligned} P(n, k) = P(n+h+1, k+h+1) + P(n-k, k) + P(n-k, k+1) + \cdots \\ + P(n-k, k+h). \end{aligned}$$

Let

$$\begin{aligned} n &= (r+1)k + t, \quad 0 \leqslant t < k, \\ t &= rh + s, \qquad 0 \leqslant s < r. \end{aligned}$$

Then

$$P(rk+k+t, k) = P(rk+k+t+h+1, k+h+1) + \sum_{j=0}^{h} P(rk+t, k+j). \tag{2}$$

Since

$$rk + k + t + h + 1 = (r+1)(k+h) + s + 1 < (r+1)(k+h+1),$$

and

$$rk \leqslant rk + t \geqslant r(k+h) \quad \text{but} \quad < (r+1)k,$$

(2) enables us to express $P(rk + k + t, k)$ as a sum of terms of the form

$$P(rk' + t', k'), \quad t' < k';$$

it thus serves as a reduction formula.

Example. Let $k > 12, r = 5, t = 11$, then we have

$$\begin{aligned} P(6k+11, k) = P(6k+14, k+3) + P(5k+11, k) + P(5k+11, k+1) \\ + P(5k+11, k+2). \end{aligned}$$

In particular, for $k = 15$, this gives

$$\begin{aligned} P(101, 15) &= P(104, 18) + P(86, 15) + P(86, 16) + P(86, 17) \\ &= 176 + 119 + 82 + 57 = 434. \end{aligned}$$

This tallies with the value given in the Tables of Partitions by H. Gupta (Indian Mathematical Society, Madras, 1939). We might state here that

$$P(k+t, k) = 1 \quad \text{or} \quad 0$$

according as $t = 0$ or $0 < t < k$.

If we can obtain formulae for $P(rk+t, k)$, $0 \leqslant t < k$, even for a few small values of r, the work of computing values of $P(n, k)$ for all n up to a preassigned limit will be greatly reduced. We will obtain such formulae shortly.

5. A relation between $P(n, k)$ and $p(n, k)$

Let n be partitioned into j summands of which k is the least. If we diminish each of the j parts by $(k-1)$, we are left with a partition of $(n - jk + j)$ into exactly j parts of which at least one of the parts is 1. Removing this part we get a partition of $(n - jk + j - 1)$ into exactly $(j-1)$ parts with no restriction on the size of these parts. Evidently the number of such partitions is $p(n - jk + j - 1, j - 1)$. This is non-zero if and only if

$$n - jk \geqslant 0.$$

Hence

$$P(n, k) = \sum_{j=1}^{[n/k]} p(n - jk + j - 1, j - 1). \tag{1}$$

In particular, we have

$$P(k, k) = p(0, 0) = 1,$$

$$P(k+t, k) = p(t, 0) = 0, \quad 0 < t < k.$$

$$P(2k+t, k) = p(k+t, 0) + p(t+1, 1) = 1.$$

5.1. $p(n, k)$ for small k

From the recurrence relation for $p(n, k)$ proved earlier, we have for $m \geqslant 1$,

$$p(2m, 2) = p(1, 1) + p(3, 1) + p(5, 1) + \cdots + p(2m-1, 1) = m,$$

$$p(2m+1, 2) = p(2, 1) + p(4, 1) + p(6, 1) + \cdots + p(2m, 1) = m.$$

Hence

$$p(n, 2) = [n/2], \quad n \geqslant 0. \tag{2}$$

Again for $m \geqslant 1$,

$$\begin{aligned} p(6m, 3) &= p(2, 2) + p(5, 2) + p(8, 2) + \cdots + p(6m-1, 2) \\ &= 1 + 2 + 4 + 5 + \cdots + (3m-2) + (3m-1) \\ &= 3 + 9 + 15 + \cdots + (6m-3) = 3m^2. \end{aligned}$$

Similarly it can be shown that

$$p(6m+1,3)=3m^2+m, \qquad p(6m+2,3)=3m^2+2m,$$

$$p(6m+3,3)=3m^2+3m+1, \quad p(6m+4,3)=3m^2+4m+1,$$

$$p(6m+5,3)=3m^2+5m+2.$$

These results hold even when $m=0$, and are contained in the single formula:

$$p(n,3)=[(n^2+3)/12]. \tag{3}$$

Formulae for $p(n,k)$, $k \leqslant 12$, are given in the Royal Society Tables of Partitions referred to earlier, but they are fairly intricate and we do not give them here.

Using (2) and (3), we now readily get from (1) the following results concerning the function $P(n,k)$. For $0 \leqslant t < k$,

$$\begin{aligned} P(3k+t,k) &= p(2k+t,0)+p(k+t+1,1)+p(t+2,2) \\ &= 2+[t/2], \end{aligned}$$

$$\begin{aligned} P(4k+t,k) &= p(3k+t,0)+p(2k+t+1,1)+p(k+t+2,2)+p(t+3,3) \\ &= 3+[(k+t)/2]+[(t^2+6t)/12]. \end{aligned}$$

Exercises

1. Use the reduction formula of section 4.1 to get the formulae for $P(rk+t,k)$, $r=2,3,4; 0 \leqslant t < k$.
2. Evaluate $P(104, 18)$ and $P(86, 15)$.
3. Assuming that

$$\prod_{j \geqslant 1} (1-zx^j)^{-1} = 1 + \sum_{t \geqslant 1} z^t \phi_t(x)$$

establish the relation between $P(n,k)$ and $p(n,k)$.
4. Show that $P(n+1,1)=p(n)$.
5. Using the formulae for $P(rk+t,k)$, $r=2,3,4$, evaluate completely $P(173, 25)$.
6. Prepare tables of values of $p(n,3)$ and $P(n,4)$ for values of $n \leqslant 25$. Show how these tables can be utilized for finding the values of $p(m)$ for $m \leqslant 21$. Using your method find $p(20)$.
7. Extend the idea suggested by exercise 6.

6. Graphs of partitions

The most common way of representing a partition of a natural number n, is by a two-dimensional array of dots (also called nodes), one dimension indicating the size of each part and the other the number of parts. The array is called a Ferrars graph of the partition. The summands are usually arranged in a non-ascending (or non-descending) order of magnitude from top down. Graph paper can be used with advantage for the purpose.

For example the partition

$$4+3+3+1+1$$

of 12 will be represented by the array

```
o o o o
o o o
o o o
o
o
```

There can be several other ways of representing a partition of n graphically, depending on the nature of the problem and the ingenuity of the person handling it.

In this section, we offer proofs of some theorems in partition theory using graphs.

THEOREM 4. *The number of partitions of n into exactly k summands is the same as the number of partitions of n into parts the largest of which is k.*

Proof. Take any partition of n into exactly k summands and represent it by k rows of dots as explained above. Then reading the graph by columns instead of rows, we get a partition of n in which the largest part is k. Since this correspondence is one–one and onto, the two types of partitions of n are equal in number.

In the graph above, the partition $4 + 3 + 3 + 1 + 1$ of 12 corresponds to the partition $5 + 3 + 3 + 1$ of 12.

THEOREM 5. *The number of partitions of n into exactly k summands is the same as the number of partitions of $(n - k)$ into at most k summands.*

Proof. Let n be greater than k. Then removing the left hand column of k dots from the Ferrars graph of any partition of n into k summands, we get a graph of a partition of $(n - k)$ with k or a smaller number of rows. The process being reversible the theorem follows.

THEOREM 6. *The number $p(n, k)$ of partitions of n into exactly k parts, is the same as the number $q(m, k)$ of partitions of m into exactly k distinct parts, for $m = n + k(k - 1)/2$.*

Proof. Take any partition

$$n = a_1 + a_2 + a_3 + \cdots + a_k, \quad a_j \leqslant a_{j+1}, \quad j \leqslant k - 1,$$

of n into exactly k summands. Represent it graphically, putting a_j dots in the jth row, $j = 1, 2, 3, \ldots, k$. Add one dot on the right of the second row, two dots on the right of the third, three on the right of the fourth, and finally $(k - 1)$ dots on the right of the kth row. This gives a graph of a partition of m into k unequal rows. The process is easily reversed. Hence the theorem.

Here it must be clearly understood that what is meant by the process being reversible is that given *any* partition of $m = n + k(k - 1)/2$ into k distinct parts one can get by the reverse process a partition of n into exactly k summands.

Question. Would the process have been reversible if we had added j dots on the right of the jth row, $j = 1, 2, 3, \ldots, k$?

THEOREM 7. *For* $1 \leqslant k \leqslant n$,

$$p(n, k) = p(n-1, k-1) + p(n-k, k)$$

with $p(0, 0) = 1$.

Proof. The partitions of n into exactly k summands can be placed in two classes:

1. those that contain 1 as a part;
2. those in which every part exceeds 1.

Removing a part equal to 1 from any partition of the first kind, we get a partition of $(n-1)$ into exactly $(k-1)$ parts and conversely. Hence there must be exactly $p(n-1, k-1)$ partitions of n in class one. Again, removing the left hand column of k dots from the graph of a partition of n of class two, we get a partition of $(n-k)$ the graph of which has exactly k rows. Hence there must be exactly $p(n-k, k)$ partitions of n in class two and the theorem follows.

THEOREM 8. *For* $n \geqslant 2$,

$$p(n) = p(n-1) + \sum_{k=1}^{[n/2]} p(n-k, k).$$

Proof. We offer a combinatorial argument which does not appear to have been employed before.

The partitions of n are of two types:

(i) Those in which the largest part is not repeated.
(ii) Those in which the largest part is repeated.

Adding 1 to the largest part, we can obtain from the set of partitions of $(n-1)$ all the partitions of n of type (i) and each only once.

Again, if k is the largest part in a partition of n of type (ii), then removing one such part, we get a partition of $(n-k)$ in which the largest part is k and conversely. The contribution of such partitions to partitions of n is, therefore, given by $p(n-k, k)$. Letting k vary now, we get the result in the theorem.

THEOREM 9. *The number of partitions of n into odd summands is the same as the number of partitions of n into distinct parts.*

Proof. Let

$$a_1, a_2, \ldots, a_k; \quad a_j \leqslant a_{j+1}, \quad j = 1, 2, \ldots, k-1,$$

be a partition of n into odd parts. Since each one of the a's is odd, we get a graph of the partition by placing a_1 dots in the first row, a_2 dots in the second row, and so on, and finally a_k dots in the kth row, in such a way that the middle dot of any row falls just below the middle dot of the row above it, as in the following diagram drawn with

$$n = 45, \quad a_j = 1, 1, 3, 5, 5, 5, 5, 7, 13.$$

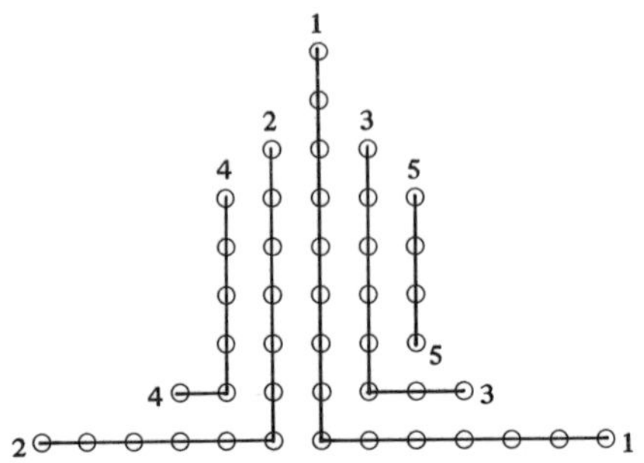

Connect the dots on the central vertical axis with those on its right in the bottom row, and then alternately on the left and right of this axis, as shown in the graph by lines numbered 1, 2, 3, 4 and 5. Let b_j denote the number of points on the line numbered j. Then, since the number of points on any line is at least one less than the number of nodes on the preceding line, the b's provide a partition of n into distinct parts. In our diagram this partition is $15 + 12 + 8 + 6 + 4$.

Thus, to every partition of n into odd parts corresponds a unique partition of n into distinct parts. This, however, does not establish a one–one correspondence between the partitions of the first kind and those of the second. We must show that, given any partition of the second kind, there is a unique partition of the first kind from which it can arise. To prove this, we make the following observations:

(i) The horizontal sections of paths 1 and 2 are on the same level, those of 3 and 4 one level above, and so on.

(ii) There is one more node on the horizontal section of 1 than on that of 2, one more on the horizontal section of 3 than on that of 4, and so on.

(iii) The vertical sections of paths 2 and 3 start at the same level, those of 4 and 5 at the same level, and so on.

(iv) The last path consists of a vertical line of dots if the number of paths in the graph is odd and of a horizontal line of dots if the total number of paths is even. It may, however, consist of a single node.

These structural properties of the graph enable us to reconstruct the graph from the given values of the b's. Take the central vertical axis as the axis of Y, positive upwards, and the bottom row as the axis of X, positive to the right. Then, the corner stones of paths 1, 2, 3, etc., are the points

$$(0, 0), (-1, 0), (1, 1), (-2, 1), (2, 2), (-3, 2), \ldots$$

We explain the process of reconstruction with the help of an example.

Consider the partition

$$12 + 9 + 7 + 6 + 5 + 2$$

of 41 into six distinct parts. Evidently, the number of paths in our graph will be six (an even number). Hence, the last path must consist of a row with two nodes. The previous path must have three nodes on the horizontal section. The vertical section of the fourth path starts at the same level as that of the fifth. The paths are, thus, completed from the end backwards and the theorem is established. The reconstructed graph is:

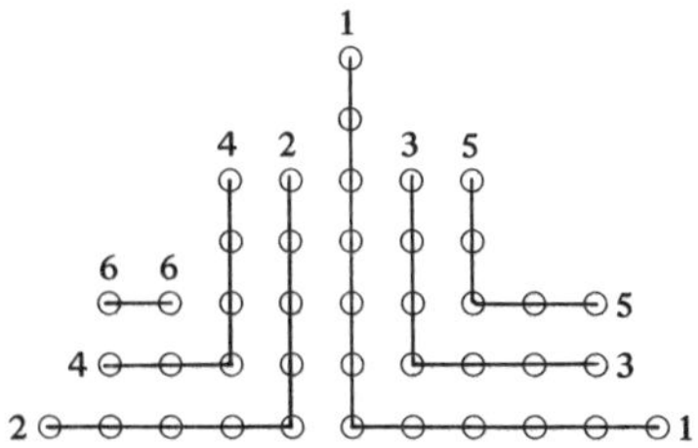

This means that the corresponding partition of 41 into odd parts is

$$1 + 1 + 5 + 5 + 9 + 9 + 11.$$

There is another method of setting up a one–one correspondence between the partitions of n into distinct summands and those into odd parts. This can be described as follows.

Given a partition of n into distinct summands, we replace each even summand by its two halves. The process is repeated till no even summands are left. Take, for example, the partition

18, 12, 9, 6, 5, 3

of 53 into distinct parts. This is successively replaced by the partitions:

9, 9, 6, 6, 9, 3, 3, 5, 3

9, 9, 3, 3, 3, 3, 9, 3, 3, 5, 3.

We thus arrive at the partition

9, 9, 9, 5, 3, 3, 3, 3, 3, 3, 3

of 53 into odd parts. In the reverse direction, we get

18, 9, 5, 6, 6, 6, 3

18, 9, 5, 12, 6, 3

thus retrieving the original partition.

Exercise

To what partition of 53 into odd parts does the partition

18, 12, 9, 6, 5, 3

correspond in the graphical method?

6.1. Euler's identity

The remarkable identity due to Euler states that

$$\prod_{r \geqslant 1} (1 - x^r) = 1 + \sum_{j \geqslant 1} (-1)^j \{x^{j(3j-1)/2} + x^{j(3j+1)/2}\}$$

which implies and is in fact equivalent to

THEOREM 10. *The number of partitions of n into an even number of distinct natural numbers is the same as the number of partitions of n into an odd number of distinct parts, except when n is of one or other of the forms:*

$$j(3j-1)/2 \quad \text{or} \quad j(3j+1)/2, \quad j > 0.$$

In these cases there is just one more partition of the even or odd kind, according as j is even or odd.

Proof. The proof we give here is due to F. Franklin [*Comptes Rendus,* **92** (1881), 448–450] and is the most elegant of the proofs available at present.

Let

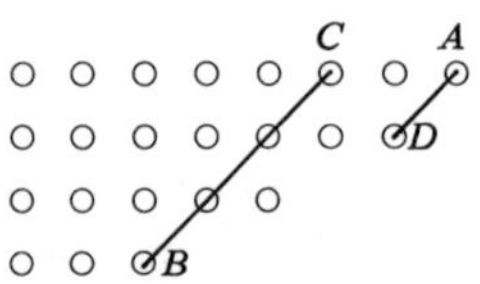

be the graph of any partition of n into distinct parts. Let A and B denote the points to the extreme right on the first and the last rows. Evidently, $A = B$, i.e. A and B coincide, if and only if there is only one row in the graph. Through B draw a line with slope 1 meeting the first row at a point C of the graph. If $A \neq C$, draw through A a line AD parallel to CB. Call CB and AD the left and right slope lines of the graph. When $A = C$, the two slope lines merge into one.

Let r denote the number of nodes on the last row, s the number of those on the right slope and t of those on the left slope. Evidently, if $A = C$, then $s = t$, and if $A \neq C$ then $s < t$. From the given graph of a partition of n of one kind, we will now try to construct a graph of a partition of n of the other kind. For this purpose, we first take the points on the last row and place them on a line parallel to AD and to the right of it. If this does not give a proper graph, then we take the nodes on AD and place them on a row below the last row in our graph. There can be no graph which will permit both these operations, giving a proper graph in each case, but there may be some graphs which will not admit any of the two operations.

In our graph, for example, only the second operation is possible, and it gives

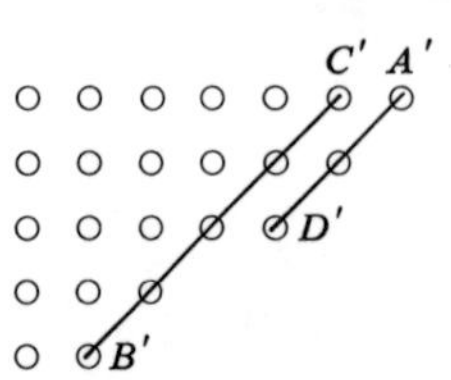

Note that the original graph was of the even kind and the new graph is of the odd kind. Note further that on this new graph only the first operation can be performed and this takes us back to the original graph. The first operation reduces the number of rows by one while the second increases the number of rows by one.

The problem can be analysed as follows.

(i) If $A \neq C, s < t$.

The first operation is possible if $r \leqslant s$, the second if $r > s$.

(ii) If $A = C$, we have $s = t$.

In this case, the first operation is possible when $r < s$, and the second when $r > s + 1$. When, however, $r = s$ or $s + 1$, none of the two operations can be performed.

When $r = s\ (= t)$, the numbers of nodes on the rows starting from the bottom are

$$r, r+1, r+2, \ldots, r+t-1;$$

so that

$$n = t(2r + t - 1)/2 = t(3t - 1)/2.$$

When $r = s + 1 = t + 1$, we get in the same manner

$$n = t(2r + t - 1)/2 = t(3t + 1)/2.$$

Now, we assert that no natural number can be expressed in both of the forms

$$t(3t - 1)/2 \quad \text{and} \quad t(3t + 1)/2, \quad t > 0.$$

If this were possible then for some positive integers t_1 and t_2 we will have

$$t_1(3t_1 - 1)/2 = n = t_2(3t_2 + 1)/2.$$

This can hold if and only if

$$(t_1 + t_2)(3t_1 - 3t_2 - 1) = 0.$$

Since

$$t_1 + t_2 > 0 \quad \text{and} \quad 3(t_1 - t_2) \neq 1$$

our assertion is proved.

Thus, we have shown that except when n is of one or the other of the forms $t(3t - 1)/2$ or $t(3t + 1)/2$, the partitions of n into distinct parts can be paired off so that one partition in each pair has an even number of parts and the other an odd number of parts. In the exceptional cases, after the pairing off, we are left with just one partition of the even or the odd kind according as t is even or odd. Euler's identity is thus proved.

Explicitly speaking, we have shown that

$$\prod_{r \geqslant 1} (1 - x^r) = 1 + \sum_{t \geqslant 1} (-1)^t (x^{t(3t-1)/2} + x^{t(3t+1)/2})$$

$$= 1 - x - x^2 + x^5 + x^7 - x^{12} - x^{15} + x^{22} + x^{26} - \cdots$$

6.2. A recursion formula for $p(n)$

As a direct application of Euler's identity, we have

$$\prod_{r\geq 1} (1-x^r)^{-1} = 1/(1-x-x^2+x^5+x^7-x^{12}-x^{15}+x^{22}+x^{26}-\cdots)$$

But the left side in this relation is the generating function of $p(n)$, the number of unrestricted partitions of n. Hence

$$(1-x-x^2+x^5+x^7-x^{12}-x^{15}+x^{22}+x^{26}-\cdots)\sum_{j\geq 0} p(j)x^j = 1.$$

Comparing the coefficients of x^n, $n\geq 0$ in this identity, we get

$$p(0)=1;$$

$$p(1)-p(0)=0;$$

$$p(2)-p(1)-p(0)=0;$$

$$\vdots$$

so that

$$p(n)=p(n-1)+p(n-2)-p(n-5)-p(n-7)+p(n-12)+p(n-15) \\ -\cdots, \tag{1}$$

where we stop just before the argument of p becomes negative. Formula (1) enables us to compute the values of $p(n)$ successively, without the intervention of the functions $p(n, k)$ or $P(n, k)$. The best way of presenting the work of computation is the following.

In a six-column table with column headings $A, B, \ldots, F$ insert the values of $t(3t-1)/2$ and $t(3t+1)/2$, $t\geq 1$, for odd values of t in the column headed A and for even values of t in the column F. Assume that the number $p(j)$ has already been computed for values of $j\leq n-1$. Then to find the value of $p(n)$ enter in columns B and E numbers ≥ 0 obtained by subtracting from n the numbers already inserted in columns A and F. Now put down in columns C and D, the numbers of partitions of the numbers in columns B and E. Then, the difference between the sum of the entries in column C and those in column D, is the required value of $p(n)$. Thus for $n=35$ we have the following table.

A	B	C	D	E	F
1	34	12310	5604	30	5
2	33	10143	3718	28	7
12	23	1255	101	13	22
15	20	627	30	9	26
35	0	1			51
		24336	– 9453		
		$p(35) = 14883$			

It is essentially in this manner that H. Gupta computed his table of values of $p(n)$ for values of n up to 1000. (Royal Society Tables of Partitions.)

We give here values of $p(n)$ for $n = 100(100)1000$.

n	$p(n)$
100	1905 69292
200	397 29990 29388
300	9 25308 29367 23602
400	6727 09005 17410 41926
500	23 00165 03257 43239 95027
600	4580 04788 00814 43085 53622
700	6 03782 85202 83447 46110 28659
800	573 30521 72321 42250 44569 11979
900	41587 36811 90459 05478 41143 65430
1000	24 06146 78640 32622 47369 21497 27991

For ready reference, values of $p(n)$ for $n \leqslant 100$ are given at the end of this chapter.

7. Durfee's identity

The identity may be stated as

THEOREM 11. *$p(n)$ is the coefficient of x^n in*

$$\sum_{r \geqslant 1} \frac{x^{r^2}}{(1-x)^2(1-x^2)^2 \ldots (1-x^r)^2}$$

for every natural number n.

Proof. The Ferrars graph of a partition of n contains a maximal square with r nodes on each side. The remaining part of the graph consists of two tails. The tail to the right of the square represents a partition of some number into at most r parts. The tail below the square is a partition in which the largest part does not exceed r. Any or both of these tails may, of course, be empty.

Take, for example, the partition graph

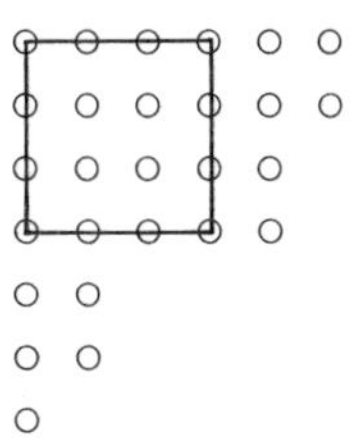

representing a partition of 27. The maximal square has four nodes on each side. The tail on the right of the square represents a partition of 6 into at most four parts, while the tail below the square is a partition of 5 in which the largest part does not exceed four. Thus a one–one correspondence is set up between the partitions of n and the two tailed Durfee square.

If m_1 is the number of nodes in one of the tails and m_2 in the other, then we have

$$n = r^2 + m_1 + m_2.$$

Since the generating function for the number of partitions of m into at most r parts or into parts the largest of which does not exceed r is

$$1/(1-x)(1-x^2)\dots(1-x^r)$$

the identity follows.

7.1. Gordon's identity

Similar to the Durfee identity, is the identity

$$\sum_{n\geqslant 0} p(n)x^n = \sum_{r\geqslant 1} \frac{x^{r(r-1)}}{(1-x)^2(1-x^2)^2\dots(1-x^{r-1})^2(1-x^r)}$$

which B. Gordon gave in 1962. The proof runs on the same lines as that for Durfee's identity, the only point of departure being that we select the maximal rectangle with r nodes on one side and $(r-1)$ nodes on the other, in place of the square.

8. The Jacobi identity

In the following, unless indicated otherwise, j runs over integers from $-\infty$ to ∞, i runs over integers from 0 to ∞, and n over all natural numbers from 1 to ∞. In accordance with our convention x will continue to denote a real variable and $|x|$ will be <1, so that all products will be convergent.

Our object here will be to expand the infinite product

$$F(a,x) = \prod_n \{(1-ax^{n-1})(1-a^{-1}x^n)(1-x^n)\} \tag{1}$$

where a is an arbitrary real number, not equal to zero.

Let

$$F(a,x) = \sum G_j a^j \tag{2}$$

where G's are independent of a and depend on x alone. We observe that

$$F(ax,x) = \frac{1-a^{-1}}{1-a} F(a,x) = -a^{-1}F(a,x) \tag{3}$$

$$F(a^{-1},x) = -a^{-1}F(a,x). \tag{4}$$

Hence

$$\sum G_j a^j x^j = -\sum G_j a^{j-1} = \sum G_j a^{-j},$$

so that for all integral values of j, we have

$$G_{j+1} = -G_j x^j, \tag{5}$$

$$G_{-j} = -G_{j+1}. \tag{6}$$

From (5), we get

$$G_j = (-1)^j G_0 x^{j(j-1)/2}. \tag{7}$$

Hence, we must have

$$F(a,x) = G_0 \sum (-1)^i (a^{-i} - a^{i+1}) x^{i(i+1)/2}. \tag{8}$$

To evaluate G_0, we replace a by x^2 and x by x^3 simultaneously in (1) and obtain

$$\begin{aligned}\prod(1-x^{3n-1})(1-x^{3n-2})(1-x^{3n})\\ &= G_0 \sum (-1)^i (x^{(3i^2-i)/2} - x^{(3i^2+7i+4)/2})\\ &= G_0 \sum (-1)^i (x^{i(3i-1)/2} - x^{(i+1)(3i+4)/2})\\ &= G_0 \sum (-1)^j x^{j(3j-1)/2}\\ &= G_0 \sum (1-x^n), \quad \text{by Euler's identity.}\end{aligned}$$

This gives $G_0 = 1$. We have thus shown that

$$\prod(1-ax^{n-1})(1-a^{-1}x^n)(1-x^n) = \sum (-1)^i (a^{-i} - a^{i+1}) x^{i(i+1)/2}. \tag{9}$$

Replacing a by $-ax$ and x by x^2 simultaneously in (9), we obtain the Jacobi identity

$$\prod(1-x^{2n})(1+ax^{2n-1})(1+a^{-1}x^{2n-1}) = \sum a^j x^{j^2}.$$

Again, dividing the two sides of (9) by $(1-a)$, we get

$$\prod(1-ax^n)(1-a^{-1}x^n)(1-x^n) = \sum (-1)^i \frac{a^{-i} - a^{i+1}}{1-a} x^{i(i+1)/2}.$$

Now letting a tend to 1, we get another famous identity of Jacobi:

$$\prod(1-x^n)^3 = \sum (-1)^i (2i+1) x^{i(i+1)/2}.$$

This will be particularly useful to us in the next chapter.

8.1. A combinatorial equivalent of the Jacobi identity due to Carlitz

In the Jacobi identity

$$\prod(1-x^{2n})(1+x^{2n-1}z)(1+x^{2n-1}z^{-1}) = \sum x^{j^2} z^j, \tag{1}$$

replace z by x/y and then x^2 by xy. This gives

$$\prod(1-x^ny^n)(1+x^{n-1}y^n)(1+x^ny^{n-1})=\sum x^{j(j+1)/2}y^{j(j-1)/2}. \tag{2}$$

We assert that (1) and (2) are equivalent. This is shown as follows. In (2), replace y by x/z and get

$$\prod(1-x^{2n}z^{-n})(1+x^{2n-1}z^{-n})(1+x^{2n-1}z^{-n+1})=\sum x^{j^2}z^{-j(j-1)/2}. \tag{3}$$

In (3), replace x by $xz^{\frac{1}{2}}$. This gives

$$\prod(1-x^{2n})(1+x^{2n-1}z^{-\frac{1}{2}})(1+x^{2n-1}z^{\frac{1}{2}})=\sum x^{j^2}z^{\frac{1}{2}j}. \tag{4}$$

Replacing z by z^2 in (4), we get (1). This proves our assertion.

Changing the signs of x and y in (2), we have

$$\prod(1-x^ny^n)(1-x^{n-1}y^n)(1-x^ny^{n-1})=\sum(-1)^jx^{\frac{1}{2}j(j+1)}y^{\frac{1}{2}j(j-1)}. \tag{5}$$

This yields

THEOREM 12. *The excess of the number of partitions of the bipartite number* (n, m) *into an even number of distinct parts of the type*

$$(a, a-1), (b-1, b), (c, c)$$

over those into an odd number of distinct parts of the same type, is $(-1)^j$ *or* 0, *according as* (n, m) *is or is not of one or the other of the forms*

$$(j(j+1)/2, j(j-1)/2) \quad \textit{and} \quad (j(j-1)/2, j(j+1)/2), \quad j \geqslant 0.$$

An equivalent of theorem 12, also due to Carlitz, is

THEOREM 13. *Let* $C(n, m)$ *denote the number of partitions of* (n, m) *into distinct parts of the type* $(a, a-1), (b-1, b)$. *Then*

$$C(n, m)=p(n-k(k+1)/2)$$

where $k=n-m$ *and we take as usual* $p(-t)=0$ *for* $t \geqslant 1$.

Proof. From the definition of $C(n, m)$, we have

$$\sum_{n,m\geqslant 0} C(n,m)x^ny^m=\prod_{r\geqslant 1}(1+x^ry^{r-1})(1+x^{r-1}y^r). \tag{6}$$

Now (5) shows that the right side of (6)

$$=\prod_{r\geqslant 1}(1-x^ry^r)^{-1}\cdot\sum_{j=-\infty}^{\infty}x^{\frac{1}{2}j(j+1)}y^{\frac{1}{2}j(j-1)}$$

$$=\sum_{t\geqslant 0}p(t)x^ty^t\cdot\left\{1+\sum_{r\geqslant 1}(x^{\frac{1}{2}r(r+1)}y^{\frac{1}{2}r(r-1)}+x^{\frac{1}{2}r(r-1)}y^{\frac{1}{2}r(r+1)})\right\}. \tag{7}$$

Equating the coefficients of x^ny^m in (6) and (7), we get the result stated in the theorem.

Notice that $C(n, m) = C(m, n)$; we therefore assume that $n \geqslant m$. Further, we take $C(0, 0) = 1$. If $n - m = k, k > 0$, then (n, m) cannot be expressed as a sum of fewer than k numbers of the type $(a, a - 1)$, $(b - 1, b)$ where a and b are each $\geqslant 1$. Since the summands are required to be distinct, $(n, n - k)$ has no partition into numbers of the said type if $n < k(k + 1)/2$. Hence

$$C(n, n - k) = 0, \quad 0 < n < k(k + 1)/2. \tag{8}$$

Again, for $n = k(k+1)/2$, $m = k(k - 1)/2$, the only partition of (n, m) into distinct numbers of the type $(a, a - 1)$, $(b - 1, b)$ is

$$(1, 0), (2, 1), (3, 2), \ldots, (k, k - 1).$$

Hence

$$C(k(k + 1)/2, k(k - 1)/2) = 1 = C(k(k - 1)/2, k(k + 1)/2). \tag{9}$$

In the following subsection, we therefore consider only the case

$$n = u + k(k + 1)/2, \quad m = u + k(k - 1)/2, \quad u \geqslant 1, \quad k \geqslant 0.$$

8.2. Cheema's graphical proof of theorem 13

Take a Ferrars graph of a partition of u. Place a triangular array of $k(k + 1)/2$ crosses above the graph. This gives a graph of a partition of n (an example of ingenuity). Separate the graph into two sections by drawing a line parallel to and just above the principal diagonal as in the following diagram:

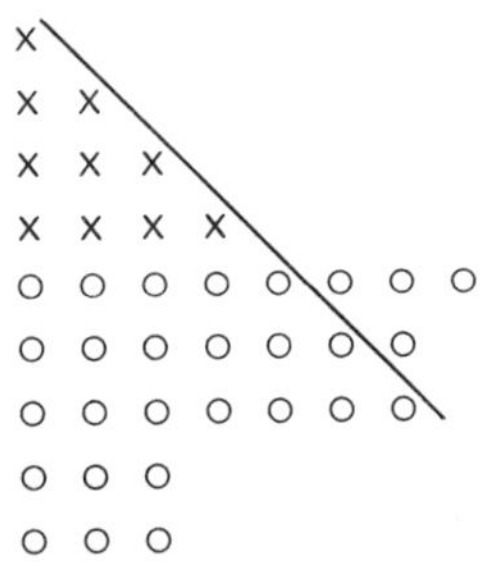

Suppose there are $k + h, h \geqslant 0$, marks (crosses and dots) on the principal diagonal, and the $k + h$ columns below the line, have in order

$$a_1, a_2, a_3, \ldots, a_{k+h}$$

marks on them. Further suppose that the h rows on the right of the line have in order

$$b_1, b_2, b_3, \ldots, b_h$$

nodes. (Here some or all the rows could be empty and the corresponding b's are taken to be zero.) Then the parts of the type $(a, a - 1)$ in the partition of (n, m) are

$$(a_1, a_1 - 1), (a_2, a_2 - 1), (a_3, a_3 - 1), \ldots, (a_{k+h}, a_{k+h} - 1)$$

and those of the type $(b-1, b)$ in the same partition are

$$(b_1, b_1+1), (b_2, b_2+1), \ldots, (b_h, b_h+1).$$

Thus, a one–one correspondence is established between the partitions of u and those of (n, m) into distinct parts of the type $(a, a-1)$ and $(b-1, b)$. Hence

$$C(n, m) = p(u). \tag{10}$$

Evidently when $h = 0$, the partition of (n, m) has no parts of the second type, that is of the type $(b-1, b)$. If $h > 0$, it is clear from the graph that there is only one b that can be zero, and that is b_h.

In our diagram, we have

$$k = 4, \quad u = 28, \quad n = 38, \quad m = 34$$

and the graph gives the partition

$$(9, 8), (8, 7), (7, 6), (4, 3), (3, 2), (2, 1), (1, 0); (3, 4), (1, 2), (0, 1).$$

Remark

As a particular case of (10), we have

$$C(u, u) = p(u).$$

8.3. Cheema's proof of theorem 12

M. S. Cheema's proof of the triple-product identity of Jacobi can now be presented as follows.

Any partition of (n, m) into distinct parts of the type

$$(a, a-1), (b-1, b), (c, c)$$

can be obtained by associating a partition of $(n-j, m-j)$ into distinct parts of the type $(a, a-1)$, $(b-1, b)$ with a partition of (j, j) into distinct parts of the type (c, c). Letting j run over the values $0, 1, 2, \ldots, u$ we get all the partitions of (n, m) of the desired kind.

If the partition of (n, m) has $(k+h)$ parts of the type $(a, a-1)$, h parts of the type $(b-1, b)$ and t parts of the type (c, c), then the total number of parts in the partition is odd or even according as k and t are, or are not, of opposite parity. Evidently, (j, j) has as many partitions into distinct parts of the type (c, c) as j itself has into distinct parts. Now, we know that j has as many partitions into an odd number of distinct parts as it has into an even number of such parts, except when j is of the form $t(3t-1)/2$ or $t(3t+1)/2$, $t > 0$. In the exceptional case, there is one more partition of the even or odd kind according as t is even or odd.

Hence, the excess of the number of partitions of (n, m) into an even number of distinct parts of the type $(a, a'-1)$, $(b-1, b)$, (c, c) over the number of those into an odd number of the same type of parts, all distinct, is given by

$$\sum_j (-1)^{k+t} C(n-j, m-j) = \sum_j (-1)^{k+t} p(u-j), \tag{11}$$

where j runs over numbers of the form $t(3t-1)/2$ and $t(3t+1)/2$, $t \geqslant 0$. From the formula (1) of section 6.2, the sum on the right of (11) is $(-1)^k$ or zero according as u is, or is not, zero, and the theorem follows.

8.4. An identity of Gauss

In 1808, Gauss proved the identity

$$\prod_{n\geqslant 1} (1-x^{2n})(1+x^n) = \sum_{j\geqslant 0} x^{j(j+1)/2}.$$

Here we derive it from the identity

$$\prod_{n\geqslant 1} (1-ax^{n-1})(1-a^{-1}x^n)(1-x^n) = \sum_{i\geqslant 0} (-1)^i(a^{-i}-a^{i+1})x^{i(i+1)/2}$$

proved earlier in this section.

Replace a by -1 in the latter formula, and we get the desired result as follows. The identity can be written in the form

$$(1-a)\prod_{n\geqslant 1} (1-ax^n)(1-a^{-1}x^n)(1-x^n) = \sum_{i\geqslant 0} (-1)^i(a^{-i}-a^{i+1})x^{i(i+1)/2}.$$

Now substitute -1 for a and the identity of Gauss is obtained immediately.

9. Winquist's identity

In the following a and b are non-zero real numbers, n runs over the set of natural numbers, i over the non-negative integers and j over all integers unless indicated otherwise. In 1969, using the Euler and Jacobi identities, Winquist proved the identity:

$$\begin{aligned}\prod &(1-ax^{n-1})(1-bx^{n-1})(1-ab^{-1}x^{n-1})(1-abx^{n-1})\\ &\quad .(1-a^{-1}x^n)(1-b^{-1}x^n)(1-a^{-1}bx^n)(1-a^{-1}b^{-1}x^n).(1-x^n)^2\\ &= \sum(-1)^i(a^{-3i}-a^{3i+3})x^{3i(i+1)/2}\\ &\quad .\sum(-1)^j(b^{-3j}-b^{3j+1})x^{j(3j+1)/2}\\ &\quad -\sum(-1)^i ab^{-1}(b^{-3i}-b^{3i+3})x^{3i(i+1)/2}\\ &\quad .\sum(-1)^j(a^{-3j}-a^{3j+1})x^{j(3j+1)/2}. \end{aligned} \tag{1}$$

In 1972, L. Carlitz and M. V. Subbarao used Jacobi's triple product identity to prove (1). Here we give a modification of Winquist's proof which leads to the identity more directly and uses only Euler's identity.

Proof of Winquist's identity

Denote the product on the left of the identity by $F(a, b, x)$. Then, as Winquist observed, we have

$$F(ax, b, x) = -a^{-3}F(a, b, x), \tag{2}$$

$$F(a^{-1}, b, x) = -a^{-3}F(a, b, x). \tag{3}$$

Besides these two relations, we further find that

$$F(b, a, x) = -ba^{-1}F(a, b, x). \tag{4}$$

Let

$$F(a, b, x) = \sum G_{r,s}a^r b^s, \tag{5}$$

where r and s run independently over all integers from $-\infty$ to ∞, and G's are functions of x alone, that is they are independent of a and b. Then, from (2), (3) and (4), we have for all integral values of r and s,

$$G_{r,s} = -G_{r-3,s}x^{r-3}, \tag{6}$$

$$G_{-r,s} = -G_{r+3,s}, \tag{7}$$

$$G_{r,s} = -G_{s+1,r-1}. \tag{8}$$

Hence

$$\begin{aligned} G_{-r,-s} &= -G_{r+3,-s} = G_{-s+1,r+2} = -G_{s+2,r+2} = G_{r+3,s+1} \\ &= -G_{-r,s+1} = G_{s+2,-r-1} = -G_{-s+1,-r-1}. \end{aligned} \tag{9}$$

In particular, taking $r = 0 = s$, we get

$$G_{0,0} = G_{1,2} = -G_{2,2} = -G_{0,1}. \tag{10}$$

Again, taking $r = -1 = s$ and $r = 0$, $s = -2$, it is easy to show that

$$G_{1,0} = G_{1,1} = G_{2,0} = G_{2,1} = G_{0,2} = 0. \tag{11}$$

Using results (6), (10) and (11), we can express each non-zero G in terms of $G_{0,0}$ and a power of x. Replacing r by $3i$ and s by $3j$ in (9), we obtain

$$\begin{aligned} G_{-3i,-3j} &= -G_{3i+3,-3j} = G_{-3j+1,3i+2} = -G_{3j+2,3i+2} = G_{3i+3,3j+1} \\ &= G_{-3i,3j+1} = G_{3j+2,-3i-1} = -G_{-3j+1,-3i-1}. \end{aligned}$$

We use (6) to evaluate one of the G's, preferably the one with both subscripts positive (there will be two such in any case). Taking $i, j > 0$, we thus have

$$G_{3i+3,3j+1} = -G_{3i,3j+1}x^{3i} = \cdots = (-1)^i G_{3,3j+1}x^{3i(i+1)/2}.$$

Further

$$\begin{aligned} G_{3,3j+1} &= -G_{3j+2,2} = G_{3(j-1)+2,2}x^{3(j-1)+2} \\ &= \cdots = (-1)^{j+1}G_{2,2}x^{2j+3j(j-1)/2} \\ &= (-1)^j G_{0,0}x^{j(3j+1)/2}. \end{aligned} \tag{12}$$

These results hold for all integral values of i, j and it is not necessary any longer to regard i and j as positive.

Since $i(i+1)/2$ has the same set of values as i runs over the set of negative integers as it has when i runs over the set of non-negative integers, we need only let i run over the set of integers from 0 to ∞. But j will have to be allowed to run over all the integers from $-\infty$ to ∞, because $j(3j+1)/2$ has distinct sets of values for negative and non-negative integral values of j.

It is now easy to verify that putting together terms having the same power of x in their expressions, we find that the expression on the right of (1) differs from $F(a, b, x)$ only by the factor $G_{0,0}$. All that we have to do now is to evaluate $G_{0,0}$. The fact that it is independent of a and b is of great help to us in this. Let x be different from zero. Then replacing a by x^3, b by x and x by x^9 simultaneously in $F(a, b, x)$, we find that it becomes

$$\prod(1-x^n)\,.\prod(1-x^{9n}).$$

On the other hand it will be

$$\begin{aligned}&= \sum(-1)^i(x^{9i(3i+1)/2} - x^{9(i+1)(3i+2)/2})\\ &\quad .\sum(-1)^j(x^{3j(9j+1)/2} - x^{(3j+1)(9j+2)/2})\\ &\quad - \sum(-1)^i(x^{(3i+1)(9i+4)/2} - x^{(3i+2)(9i+5)/2})\\ &\quad .\sum(-1)^j(x^{9j(3j-1)/2} - x^{3(3j+1)(3j+2)/2}).\end{aligned} \tag{13}$$

Since

$$\sum(-1)^j x^{3(3j+1)(3j+2)/2} = 0,$$

Euler's identity yields

$$\begin{aligned}\sum(-1)^i(x^{9i(3i+1)/2} - x^{9(i+1)(3i+2)/2}) &= \prod(1-x^{9n})\\ &= \sum(-1)^j x^{9j(3j-1)/2}\\ &= \sum(-1)^j(x^{9j(3j-1)/2} - x^{3(3j+1)(3j+2)/2}).\end{aligned}$$

This means that of the four sums on the right of (13), the first and the fourth are each equal to $\prod(1-x^{9n})$. Again, with all congruences modulo 3, the second sum is

$$\sum_{j\equiv 0}(-1)^j x^{j(3j+1)/2} + \sum_{j\equiv 1}(-1)^j x^{j(3j-1)/2}$$

and the third sum is

$$-\sum_{i\equiv 1}(-1)^i x^{i(3i+1)/2} - \sum_{i\equiv 2}(-1)^i x^{i(3i-1)/2}$$

so that the second sum, minus the third sum, is

$$\sum_{j\equiv 0} (-1)^j x^{j(3j+1)/2} + \sum_{j\equiv 1} (-1)^j x^{j(3j-1)/2} + \sum_{i\equiv 1} (-1)^i x^{i(3i+1)/2}$$
$$+ \sum_{i\equiv 2} (-1)^i x^{i(3i-1)/2}. \tag{14}$$

Now recall that i runs over non-negative integers while j runs over the whole set of integers. Then the sum on the right of (14)

$$= \prod (1 - x^n).$$

This gives $G_{0,0} = 1$ and we are through.

Exercise

Show how the four sums on the right of (14) take care of all the exponents of x of the form $t(3t - 1)/2$ and $t(3t + 1)/2$.

Winquist's identity will be useful to us in the next chapter.

10. Two-line partitions

There are so many types of partition problems that it is not even possible to make a mention of them in this chapter. We will, however, take up for consideration the problem of two-line partitions which is simple and interesting.

Consider a partition of a natural number n in which no summand occurs more than twice. Place in set A the summands which appear only once and in set B those that appear twice in the partition.

Let

$$A = \{a_1, a_2, a_3, \ldots, a_h\}$$
$$B = \{b_1, b_2, b_3, \ldots, b_k\}$$

where we assume that

$$a_1 > a_2 > a_3 > \cdots > a_h;$$
$$b_1 > b_2 > b_3 > \cdots > b_k.$$

Evidently, at least one of the two sets is non-empty and the sets A and B are disjoint.

Let the elements

$$a_1, a_2, \ldots, a_h; \quad b_1, b_1, b_2, b_2, \ldots, b_k, b_k$$

which constitute the partition of n presently being considered, be arranged in the form

$$\begin{matrix} c_1 & c_2 & c_3 & \ldots & c_v & \ldots & c_u \\ d_1 & d_2 & d_3 & \ldots & d_v & & \end{matrix} \tag{1}$$

where $u + v = h + 2k$. Then the arrangement in (1) is said to give a two-line partition of n if the following conditions are satisfied:

(i) $c_1 > c_2 > c_3 > \cdots > c_u$,

$d_1 > d_2 > d_3 > \cdots > d_v$.

(ii) For each $t \leqslant v$, $d_t \leqslant c_t$.

Evidently, in an acceptable arrangement, that is an arrangement satisfying conditions (i) and (ii), the first line cannot have fewer elements than the second. Also each member of B must occur once on the first line and once on the second. Further, condition (ii) implies that an element b_j of B cannot appear on the first line earlier than it appears on the second.

We first determine the number $R(A, B)$ of two-line partitions of n that can arise from the given partition of n. We might state here that we include among the two-line partitions of n even those in which the second line is empty.

Example. The seven two-line partitions of 5 are

$$\begin{matrix} 5; & 4 & 1; & 4; & 3 & 2; & 3; & 3 & 1; & 2 & 1. \\ & & & 1 & & & 2 & 1 & & 2 & \end{matrix}$$

As an exercise write out the two-line partitions of 4, 6 and 7.

THEOREM 14. *The number $R(A, B)$ depends only on A and n.*

Proof. If B is empty, there of course is nothing to prove. If A is empty, the only two-line partition of n arising from the given partition is

$$\begin{matrix} b_1 & b_2 & b_3 & \ldots & b_k \\ b_1 & b_2 & b_3 & \ldots & b_k \end{matrix}$$

so that $R(A, B) = 1$, when A is empty. If A and B are both non-empty, let (1) be a two-line partition of n arising from the sets A and B. If $b_1 = c_j$, then for some value of $t \leqslant j$, we have

$$d_t = b_1.$$

Delete b_1 from each of the two rows of (1), and close up the gaps by moving the elements on the right of c_j and d_t one step to the left, thus getting the arrangement

$$\begin{matrix} c_1 \ldots c_{t-1} & c_t & \ldots c_{j-1} & c_{j+1} \ldots c_v \ldots c_u \\ d_1 \ldots d_{t-1} & d_{t+1} & \ldots d_j & d_{j+1} \ldots d_v \end{matrix} \tag{2}$$

This is an acceptable arrangement because

$$c_t \geqslant d_t > d_{t+1}; \ldots \quad c_{j-1} \geqslant d_{j-1} > d_j.$$

Removing $b_2, b_3, \ldots, b_k$ one by one from (2) in this manner, we get a unique two-line partition of

$$n - 2(b_1 + b_2 + \cdots + b_k)$$

formed from the elements of A alone. Insertion of b's on the two lines of this partition reproduces the two-line partition (1) of n. This one–one correspondence between the two sets of partitions proves the theorem.

10.1. The formula for $R(A, B)$

We can now assume that the set A is non-empty and find the number of two-line partitions that can arise from the elements of A alone.

Suppose in any such partition, r of the h elements of A lie on the first line and s on the second. Denote the contribution of this type of partitions to $R(A, B)$ by $T(r, s)$. Evidently, $r + s = h$, and $r \geqslant s$. We take $T(r, s) = 0$ if $r < s$ or if either of them is negative.

If a_h, the least element of A, lies on the first line, then there can be no entry on the second line either below or to the right of a_h. We are thus left to arrange $(r - 1)$ elements out of the remaining $(h - 1)$ on the first line and s on the second. This can be done in $T(r - 1, s)$ ways. On the other hand, if a_h appears on the second line, then we have to fill up the remaining $(s - 1)$ places on the second line and all the r on the first line. This can be done in $T(r, s - 1)$ ways. Hence, we have the recurrence relation

$$T(r, s) = T(r - 1, s) + T(r, s - 1). \tag{3}$$

When $s = 0$, the elements $a_1, a_2, \ldots, a_h$ have all to be placed on the first line and in this order. Hence

$$T(h, 0) = 1. \tag{4}$$

Again, if $s = 1$, any one of the elements of A other than a_1 can be placed on the second line. Hence

$$T(h - 1, 1) = h - 1. \tag{5}$$

Using inductive reasoning and relations (3)–(5), the reader can now prove for himself, that

$$T(h - s, s) = \binom{h}{s} - \binom{h}{s - 1}, \quad s \leqslant [h/2]. \tag{6}$$

Adding up all these results, we get

$$R(A, B) = \binom{h}{[\frac{1}{2}h]}. \tag{7}$$

10.2. Gordon's Theorem

In 1962, B. Gordon proved the interesting

THEOREM 15. *The number $t(n)$ of two-line partitions of n is the same as the number $p(n)$ of unrestricted partitions of n.*

Proof. The formula for $R(A, B)$ implies that the contribution to $t(n)$ of any partition of n in which no part occurs more than twice is the same as the number of ways in which the h unrepeated elements in the partition can be divided into two subsets of $[\frac{1}{2}(h+1)]$ and $[\frac{1}{2}h]$ elements respectively. The elements of the first subset, along with the elements of B, give a partition of some integer m $(<n)$ into $[\frac{1}{2}(h+1)+k]$ distinct parts. So also, the elements of the second subset along with the elements of B, provide a partition of $(n-m)$ into $[\frac{1}{2}h+k]$ distinct parts. The two together give the original partition of n.

Hence, writing g for $(h+2k)$, we get

$$t(n) = \sum_{m=0}^{n} \sum_{g=0}^{\infty} q(m, [\tfrac{1}{2}(g+1)])q(n-m, [\tfrac{1}{2}g])$$

$$= \sum_{m=0}^{n} \sum_{j=0}^{\infty} \{q(m, j+1)q(n-m, j) + q(m, j)q(n-m, j)\}, \tag{8}$$

where $q(u, j)$ denotes, as usual, the number of partitions of u into exactly j distinct parts, and we take $q(0, 0) = 1$. Recall that $q(u, j)$ is the coefficient of x^u in

$$x^{j(j+1)/2}/(1-x)(1-x^2)(1-x^3)\dots(1-x^j).$$

Hence (8) implies that

$$\sum_{n\geqslant 0} t(n)x^n = \sum_{j\geqslant 0} \left\{ \frac{x^{\frac{1}{2}(j+1)(j+2)+\frac{1}{2}j(j+1)}}{(1-x)^2(1-x^2)^2\dots(1-x^j)^2(1-x^{j+1})} + \frac{x^{j(j+1)}}{(1-x)^2(1-x^2)^2\dots(1-x^j)^2} \right\}$$

$$= \sum_{j\geqslant 0} \frac{x^{j(j+1)}}{(1-x)^2(1-x^2)^2\dots(1-x^j)^2(1-x^{j+1})}$$

$$= \sum_{n\geqslant 0} p(n)x^n,$$

by Gordon's identity. Hence the theorem follows.

THEOREM 16. *The number of two-line partitions of n when the summands are odd is $p([\frac{1}{2}n])$.*

The proof proceeds on the same lines as that of theorem 15 and is left to the reader.

Some remarks. It is noteworthy that $R(A, B)$ does not depend on the size of the members of A but only on their number.

Two-line partitions considered in this section are a special case of plane partitions where there can be any number of lines. Even partitions in three-dimensional space can be considered.

11. Equinumerous partitions and Euler's pairs

Let

$$A = \{a_1, a_2, a_3, \ldots, a_r, \ldots\}, \quad a_r < a_{r+1}, \quad r \geqslant 1,$$

$$B = \{b_1, b_2, b_3, \ldots, b_r, \ldots\}, \quad b_r < b_{r+1}, \quad r \geqslant 1,$$

etc., be subsets of natural numbers, with $A - B$ and tA defined as in chapter 4. Also let $p(n, S)$ denote the number of partitions of n into the elements of S, where the parts can be repeated, and let $q(n, S)$ be the number of partitions of n into distinct elements of S. Then, we have the following interesting theorem due to George E. Andrews (1969).

THEOREM 17. *If for all natural numbers n, $p(n, A) = p(n, B)$, then $A = B$.*

Proof. If $A \neq B$, let m be the least element of $A - B$. Then $p(m, A)$ is the number of partitions of m into elements of A which are $\leqslant m$. Thus $p(m, A)$ is 1 plus the number of partitions of m into elements of A which are $\leqslant m - 1$, or 1 plus the number of partitions of m into the elements of B which do not exceed $(m - 1)$. Hence $p(m, A) = 1 + p(m, B)$. This contradicts the hypothesis that $p(n, A) = p(n, B)$ for all n. The contradiction proves the theorem.

THEOREM 18. *$q(n, A)$ and $p(n, B)$ are equal for all natural numbers n if and only if $A - B = 2A$.*

Proof. First assume that $2A \subseteq A$ and $B = A - 2A$. Then, we have

$$\prod_{a \in A} (1 + x^a) = \prod_{a \in A} (1 - x^{2a}) \Big/ \prod_{a \in A} (1 - x^a)$$

$$= 1 \Big/ \prod_{c \in A - 2A} (1 - x^c) = 1 \Big/ \prod_{b \in B} (1 - x^b).$$

Hence $q(n, A) = p(n, B)$ for all n. Next, let $q(n, A) = p(n, B)$ for all n, so that

$$\prod_{a \in A} (1 + x^a) = 1 \Big/ \prod_{b \in B} (1 - x^b),$$

that is

$$\prod_{a \in A} (1 - x^{2a}) \Big/ \prod_{a \in A} (1 - x^a) = \prod_{b \in B} (1 - x^b)^{-1}. \tag{1}$$

If $2A \not\subseteq A$, let

$$C = A - (2A \cap A), \quad D = 2A - (2A \cap A).$$

Evidently then, we have

$$\prod_{d\in D}(1-x^d)\Big/\prod_{c\in C}(1-x^c)=\prod_{b\in B}(1-x^b)^{-1},$$

or what is the same thing

$$\prod_{c\in C}(1-x^c)^{-1}=\prod_{b\in B}(1-x^b)^{-1}\cdot\prod_{d\in D}(1-x^d)^{-1}. \tag{2}$$

Since the elements of C are necessarily distinct, B and D must be disjoint and in view of theorem 17, we must have

$$C=B\cup D. \tag{3}$$

This implies that

$$D\subset C. \tag{4}$$

But by definition C and D are disjoint. D must therefore be a null set. Hence $B=C=A-2A$ and the theorem is proved.

Remark

This theorem of Andrews gives a complete characterization of sets A and B for which $q(n,A)=p(n,B)$ for all natural numbers n. A generalization of the theorem was given by M. V. Subbarao in 1971.

11.1. Some applications of theorem 18

(i) Let

$$A=\{1,2,2^2,2^3,\ldots,2^r,\ldots\}$$

then

$$B=A-2A=\{1\}.$$

Hence the number of partitions of any natural number n into distinct powers of 2 equals the number of partitions of n into 1's. Hence

$$q(n,A)=p(n,B)=1.$$

This means that every natural number has a unique representation as a sum of distinct powers of 2. This result is due to Euler.

(ii) Let

$$A=\{1,2,3,\ldots,r,\ldots\}$$

then

$$B=A-2A=\{1,3,5,\ldots,2r-1,\ldots\}$$

and we have the interesting theorem:

The number of partitions of n into distinct parts equals the number of partitions of n into odd parts.

This theorem is also due to Euler. A proof of the theorem was given earlier in this chapter with the help of graphs.

(iii) Let

$$A = \{a : a = x^2 + y^2; \quad x, y \text{ integers such that} \quad 0 \leqslant x < y\}$$

Then, since

$$2a = (x+y)^2 + (x-y)^2,$$

it follows that $2a \in A$. Hence

$$B = A - 2A = \{b : b = x^2 + y^2, \quad 0 \leqslant x < y, \quad b \text{ odd}\}.$$

This gives the theorem:

The number of partitions of n into distinct summands each of which is representable as a sum of two squares equals the number of partitions of n into odd parts themselves representable as sums of two squares each.

Remark

George E. Andrews and Basil Gordon have done a lot of work on equinumerous partitions which we cannot even touch upon here.

12. Highly restricted partitions

Earlier it had been stated that the restrictions placed on the partitions of numbers n may sometimes be quite stringent. Here we present such an example.

The problem is to find the number $g(n, m, h, k)$ of partitions of n into exactly k summands, just h (and any h) of the natural numbers not exceeding m being used in any partition as summands.

For the conditions to be meaningful, we must have

$$h \leqslant k, \quad h \leqslant m \quad \text{and} \quad n \leqslant mk.$$

Thus, for $n = 10, m = 3, h = 2, k = 4$, there are just two acceptable partitions, namely

$$1, 3, 3, 3; \quad \text{and} \quad 2, 2, 3, 3.$$

THEOREM 19. *The number $g(n, m, h, k)$ is the coefficient of $x^n z^k t^h$ in the expansion of the product*

$$\prod_{j=1}^{m} \{1 + (x^j z + x^{2j} z^2 + x^{3j} z^3 + \cdots) t\}.$$

Proof. If a partition of n of the desired type uses the summands

$$j_1, j_2, j_3, \ldots, j_h$$

then, from h of the m factors in the above product, viz. those with $j = j_1, j_2, j_3, \ldots, j_h$, take the terms

$$(x^{j_r}z + x^{2j_r}z^2 + x^{3j_r}z^3 + \cdots)t, \quad r = 1, 2, 3, \ldots, h$$

and take 1 from the rest of them. In taking the product of these h factors, select from the rth factor a term

$$x^{s_r j_r} z^{s_r}, \quad r = 1, 2, 3, \ldots, h,$$

such that

$$s_1 + s_2 + s_3 + \cdots + s_h = k, \tag{1}$$

and

$$s_1 j_1 + s_2 j_2 + s_3 j_3 + \cdots + s_h j_h = n. \tag{2}$$

Then a partition of n of the required type is obtained by taking the summand j_1 with the frequency s_1, j_2 with the frequency s_2, and so on. Relation (1) ensures that any such partition has k parts, the appearance of t^h in the product ensures that just h and any h of the numbers $j \leqslant m$ occur in any partition. The correspondence between the solutions of (1) and (2) and the acceptable partitions of n being one–one, the theorem follows.

12.1. Recurrence relations for $g(n, m, h, k)$

We have shown that $g(n, m, h, k)$ is the number of solutions of the equations

$$\begin{aligned} a_1x_1 + a_2x_2 + a_3x_3 + \cdots + a_hx_h &= n \\ x_1 + x_2 + x_3 + \cdots + x_h &= k \end{aligned} \tag{3}$$

in positive integers a and x, with

$$1 \leqslant a_1 < a_2 < a_3 < \cdots < a_h \leqslant m.$$

The case $h = 1$ is trivial. For $h > 1$, equations (3) imply that

$$\begin{aligned} (a_2 - a_1)x_2 + (a_3 - a_1)x_3 + \cdots + (a_h - a_1)x_h &= n - a_1k \\ x_2 + x_3 + \cdots + x_h &= k - x_1 \end{aligned} \tag{4}$$

Hence

$$g(n, m, h, k) = \sum_{a_1} \sum_{x_1} g(n - a_1k, m - a_1, h - 1, k - x_1), \tag{5}$$

where evidently a_1 runs over integers $1, 2, \ldots, (n - h + 1)/k$ and x_1 runs over positive integers from 1 to $(k - h + 1)$. Replacing n by $(n - k)$ and m by $(m - 1)$ in (5), we get

$$g(n - k, m - 1, h, k) = \sum_{a_1 \geqslant 2} \sum_{x_1} g(n - a_1k, m - a_1, h - 1, k - x_1). \tag{6}$$

Subtracting (6) from (5), we obtain

$$g(n, m, h, k) - g(n-k, m-1, h, k) = \sum_{x_1} g(n-k, m-1, h-1, k-x_1). \tag{7}$$

Replacing n by $(n-1)$ and k by $(k-1)$ in (7), we get

$$\begin{aligned} g(n-1, m, h, k-1) &- g(n-k, m-1, h, k-1) \\ &= \sum_{x_1 \geqslant 2} g(n-k, m-1, h-1, k-x_1). \end{aligned} \tag{8}$$

Subtracting (8) from (7), we get the reduction formula

$$\begin{aligned} g(n, m, h, k) = g(n-k, m-1, h, k) &+ g(n-k, m-1, h-1, k-1) \\ &+ g(n-1, m, h, k-1) - g(n-k, m-1, h, k-1). \end{aligned} \tag{9}$$

Also for $h = 1$, we have

$$g(n, m, h, k) = \begin{cases} 1, & k \mid n \quad \text{and} \quad n \leqslant km, \\ 0, & \text{otherwise.} \end{cases} \tag{10}$$

Another recurrence formula for $g(n, m, h, k)$ is obtained as follows. The g-type partitions of n can be placed in two classes:

(i) those in which m does not occur as a part,
(ii) those in which m occurs as a part.

The number of partitions in class (i) is $g(n, m-1, h, k)$. The number of those in class (ii) is given by

$$\sum_j g(n-jm, m-1, h-1, k-j),$$

where j runs over positive integers from 1 to $[(n-1)/m]$. Hence

$$g(n, m, h, k) - g(n, m-1, h, k) = \sum_j g(n-jm, m-1, h-1, k-j). \tag{11}$$

Replacing n by $n - m$ and k by $k - 1$ in (11), we get

$$\begin{aligned} g(n-m, m, h, k-1) &- g(n-m, m-1, h, k-1) \\ &= \sum_{j \geqslant 2} g(n-jm, m-1, h-1, k-j). \end{aligned} \tag{12}$$

Finally, subtracting (12) from (11), we get the recurrence

$$\begin{aligned} g(n, m, h, k) = g(n, m-1, h, k) &+ g(n-m, m-1, h-1, k-1) \\ &+ g(n-m, m, h, k-1) - g(n-m, m-1, h, k-1). \end{aligned}$$

The two recurrence relations, being independent, serve as a good check upon one another.

For tables of $g(n, m, h, k)$, see H. Gupta: Highly restricted partitions. *J. Res. Nat. Bur. Standards,* **73B** (1969) 329–350.

12.2. Two theorems concerning $g(n, m, h, k)$

The partition function $g(n, m, h, k)$ has two interesting properties which admit of graphical proofs. These are presented in the following theorems.

THEOREM 20. *For positive integers n, m, h, k such that*

$$h \leqslant k, \quad h \leqslant m, \quad n \leqslant km,$$

we have

$$g(n, m, h, k) = g(k(m + 1) - n, m, h, k).$$

Proof. Let

```
o o
o o
o o
o o o
o o o
o o o o o
```

be the graph of a g-type partition of n. (In the above graph, $n = 17, h = 3, k = 6$, and we take $m = 7$.) Place crosses in each row so that the total number of marks in every row will be $(m + 1)$. The above graph will then take the form:

```
o o x x x x x x
o o x x x x x x
o o x x x x x x
o o o x x x x x
o o o x x x x x
o o o o o x x x
```

The crosses provide an acceptable partition of $k(m + 1) - n$. The correspondence being one–one, the theorem is proved.

THEOREM 21. *For fixed m, h, k with $h \leqslant k, h \leqslant m$,*

$$\sum_{n=1}^{\infty} g(n, m, h, k) = \binom{k-1}{h-1}\binom{m}{h}.$$

Proof. Let

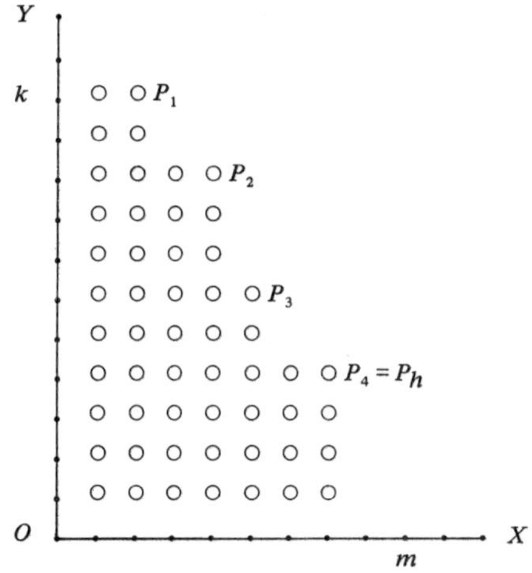

be the graph of a g-partition of n. This k-row graph determines uniquely, and is uniquely determined by, its h corner points

$$P_1, P_2, \ldots, P_h.$$

These in turn are uniquely determined by and uniquely determine two sets of h points each, namely their projections on the axis of X and their projections on the axis of Y. (Mark that the X-axis is drawn one unit of length below the last row and the axis of Y one unit of length to the left of the first column.) The h points on the axis of X are distinct and lie strictly between the origin O and the point $(m + 1, 0)$. Of the h points on the Y-axis, the first lies at $(0, k)$, and the remaining $(h - 1)$ are distinct and lie strictly between O and the point $(0, k)$. The total number of ways in which the h points on the X-axis can be located is $\binom{m}{h}$, and the number of ways in which the $(h - 1)$ points on the Y-axis can be arranged is $\binom{k-1}{h-1}$. From these results the theorem follows immediately.

Problems

1. In extending a table of values of $p(n)$ from $n = 300$ to $n = 400$, using the method described in section 6.2, a tabulator committed a mistake at $n = 328$ but this was realized only after the calculations had been carried on up to $n = 362$. Assuming that no further mistakes had been made, suggest a formula which will enable the tabulator to correct his results without having to recalculate all the wrong answers.
2. The generating function

$$x^k/(1 - x)(1 - x^2) \ldots (1 - x^k)$$

of $p(n, k)$ can be written in the form

$$G(x)/(1 - x^{L_k})^k$$

where L_k denotes the l.c.m. of the first k natural numbers and $G(x)$ is a polynomial in x with integral coefficients and of degree $k(2L_k - k + 1)/2$ in x. For $n \equiv 5 \pmod{12}$, express $p(n, 4)$ in terms of combinatorial functions.

3. If

$$A = \{a : a \not\equiv 0 \pmod 5,\ a\text{'s natural numbers}\}$$

then find a set B of natural numbers, such that for all natural numbers n, we shall have

$$q(n, A) = p(n, B).$$

4. If p is a prime of the form $8j + 1$ or $8j - 1$, then show that given $A = \{a : (a/p) = 1$ where a's are natural numbers$\}$ a set B can be found such that $q(n, A) = p(n, B)$ for all n.

5. Show that the number of partitions of n in which 1 occurs at most once, 2 at most twice, . . . , j at most j times, . . . is the same as the number of partitions of n in which numbers of the form $k(k + 1)$, $k \geqslant 1$, do not appear.

6. (L. Carlitz, 1965.) Let $L(n, m)$ denote the number of arrays

$$\begin{matrix} n_1 & n_2 & n_3 & \dots \\ m_1 & m_2 & m_3 & \dots \end{matrix}$$

such that

$$\min(n_j, m_j) \geqslant \max(n_{j+1}, m_{j+1}), \qquad j = 1, 2, 3, \dots,$$

and

$$n = n_1 + n_2 + n_3 + \cdots$$
$$m = m_1 + m_2 + m_3 + \cdots$$

Then, prove that $L(n, m)$ is equal to the number of partitions of (n, m) into parts of the type $(a, a - 1)$, $(b - 1, b)$, $(2c, 2c)$ where a, b, c are positive integers.

7. (Sylvester's Theorem.) If $A_k(n)$ denotes the number of partitions of n into odd parts, repetitions allowed, just k distinct odd numbers being used as summands in any partition, and $B_k(n)$ the number of partitions of n into distinct parts such that exactly k sequences of consecutive integers appear in each partition, then

$$A_k(n) = B_k(n).$$

Prove the above theorem graphically or otherwise.

8. Let $h(n)$ denote the number of partitions of n in which each part is at least as large as the number of parts. Then express $h(n)$ in terms of the function $p(m, k)$.

Table of values of $p(n)$

n	$p(n)$	n	$p(n)$	n	$p(n)$	n	$p(n)$
1	1	26	2436	51	2 39943	76	92 89091
2	2	27	3010	52	2 81589	77	106 19863
3	3	28	3718	53	3 29931	78	121 32164
4	5	29	4565	54	3 86155	79	138 48650
5	7	30	5604	55	4 51276	80	157 96476
6	11	31	6842	56	5 26823	81	180 04327
7	15	32	8349	57	6 14154	82	205 06255
8	22	33	10143	58	7 15220	83	233 38469
9	30	34	12310	59	8 31820	84	265 43660
10	42	35	14883	60	9 66467	85	301 67357
11	56	36	17977	61	11 21505	86	342 62962
12	77	37	21637	62	13 00156	87	388 87673
13	101	38	26015	63	15 05499	88	441 08109
14	135	39	31185	64	17 41630	89	499 95925
15	176	40	37338	65	20 12558	90	566 34173
16	231	41	44583	66	23 23520	91	641 12359
17	297	42	53174	67	26 79689	92	725 33807
18	385	43	63261	68	30 87735	93	820 10177
19	490	44	75175	69	35 54345	94	926 69720
20	627	45	89134	70	40 87968	95	1046 51419
21	792	46	1 05558	71	46 97205	96	1181 14304
22	1002	47	1 24754	72	53 92783	97	1332 30930
23	1255	48	1 47273	73	61 85689	98	1501 98136
24	1575	49	1 73525	74	70 89500	99	1692 29875
25	1958	50	2 04226	75	81 18264	100	1905 69292

A specimen of the table of values of $P(n, k)$

n \ k	1	2	3	4	5	6	7	$>n/4$
24	1255	210	60	24	10	6		10
25	1575	253	73	27	13	6		11
26		320	88	34	15	8		13
27			110	39	18	9		15
28				50	21	11	6	12
29					26	12	7	13
30						16	8	16
31							10	17

Tables of values of $q(n)$ and $q_o(n)$

n	q	q_o	n	q	q_o	n	q	q_o	n	q	q_o
1	1	1	26	165	12	51	4097	107	76	53250	609
2	1	0	27	192	14	52	4582	117	77	58499	647
3	2	1	28	222	16	53	5120	125	78	64234	686
4	2	1	29	256	17	54	5718	133	79	70488	732
5	3	1	30	296	18	55	6378	144	80	77312	784
6	4	1	31	340	20	56	7108	157	81	84756	833
7	5	1	32	390	23	57	7917	168	82	92864	881
8	6	2	33	448	25	58	8808	178	83	101698	939
9	8	2	34	512	26	59	9792	192	84	111322	1004
10	10	2	35	585	29	60	10880	209	85	121792	1065
11	12	2	36	668	33	61	12076	223	86	133184	1126
12	15	3	37	760	35	62	13394	236	87	145578	1199
13	18	3	38	864	37	63	14848	255	88	159046	1279
14	22	3	39	982	41	64	16444	276	89	173682	1355
15	27	4	40	1113	46	65	18200	294	90	189586	1433
16	32	5	41	1260	49	66	20132	312	91	206848	1523
17	38	5	42	1426	52	67	22250	335	92	225585	1621
18	46	5	43	1610	57	68	24576	361	93	245920	1717
19	54	6	44	1816	63	69	27130	385	94	267968	1814
20	64	7	45	2048	68	70	29927	408	95	291874	1925
21	76	8	46	2304	72	71	32992	437	96	317788	2048
22	89	8	47	2590	78	72	36352	471	97	345856	2166
23	104	9	48	2910	87	73	40026	501	98	376256	2286
24	122	11	49	3264	93	74	44046	530	99	409174	2425
25	142	12	50	3658	98	75	48446	568	100	444793	2574

$q_o(n)$ denotes the number of partitions of n into distinct odd parts.

References

Andrews, George E.: Two theorems of Euler and a general partition theorem. *Proc. Amer. Math. Soc.* **20** (1969), 499–502.

Carlitz, L.: A note on the Jacobi theta formula. *Bull. Amer. Math. Soc.* **68** (1962), 591–592.

——: Generating functions and partition problems. *Proc. Sympos. Pure Math. Vol. VIII, Amer. Math. Soc.*, Providence, R.I. 1965, 144–169.

——; Subbarao, M. V.: On a combinatorial identity of Winquist and its generalization. *Duke Math. J.* **39** (1972), 165–172.

——: Rectangular arrays and plane partitions. *Acta Arith.* **13** (1967/68), 29–47.

Cheema, M. S.: Vector partitions and combinatorial identities. *Math. Comp.* **18** (1964), 414–420.

Gordon, Basil: Two new representations of the partition function. *Proc. Amer. Math. Soc.* **13** (1962), 869–873.

Subbarao, M. V.: Partition theorem for Euler pairs. *Proc. Amer. Math. Soc.* **28** (1971), 330–336.

Winquist, Lasse: An elementary proof of $p(11m+6) \equiv 0 \pmod{11}$. *J. Combinatorial Theory* **6** (1969), 56–59.

CHAPTER 8

PARTITIONS: INEQUALITIES AND CONGRUENCES

1. An inequality for $p(n, k)$

Denoting by $p(n, k)$ – the number of partitions of n into exactly k summands, or equivalently, into summands the greatest of which is k, in the last chapter we saw that

$$p(n, k) = p(n-k, k) + p(n-1, k-1), \tag{1}$$

with the conventions that

$$p(n, k) = \begin{cases} 0, & 0 < n < k, \\ 0, & k < 0, \\ 1, & k = 0 = n. \end{cases} \tag{2}$$

From (1), we readily obtain

$$\begin{aligned} &p(n, k) + p(n-1, k) + p(n-2, k) + \cdots + p(n-k+1, k) \\ &\quad = p(n-1, k-1) + p(n-2, k-1) + p(n-3, k-1) + \cdots + p(k-1, k-1) \\ &\quad = \sum_{m=k}^{n} p(m-1, k-1). \end{aligned} \tag{3}$$

We can now proceed to prove

THEOREM 1. *For* $1 \leqslant j \leqslant n$,

$$\binom{n-1}{j-1} \leqslant j!\, p(n, j) \leqslant \binom{n-1+c_j}{j-1} \tag{4}$$

where c_j depends on j alone.

Proof. The theorem holds for $j = 1$ with $c_j = 0$. Assuming that it holds for each $j \leqslant k-1$, we notice that the right side of (3) is

$$\leqslant \frac{1}{(k-1)!} \sum_{m=k}^{n} \binom{m-2+c_{k-1}}{k-2} \leqslant \frac{1}{(k-1)!} \binom{n-1+c_{k-1}}{k-1},$$

while the left side of (3) is $\geqslant k\, p(n-k+1, k)$. Hence

$$k!p(n-k+1,k) \leqslant \binom{n-1+c_{k-1}}{k-1}$$

so that

$$k!p(n,k) \leqslant \binom{n+k-2+c_{k-1}}{k-1} = \binom{n-1+c_k}{k-1},$$

where

$$c_k = k-1+c_{k-1}. \tag{5}$$

Since $c_1 = 0$, (5) readily gives

$$c_k = \tfrac{1}{2}k(k-1). \tag{6}$$

This is of course not the best value for c_k. The right half of the inequality in the theorem now follows by induction. Again, of the k terms on the left side of (3), $p(n, k)$ is obviously the greatest, and the right side is

$$\geqslant \frac{1}{(k-1)!} \sum_{m=k}^{n} \binom{m-2}{k-2} = \frac{1}{(k-1)!}\binom{n-1}{k-1}.$$

Hence

$$k\,p(n,k) \geqslant \frac{1}{(k-1)!}\binom{n-1}{k-1}.$$

This proves the left half of the inequality in the theorem. The inequalities of theorem 1 provide a simple proof of a theorem of Erdös and Lehner (1941). This may be stated as

THEOREM 2. *For a fixed k and a large n,*

$$k!\,p(n,k) \sim \binom{n-1}{k-1}.$$

Proof. From (4), we have

$$1 \leqslant \frac{k!\,p(n,k)}{\binom{n-1}{k-1}} \leqslant \frac{n-1+c_k}{n-1}\cdot\frac{n-2+c_k}{n-2}\cdot\ldots\cdot\frac{n-k+1+c_k}{n-k+1}. \tag{7}$$

The largest factor on the right of (7) is

$$\frac{n-k+1+c_k}{n-k+1} = 1 + \frac{c_k}{n-k+1}.$$

Since $1 + x < e^x$, the right side of (7) is

$$< \left(\exp\frac{c_k}{n-k+1}\right)^{k-1} = \exp\frac{k(k-1)^2}{2(n-k+1)}.$$

If $n/k^3 = t$, then for large t that is for $k = o(n^{1/3})$,

$$\exp \frac{k(k-1)^2}{2(n-k+1)} \sim \exp\left(\frac{1}{2t}\right) \to 1, \quad \text{as } t \to \infty.$$

Hence the theorem.

2. A more general inequality

Let $p(n, A_k)$ denote the number of partitions of n into the members $1, a_1, a_2, \ldots, a_k$ of the set

$$A = \{1, a_1, a_2, \ldots, a_k, \ldots, a_m, \ldots\}$$

where

$$1 < a_1 < a_2 < \ldots < a_m < \ldots,$$

so that

$$\{(1-x)(1-x^{a_1})(1-x^{a_2}) \ldots (1-x^{a_k})\}^{-1} = \sum_{n=0}^{\infty} p(n, A_k)\, x^n,$$

with $p(0, A_k) = 1, k \geqslant 0$. We take

$$p(z, A_k) = 0$$

unless z is a non-negative integer. Then writing

$$A_k! = a_1 a_2 a_3 \ldots a_k, \quad S_k = a_1 + a_2 + \cdots + a_k;$$

with $A_0! = 1$, and $S_0 = 0$, we have the following generalization of theorem 1.

THEOREM 3. *For $m \geqslant 0$,*

$$\binom{n+m}{m} \leqslant A_m!\, p(n, A_m) \leqslant \binom{n+S_m}{m}$$

Proof. It is easy to show that

$$p(n, A_m) = p(n, A_{m-1}) + p(n - a_m, A_m). \tag{1}$$

Moreover

$$p(n, A_m) \geqslant p(n-1, A_m) \geqslant p(n-2, A_m) \geqslant \cdots \geqslant p(0, A_m) = 1 \tag{2}$$

because adding unity to each of the $p(j, A_m)$ partitions of j, we get some of the partitions of $j+1$, though not generally all. Now, replacing n by $n, n-1, n-2, \ldots, 0$ in succession in (1) and adding the results, we get

$$p(n, A_m) + p(n-1, A_m) + \cdots + p(n - a_m + 1, A_m) = \sum_{j=0}^{n} p(j, A_{m-1}) \tag{3}$$

with $p(n, A_0) = 1$ for each $n \geqslant 0$. Then (1) and (3) give

$$\begin{aligned} p(n, A_1) &= p(n, A_0) + p(n - a_1, A_1) \\ &= p(n, A_0) + p(n - a_1, A_0) + p(n - 2a_1, A_0) + \cdots + p(n - ta_1, A_0) \end{aligned}$$

with $t = [n/a_1]$. Hence

$$p(n, A_1) = t + 1 = [(n + a_1)/a_1]. \tag{4}$$

This shows that the theorem holds for $m = 1$. Let us assume that the theorem holds for $m \leqslant k - 1$. Then in view of the above results, we have

$$\begin{aligned} a_k\, p(n, A_k) &\geqslant \sum_{j=0}^{n} p(j, A_{k-1}) \\ &\geqslant \frac{1}{A_{k-1}!} \sum_{j=0}^{n} \binom{j + k - 1}{k - 1} = \binom{n + k}{k} \Big/ A_{k-1}! \end{aligned}$$

This proves the left half of the inequality in the theorem. On the other hand, we have

$$\begin{aligned} a_k p(n - a_k + 1, A_k) &\leqslant \sum_{j=0}^{n} p(j, A_{k-1}) \\ &\leqslant \frac{1}{A_{k-1}!} \sum_{j=0}^{n} \binom{j + S_{k-1}}{k - 1} \leqslant \binom{n + S_{k-1} + 1}{k} \Big/ A_{k-1}! \end{aligned}$$

so that

$$a_k p(n, A_k) \leqslant \frac{1}{A_{k-1}!} \binom{n + a_k - 1 + S_{k-1} + 1}{k} = \binom{n + S_k}{k} \Big/ A_{k-1}!$$

This proves the other half of the inequality in the theorem. A generalization of the theorem of Erdös and Lehner is given by

THEOREM 4. *For a fixed m and a large n such that* $mS_m = o(n)$,

$$A_m! p(n, A_m) \sim \binom{n + m}{m}.$$

Proof. We have

$$1 \leqslant A_m! p(n, A_m) \Big/ \binom{n + m}{m} \leqslant \binom{n + S_m}{m} \Big/ \binom{n + m}{m}. \tag{5}$$

Now, the expression on the extreme right of (5) lies between

$$\left\{ \frac{n + S_m}{n + m} \right\}^m \quad \text{and} \quad \left\{ \frac{n - m + 1 + S_m}{n + 1} \right\}^m,$$

that is, between

$$\left\{1+\frac{S_m-m}{n+m}\right\}^m \quad \text{and} \quad \left\{1+\frac{S_m-m}{n+1}\right\}^m.$$

For a fixed m and a large n, the greater of the two is

$$<\exp\frac{m(S_m-m)}{n+1}<\exp\frac{mS_m}{n}. \tag{6}$$

If $mS_m = o(n)$, the right side of (6) is $1+o(1)$ and the theorem follows.

3. A generalization of the partition function

Throughout this section, products shall extend from 1 to ∞ and sums from 0 to ∞, unless expressly stated otherwise, and we shall assume that $|x|<1$. We further write

$$\psi(x)=\prod(1-x^j)$$

and for all integral values of r let

$$\{\prod(1-x^j)\}^r=\sum q_r(n)x^n. \tag{1}$$

Evidently, then

$$q_{-1}(n)=p(n) \tag{2}$$

so that $q_r(n)$ is a generalization of the partition function. In the preceding chapter, we have already shown that

$$\psi(x)=\sum_{j=-\infty}^{\infty}(-1)^j x^{j(3j-1)/2}$$

and

$$\{\psi(x)\}^3=\sum(-1)^j(2j+1)x^{j(j+1)/2}.$$

For $r=24$, $q_r(n)$ is Ramanujan's function $\tau(n+1)$. We take $q_r(0)=1$ for every r, and $q_r(z)=0$ unless z is a nonnegative integer.

Differentiating (1) logarithmically with respect to x, we get

$$-r\left\{\frac{x}{1-x}+\frac{2x^2}{1-x^2}+\cdots+\frac{jx^j}{1-x^j}+\cdots\right\}\sum q_r(n)x^n=\sum nq_r(n)x^n.$$

Since the coefficient of x^n in the bracketed expression on the left is $\sigma(n)$, comparing coefficients, we obtain

$$-r\sum_{j=1}^{n}\sigma(j)q_r(n-j)=nq_r(n). \tag{3}$$

From (3), it is easy to prove by induction

THEOREM 5. (M. Newman.) *The function $q_r(n)$ is a polynomial in r of degree n.*

Thus, for example, we have

$$q_r(10) = \binom{r}{2} + 6\binom{r}{3} - 16\binom{r}{4} + 19\binom{r}{5} + 9\binom{r}{6} - 35\binom{r}{7} + 28\binom{r}{8} - 9\binom{r}{9} + \binom{r}{10}.$$

The function $q_r(n)$ has been studied at some length by M. Newman (1957), O. Kölberg (1959), J. M. Gandhi (1963), H. Gupta (1964) among several others.

In the next section, we study* some of the congruence properties of this function. Later in this chapter, we apply these to prove some congruence properties of the partition function $p(n)$ which were first noticed by S. Ramanujan.

4. Congruence properties of $q_r(n)$

To start with, (3.3) readily gives

$$q_r(n) \equiv 0 \left(\bmod \frac{r}{(n,r)}\right).$$

THEOREM 6. *For any prime p,*

$$q_{mp}(n) \equiv q_m(n/p) \pmod p,$$

where we define $q_m(n/p) = 0$ if $p \nmid n$.

Proof. We have

$$\begin{aligned}\{\psi(x)\}^{mp} &= \{(1-x)^p(1-x^2)^p(1-x^3)^p \ldots (1-x^j)^p \ldots\}^m \\ &\equiv \{(1-x^p)(1-x^{2p})(1-x^{3p}) \ldots (1-x^{jp}) \ldots\}^m \pmod p\end{aligned}$$

so that

$$\sum q_{mp}(n)x^n \equiv \sum q_m(j)x^{jp} \pmod p.$$

Hence the theorem.

In general, since

$$\{\psi(x)\}^{mp+r} = \{\psi(x)\}^{mp} \,.\, \{\psi(x)\}^r$$

we shall have

$$\sum q_{mp+r}(n)x^n \equiv \sum q_m(n/p)x^n \,.\, \sum q_r(n)x^n \pmod p.$$

*In the manner of J. M. Gandhi.

Hence

$$q_{mp+r}(n) \equiv \sum_{j=0}^{n} q_m(j/p)q_r(n-j)\,(\text{mod}\,p)$$

$$\equiv \sum_{j=0}^{[n/p]} q_m(j)q_r(n-jp)\,(\text{mod}\,p). \tag{1}$$

As special cases of (1), we have the following results:

(i) Take $r = 1 = m$, then

$$q_{p+1}(n) \equiv \sum_{j=0}^{[n/p]} q_1(j)q_1(n-jp)\,(\text{mod}\,p)$$

$$\equiv \sum_{j=-\infty}^{\infty} q_1(j)q_1(n-jp)\,(\text{mod}\,p)$$

$$\equiv \sum_{j,k} (-1)^{j+k}\,(\text{mod}\,p),$$

where j and k run over the set of all integers, such that

$$2n = j(3j+1) + k(3k+1)p. \tag{2}$$

If (2) has no solution in integers j and k, evidently then

$$q_{p+1}(n) \equiv 0\,(\text{mod}\,p). \tag{3}$$

Now (2) can have such a solution only if the congruence

$$24n+1 \equiv (6j+1)^2\,(\text{mod}\,p)$$

has a solution, that is only if $24n + 1$ is a quadratic residue of p. Hence (3) holds if $\left(\dfrac{24n+1}{p}\right) = -1$.

Proceeding in the same manner, it can be shown that

$$q_{-p+1}(n) \equiv 0\,(\text{mod}\,p) \quad \text{if} \quad \left(\frac{24n+1}{p}\right) = -1. \tag{4}$$

(ii) Next, take $r = 3$, $m = 1$. Then

$$q_{p+3}(n) \equiv \sum_{j=-\infty}^{\infty} q_1(j)q_3(n-pj)\,(\text{mod}\,p)$$

$$\equiv \sum_{j=-\infty}^{\infty} (-1)^j q_3(n - \tfrac{1}{2}j(3j-1)p - jp)\,(\text{mod}\,p)$$

$$\equiv \sum_{j=-\infty}^{\infty} (-1)^j q_3(n - \tfrac{1}{2}j(3j+1)p)\,(\text{mod}\,p).$$

Hence

$$q_{p+3}(n) \equiv 0 \pmod p$$

if the equation

$$2n = k(k+1) + j(3j+1)p$$

has no solution in integers j and k, and hence if

$$\left(\frac{8n+1}{p}\right) = -1. \tag{5}$$

Similarly, it can be shown that

$$q_{-p+3}(n) \equiv 0 \pmod p \quad \text{if} \quad \left(\frac{8n+1}{p}\right) = -1. \tag{6}$$

(iii) Some more congruences are obtained if we start with the obvious relation

$$q_{r+s}(n) = \sum_{j=0}^{n} q_r(j) q_s(n-j). \tag{7}$$

As particular cases of (7), we consider the following:

(a) Take $r = 1 = s$, then

$$q_2(n) = \sum_{j,k} (-1)^{j+k}$$

where j and k run over all the integral solutions of

$$2n = j(3j+1) + k(3k+1). \tag{8}$$

This can be written in the form

$$24n + 2 = (6j+1)^2 + (6k+1)^2. \tag{9}$$

Let $12n + 1 = PQ^2$, where P and Q are positive integers and P is square-free. Then $24n + 2$ can be expressed as a sum of two squares only if P has no prime divisor of the form $4t + 3$. If it has at least one such divisor, then

$$q_2(n) = 0. \tag{10}$$

If P has no divisor of the form $4t + 3$, but Q has a prime divisor p of this form, then we have

$$6j + 1 \equiv 0 \equiv 6k + 1 \pmod p.$$

This however does not lead us to any conclusion about the divisibility of $q_2(n)$ by p.

(b) Take $r = 1$, $s = 3$, then

$$q_4(n) = \sum_{j,k} (-1)^{j+k}(2k+1),$$

where j and k run over the solutions of the equation

$$2n = k(k+1) + j(3j+1), \quad k \geqslant 0, \quad -\infty < j < \infty. \tag{11}$$

We can write (11) in the form

$$24n + 4 = 3(2k+1)^2 + (6j+1)^2. \tag{12}$$

Let $6n + 1 = PQ^2$, where P and Q are positive integers and P is square-free. If P has a prime divisor of the form $3t + 2$, then $24n + 4$ cannot be expressed in the form (12) and we have

$$q_4(n) = 0. \tag{13}$$

If P has no prime divisor of the form $3t + 2$, but Q has a prime divisor p of this form, then we must have

$$2k + 1 \equiv 0 \pmod{p}$$

and consequently

$$q_4(n) \equiv 0 \pmod{p}. \tag{14}$$

From (13) and (14), we conclude that $q_4(n)$ is divisible by p (a prime of the form $3t + 2$) if p divides $6n + 1$. Thus for example, we have for each $m \geqslant 0$,

$$q_4(5m+4) \equiv 0 \pmod{5}. \tag{15}$$

(c) Finally, take $r = 3 = s$. Then

$$q_6(n) = \sum_{j,k} (-1)^{j+k}(2j+1)(2k+1), \tag{16}$$

where j and k run over the nonnegative integral solutions of

$$2n = j(j+1) + k(k+1),$$

or what is the same thing, of the equation

$$8n + 2 = (2j+1)^2 + (2k+1)^2.$$

If $4n + 1 = PQ^2$ where P and Q are positive integers and P is square-free, then $8n + 2$ can be expressed as a sum of two squares only when P has no prime divisor of the form $4t + 3$. If P has such a divisor, then we have

$$q_6(n) = 0. \tag{17}$$

If P has no divisor of the form $4t + 3$, but Q has a prime divisor p of this form, then we must have

$$2j + 1 \equiv 0 \pmod{p}, \quad 2k + 1 \equiv 0 \pmod{p}.$$

Hence

$$q_6(n) \equiv 0 \pmod{p^2}. \tag{18}$$

From (17) and (18), we have the important result:

If p – a prime of the form $4t + 3$ – divides $4n + 1$, then

$$q_6(n) \equiv 0 \pmod{p^2}. \tag{19}$$

For example, we have for each $m \geqslant 0$,

$$q_6(7m + 5) \equiv 0 \pmod{49}. \tag{20}$$

Also

$$q_6(30) \equiv 0 \pmod{121}.$$

5. Ramanujan's conjecture

About 1919, S. Ramanujan proved that for any nonnegative integer m,

$$p(5m + 4) \equiv 0 \pmod 5, \quad p(25m + 24) \equiv 0 \pmod{25},$$

$$p(7m + 5) \equiv 0 \pmod 7, \quad p(49m + 47) \equiv 0 \pmod{49}.$$

He further stated that

$$p(11m + 6) \equiv 0 \pmod{11}.$$

All these results are included in his famous conjecture:

If $p = 5, 7$ or 11 and $24n - 1 \equiv 0 \pmod{p^a}$, $a \geqslant 1$, then

$$p(n) \equiv 0 \pmod{p^a}.$$

This conjecture, astounding as it was, held good till, using H. Gupta's table of values of $p(n)$ for $n \leqslant 300$, S. Chowla found that the conjecture failed for $n = 243$. For this n,

$$24n - 1 = 5831 \equiv 0 \pmod{7^3}$$

while

$$\begin{aligned} p(243) &= 13397\ 82593\ 44888 \\ &\equiv 0 \pmod{7^2} \\ &\not\equiv 0 \pmod{7^3}. \end{aligned}$$

This discovery led to a lot of work, theoretical as well as computational, particularly by D. H. Lehmer (1936). Later it was found that the Ramanujan conjecture failed also when $n = 586$.

G. N. Watson (1938) proved Ramanujan's conjecture for powers of 5, while for powers of 7, he proved the following modification of the conjecture:

If $24n - 1 \equiv 0 (\mathrm{mod}\ 7^b)$, $b > 0$, then $p(n) \equiv 0 (\mathrm{mod}\ 7^d)$ where $d = [(b + 2)/2]$.

Finally, A. O. L. Atkin (1967) settled the problem by proving the conjecture for powers of 11.

Ramanujan's conjecture can now be stated in the form of

THEOREM 7. *If*

$$24n - 1 \equiv 0 \pmod{5^a\, 7^b\, 11^c}$$

then

$$p(n) \equiv 0 \pmod{5^a\, 7^d\, 11^c}$$

where $a, b, c \geqslant 0$ *and* $d = [(b+2)/2]$ *if* $b > 0, d = 0$ *if* $b = 0$.

The proof of this theorem in all its generality is beyond the scope of this book. We shall, however, prove it for $a, b, c \leqslant 1$. For this purpose we need, besides Winquist's identity, the following:

THEOREM 8. *If for a fixed* $h <$ *a given prime* p, *and each* $m \geqslant 0$,

$$q_{p-1}(pm+h) \equiv 0 \pmod{p},$$

and

$$q_{-1}(h) \equiv 0 \pmod{p},$$

then for each $m \geqslant 0$, *we have*

$$q_{-1}(pm+h) \equiv 0 \pmod{p}.$$

Proof. Evidently

$$(1-y)^p \equiv 1 - y^p \pmod{p}.$$

Therefore

$$\{\psi(x)\}^p \equiv \psi(x^p) \pmod{p}, \tag{1}$$

where

$$\psi(x) = (1-x)(1-x^2)(1-x^3)\ldots(1-x^t)\ldots.$$

From (1), we have

$$\{\psi(x)\}^{p-1} \equiv \psi(x^p)/\psi(x) \pmod{p}. \tag{2}$$

Comparing coefficients of x^n in (2), we get modulo p,

$$q_{p-1}(n) \equiv q_{-1}(n) + \sum_{j\geqslant 1} (-1)^j \left\{ q_{-1}\left(n - \frac{pj(3j-1)}{2}\right) + q_{-1}\left(n - \frac{pj(3j+1)}{2}\right) \right\}.$$

Replacing n by $p+h, 2p+h, 3p+h, \ldots$ in this result in succession and recalling that $q_{-1}(-t) = 0$ for $t > 0$, we obtain

$$q_{-1}(mp+h) \equiv 0 \pmod{p}.$$

This is the same thing as

$$p(mp+h) \equiv 0 \pmod{p}.$$

Proof of theorem 7 for a, b, c ⩽ 1

(i) For modulus 5, we have from (4.15) for each $m \geqslant 0$,

$$q_4(5m+4) \equiv 0 \pmod{5}.$$

Also

$$p(4) = 5 \equiv 0 \pmod{5}.$$

The conditions of theorem 8 are thus satisfied and we have

$$p(5m+4) \equiv 0 \pmod{5}.$$

(ii) For modulus 7, (4.20) and the fact that

$$p(5) = 7 \equiv 0 \pmod{7}$$

together give

$$p(7m+5) \equiv 0 \pmod{7}, \quad m \geqslant 0.$$

(iii) To prove the congruence

$$p(11m+6) \equiv 0 \pmod{11}$$

we need Winquist's identity, proved in the preceding chapter. Thus for $i \geqslant 0$, $-\infty < j < \infty$, we have

$$\prod_{n=1}^{\infty} \{(1-ax^{n-1})(1-bx^{n-1})(1-ab^{-1}x^{n-1})(1-abx^{n-1}) \times (1-a^{-1}x^n)(1-b^{-1}x^n)(1-a^{-1}bx^n)(1-a^{-1}b^{-1}x^n)(1-x^n)^2\}$$
$$= \sum_{i,j} (-1)^{i+j}\{(a^{-3i}-a^{3i+3})(b^{-3j}-b^{3j+1}) + (a^{-3j+1}-a^{3j+2})(b^{3i+2}-b^{-3i-1})\} \,.\, x^{\{3i(i+1)+j(3j+1)\}/2}$$

We note that the expression within curly brackets on the right is divisible by $(1-a)(1-b)(1-ab^{-1})(1-ab)$ for each $i \geqslant 0$ and each j, $-\infty < j < \infty$. In fact the expression vanishes if a is replaced by $1, b$ or b^{-1}, and also when b is replaced by 1. Now

$$\lim_{\substack{a\to 1\\ b\to 1}} \frac{(a^{-3i}-a^{3i+3})(b^{-3j}-b^{3j+1}) + (a^{-3j+1}-a^{3j+2})(b^{3i+2}-b^{-3i-1})}{(1-a)(1-b)(1-ab^{-1})(1-ab)}$$
$$= \lim_{a\to 1} \frac{(a^{-3i}-a^{3i+3})(6j+1) - (a^{-3j+1}-a^{3j+2})(6i+3)}{(1-a)^3}.$$

Differentiating the numerator and the denominator thrice with respect to a, the required limit is

$$= (2i + 1)(6j + 1)\{(3i + 1)(3i + 2) - (3j + 1)(3j)\}/2.$$

The Winquist identity now gives

$$2 \prod_{n=1} (1 - x^n)^{10} = \sum_{i,j} (-1)^{i+j}(2i + 1)(6j + 1)\{(3i + 1)(3i + 2)$$
$$- (3j + 1)(3j)\} \,.\, x^{\{3i(i+1)+j(3j+1)\}/2}.$$

Let

$$2n = 3i(i + 1) + j(3j + 1),$$

or what is the same thing

$$24n + 10 = (6i + 3)^2 + (6j + 1)^2.$$

Then for $n = 11m + 6$, we have

$$264m + 154 = (6i + 3)^2 + (6j + 1)^2. \tag{3}$$

In (3), the left side is divisible by 11 (a prime of the form $4t + 3$) and the right side is a sum of two squares, hence we must have

$$6i + 3 \equiv 0 \,(\text{mod } 11), \quad 6j + 1 \equiv 0 \,(\text{mod } 11).$$

Since $p(6) = 11 \equiv 0$ (mod 11), the conditions of theorem 8 are satisfied and we get

$$p(11m + 6) \equiv 0 \,(\text{mod } 11).$$

6. Ramanujan's function $\tau(n)$

As already hinted at in section 3, the function is defined by the relation:

$$x \prod_{j=1}^{\infty} (1 - x^j)^{24} = \sum_{n=1}^{\infty} \tau(n)x^n. \tag{1}$$

so that

$$\tau(n + 1) = q_{24}(n). \tag{2}$$

Since

$$q_r(n) \equiv 0 \,(\text{mod } r/(n, r))$$

we have, for $k = 1, 2, \ldots, 7$,

$$q_{24}(8m + k) \equiv 0 \,(\text{mod } 2).$$

Moreover, since for any prime p,

$$q_{mp}(n) \equiv q_m(n/p) \,(\text{mod } p)$$

we must have

$$q_{24}(8m) \equiv q_3(m) \pmod 2$$
$$\equiv 1 \text{ or } 0 \pmod 2$$

according as m is or is not of the form $j(j+1)/2$, $j \geqslant 0$. Hence $q_{24}(n)$ is odd if and only if

$$1 + n = 8m + 1 = (2j+1)^2.$$

This means that $\tau(n)$ is odd if and only if n is an odd square. We might here just mention that $\tau(n)$ is a multiplicative function but the proof is too intricate to be presented here. We leave it to the reader to prove that

$$\tau(3m) \equiv 0 \equiv \tau(3m+2) \pmod 3.$$

7. Parity of $p(n)$

It is rather strange that a number of congruence properties of $p(n)$ should be known and still we should know no simple method of finding whether $p(n)$ is odd or even. We do not even know explicitly an infinity of values of n for which $p(n)$ is always odd or always even.

In 1920, MacMahon gave a method of finding the parity of $p(n)$ using Jacobi's elliptic products. Here we obtain his result by a simple method. We have from (6.1),

$$\prod_{j=1}^{\infty} \{(1-x^j)^{32}\}^{-1} \cdot \sum_{n=1}^{\infty} \tau(n)x^{n-1} = \prod_{j=1}^{\infty} \{(1-x^j)^8\}^{-1}.$$

Since

$$(1-x^j)^8 \equiv (1-x^{8j}) \pmod 2$$

and

$$(1-x^j)^{32} \equiv (1-x^{32j}) \pmod 2$$

and since $\tau(n)$ is odd if and only if n is an odd square, it follows that

$$\sum_{m=0}^{\infty} p(m)x^{32m} \cdot \sum_{j=0}^{\infty} x^{4j(j+1)} \equiv \sum_{n=0}^{\infty} p(n)x^{8n} \pmod 2.$$

Comparing coefficients of powers of x, we get

$$p(n) \equiv \sum_t p(t) \pmod 2$$

where t runs over the set of positive integers given by the relation

$$8t = 2n - j(j+1), \quad j \geqslant 0.$$

Evidently j need take only those values for which

$$j(j+1) \equiv 2n \pmod 8.$$

It will be seen that the result is not of much practical use.

8. M. Newman's conjecture

In 1960, M. Newman conjectured that for all integers $m \geqslant 2$, each of the congruences

$$p(n) \equiv r \pmod m, \qquad r = 0, 1, 2, \ldots, m-1,$$

has infinitely many solutions in positive integers n. He proved his conjecture for $m = 2, 5, 13, 65$. Later, T. Klove (1968) proved it for $m = 7, 17, 19, 29, 31$. The case $m = 2$ was proved also by O. Kolberg in 1959. Here, we give the proof of the conjecture for $m = 2$.

First, assume that $p(n)$ is odd for every $n \geqslant n_0$. Since $p(2) = 2$, $n_0 \geqslant 3$. Take $j > n_0$ and n such that

$$\frac{3j(j+1)}{2} \leqslant n < \frac{(j+1)(3j+2)}{2}.$$

Evidently then

$$n - \tfrac{1}{2}j(3j+1) \geqslant \tfrac{1}{2}j(3j+3) - \tfrac{1}{2}j(3j+1) = j > n_0.$$

Now, we have

$$p(n) = \{p(n-1) + p(n-2)\} - \{p(n-5) + p(n-7)\} + \cdots$$
$$- (-1)^j\{p(n - \tfrac{1}{2}j(3j-1)) + p(n - \tfrac{1}{2}j(3j+1))\}.$$

Since each $p(\cdots)$ on the right of this relation is odd, we have

$$p(n) \equiv 0 \pmod 2.$$

This contradicts our supposition, so that n_0 does not exist.

Next assume that $p(n)$ is even for every $n \geqslant n_1$. Since $p(7) = 15$, $n_1 > 7$. Take $j > n_1$. Then for $n = \frac{1}{2}j(3j-1) + 1$,

$$n - \tfrac{1}{2}(j-1)(3j-2) = 2j > n_1.$$

This time, we have

$$p(n) = \{p(n-1) + p(n-2)\} - \{p(n-5) + p(n-7)\} + \cdots$$
$$- (-1)^{j-1}\{p(3j-1) + p(2j)\} - (-1)^j p(1).$$

Since $p(1) = 1$ is odd, while every other $p(\cdots)$ on the right is even, our $p(n)$ is odd. This shows that n_1 does not exist and the Newman conjecture is completely proved for $m = 2$.

9. Binary partitions

Denoting by $b(n)$ the number of partitions of n into powers of 2, that is into elements of the set

$$S = \{1, 2, 4, \ldots, 2^r, \ldots\}$$

R. F. Churchhouse conjectured in 1969, that

The highest power of 2 *which divides* $b(2^{k+2} . n) - b(2^k . n)$ *for* $k \geqslant 1$ *and* n *odd is given by* $q_k = [(3k+4)/2]$.

In this section, our object will be to give a proof of this conjecture which will be simpler and more direct than any of the three proofs that have so far appeared – one by Rödseth (1970) and two by Gupta (1971, 1972). To this end, we first study some properties of $b(n)$.

9.1. The generating function of $b(n)$

Let $P(x)$ denote the generating function of $b(n)$, then

$$P(x) = \prod_{r \geqslant 0} (1 - x^{2^r})^{-1} = \sum_{n \geqslant 0} b(n)x^n, \quad b(0) = 1.$$

It will be readily seen that

$$P(x^2) = (1 - x)P(x)$$

so that

$$\sum_{n \geqslant 0} b(n)x^{2n} = (1 - x) \sum_{n \geqslant 0} b(n)x^n.$$

Equating coefficients in this relation, we get

$$b(2n) = b(2n + 1), \tag{1}$$

$$b(n) = b(2n) - b(2n - 1) = b(2n + 1) - b(2n - 1). \tag{2}$$

From (2), it follows without difficulty that

$$b(2n) = b(0) + b(1) + b(2) + \cdots + b(n)$$

$$= \sum_{j=0}^{n} C_1(j)b(n - j), \tag{3}$$

with

$$C_1(j) = 1, \quad j \geqslant 0. \tag{4}$$

From (1) and (3), it follows that $b(n)$ is even except for $n = 0$ and $n = 1$.

Assuming that

$$b(2^k \cdot n) = \sum_{j=0}^{n} C_k(j) b(n - j) \tag{5}$$

for some k, we get

$$\begin{aligned} b(2^{k+1} \cdot n) = b(2^k \cdot 2n) &= \sum_{j=0}^{2n} C_k(j) b(2n - j) \\ &= \sum_{j=0}^{2n} C_k(j)\{b(0) + b(1) + \cdots + b(n - [(j+1)/2])\} \\ &= \sum_{j=0}^{n} C_{k+1}(j) b(n - j), \end{aligned}$$

where

$$C_{k+1}(j) = \sum_{h=0}^{2j} C_k(h). \tag{6}$$

In particular

$$C_{k+1}(0) = C_k(0) = \cdots = C_1(0) = 1$$

i.e.

$$C_k(0) = 1, \quad k \geqslant 1. \tag{7}$$

Starting with (4) and using (6) repeatedly, we get

$$C_2(j) = 2j + 1,$$

$$C_3(j) = 4j^2 + 4j + 1,$$

and so on. Writing

$$B_k(j) = C_{k+2}(j) - C_k(j),$$

we have

$$b(2^{k+2} \cdot n) - b(2^k \cdot n) = \sum_{j=0}^{n} B_k(j) b(n - j), \tag{8}$$

where

$$B_k(j) = \sum_{h=0}^{2j} B_{k-1}(h), \quad k \geqslant 2, \tag{9}$$

with

$$B_1(j) = C_3(j) - C_1(j) = 2^3 \binom{j+1}{2}. \tag{10}$$

9.2. Kemmer's identity

For finding explicit expressions for $B_k(j)$, we need two results. Of these one is the well-known formula

$$\sum_{s=0}^{2j} \binom{s+1}{r} = \binom{2j+2}{r+1} \tag{11}$$

and the other is Kemmer's identity

$$\binom{2j+2}{r} = \sum_{t \geq 0} \binom{r-t}{t} \binom{j+1}{r-t} 2^{r-2t}. \tag{12}$$

The following is a simple combinatorial proof of the identity.

The left side of the identity in (12) represents the number of ways of selecting r objects from a set of $2j + 2$ objects. Pair off the $2j + 2$ objects into $j + 1$ pairs in any manner whatsoever. From the $j + 1$ pairs select $r - t$ pairs. This can be done in $\binom{j+1}{r-t}$ ways. Out of these $r - t$ pairs, select t pairs. This can be done in $\binom{r-t}{t}$ ways. To select r objects take the $2t$ objects contained in the t pairs and select one object from each of the remaining $r - 2t$ pairs. This can be done in 2^{r-2t} ways and the identity follows at once.

9.3. An expression for $B_k(j)$

For convenience write

$$y_k = y_k(j) = \binom{2j+2}{k}, \quad x_k = x_k(j) = \binom{j+1}{k}, \quad B_k = B_k(j).$$

Then, (11) and (12) take the form

$$\sum_{s=0}^{2j} x_r(s) = y_{r+1}, \tag{13}$$

and

$$y_r = 2^r x_r + 2^{r-2} \binom{r-1}{1} x_{r-1} + \cdots + 2^{r-2t} \binom{r-t}{t} x_{r-t} \tag{14}$$

with $t = [r/2]$. It will be useful to record the expressions for the first few B_k's.

Thus

$$
\begin{aligned}
B_1 &= 2^3 x_2,\\
B_2 &= 2^3 y_3 = 2^5 x_2 + 2^6 x_3;\\
B_3 &= 2^5 y_3 + 2^6 y_4 = 3.2^6 x_2 + 4.2^8 x_3 + 2^{10} x_4;\\
B_4 &= 7.2^8 x_2 + 39.2^9 x_3 + 12.2^{12} x_4 + 2^{15} x_5;\\
B_5 &= 53.2^9 x_2 + 284.2^{11} x_3 + 327.2^{13} x_4 + 32.2^{17} x_5 + 2^{21} x_6.
\end{aligned} \tag{15}
$$

Observe that the expression for B_{k+1} is obtained by replacing x_h by y_{h+1} in the expression for B_k.

Assume that for some $k \geqslant 1$,

$$B_k = \sum_{h=1}^{k} a_k(h) x_{h+1}. \tag{16}$$

where $a_k(h)$'s are positive integers. Then, we have

$$B_{k+1} = \sum_{h=1}^{k} a_k(h) y_{h+2} = \sum_{h=1}^{k+1} a_{k+1}(h) x_{h+1}, \tag{17}$$

where

$$
\begin{aligned}
a_{k+1}(h) = 2^{h+1} a_k(h-1) + 2^h \binom{h+1}{1} a_k(h) + 2^{h-1} \binom{h+1}{2} a_k(h+1) + \cdots \\
+ \binom{h+1}{h+1} a_k(2h),
\end{aligned} \tag{18}
$$

with $a_m(t) = 0$ whenever t is not a positive integer $\leqslant m$. In particular, (18) gives

$$
\begin{aligned}
a_{k+1}(1) &= 4 \,.\, a_k(1) + a_k(2),\\
a_{k+1}(2) &= 8 \,.\, a_k(1) + 12 \,.\, a_k(2) + 6 \,.\, a_k(3) + a_k(4).
\end{aligned}
$$

Moreover

$$a_{k+1}(k+1) = 2^{k+2} \,.\, a_k(k)$$

so that

$$a_k(k) = 2^{\binom{k+2}{2}}. \tag{19}$$

9.4. The potency of $a_k(h)$

Denoting by pot m, the highest power of 2 which divides m, we now proceed to prove

THEOREM 9. *For* $1 \leqslant h \leqslant k$,

$$\text{pot } a_k(h) \geqslant \binom{h+2}{2} + \left[\frac{3(k-h)+1}{2}\right],$$

with equality only when either $h = 1$, or k and h have the same parity.

Proof. The proof is by induction on k. Since the theorem holds for $k \leqslant 5$ by (15), we take $k > 5$. For $h \geqslant 3$, the theorem follows from (18). For $h = 1$, we have

$$a_{k+1}(1) = 4 \,.\, a_k(1) + a_k(2).$$

By the induction hypothesis,

$$\text{pot } 4a_k(1) = 2 + \binom{3}{2} + [(3k-2)/2] = [(3k+8)/2];$$

and

$$\text{pot } a_k(2) = \binom{4}{2} + [(3k-5)/2] = [(3k+7)/2] \quad \text{if } k \text{ is even},$$

$$> (3k+7)/2 \quad \text{if } k \text{ is odd}.$$

Therefore

$$\text{pot } 4a_k(1) > \text{pot } a_k(2), \quad k \text{ even},$$

$$< \text{pot } a_k(2), \quad k \text{ odd}.$$

In either case,

$$\text{pot } a_{k+1}(1) = [(3k+7)/2] = \binom{3}{2} + [(3k+1)/2].$$

For $h = 2$, we have

$$a_{k+1}(2) = 8 \,.\, a_k(1) + 12 \,.\, a_k(2) + 6 \,.\, a_k(3) + a_k(4). \tag{20}$$

Using the induction hypothesis again, we find that

$$\text{pot } 8a_k(1) = \binom{4}{2} + [(3k-2)/2]$$

$$= \text{pot } 12a_k(2), \quad k \text{ even},$$

$$< \text{pot } 12a_k(2), \quad k \text{ odd}.$$

The potency of each of the remaining two terms in (20) is definitely greater than

$$1 + \binom{4}{2} + [(3k-2)/2].$$

Hence

$$\text{pot } a_{k+1}(2) = \binom{4}{2} + [(3k-2)/2], \quad k \text{ odd},$$

$$> \binom{4}{2} + [(3k-2)/2], \quad k \text{ even}.$$

The theorem is thus proved.

As a direct consequence of theorem 9, we now have

THEOREM 10. *If j is such that $\binom{j+1}{2}$ is odd, then*

$$\text{pot } B_k(j) = \text{pot } a_k(1) = \binom{3}{2} + [(3k-2)/2] = q_k.$$

Since for n odd, $\binom{n}{2}$ and $\binom{n+1}{2}$ are of opposite parity, and $b(0) = 1 = b(1)$, while $b(m)$ is even for each $m \geqslant 2$, the Churchhouse conjecture follows from (8).

10. A theorem of Lehmer

We end this chapter with a simple combinatorial proof of a result of D. H. Lehmer which he announced at a group meeting in August, 1974, prior to the International Congress of Mathematicians at Vancouver. Lehmer had employed generating functions in proving his result which is equivalent to the following:

THEOREM 11. (Lehmer.) *Let $p_o(n, 2)$ and $p_e(n, 2)$ denote the number of partitions of n with an odd or even number of even parts respectively. Then*

$$p_e(n, 2) = p_o(n, 2) + q_o(n), \tag{1}$$

where $q_o(n)$ is the number of partitions of n into odd and unequal parts.

Proof. Let S be the set of partitions P of n with an even part or a repeated part. For all P in S, set

$$r = r(P) = \text{largest repeated part of } P$$

$$g = g(P) = \text{largest even part of } P.$$

From our definition of S, it is evident that of the numbers g, r at least one exists for each P. If g does not exist we set $g = 0$; if r does not exist we take $r = 0$.

We now partition S into two disjoint subsets A and B, where

$$P \in A \quad \text{if} \quad 2r > g$$

$$P \in B \quad \text{if} \quad 2r \leqslant g.$$

Take any partition P of n from S. Write the parts of P in non-decreasing order, underline the last two parts with value r and enclose the last part with value g in ( ).

Let T be the transformation which replaces two parts each of value r by a single part $2r$ and T^* the transformation which replaces (g) by two parts each of value $\frac{1}{2}g$. If P belongs to A, $T(P)$ belongs to B. If P belongs to B, $T^*(P)$ is in A. For any P in A, $T^*T(P) = P$. Similarly for any P in B, $TT^*(P)$ is the same as P. In either case let

P' denote the transform of P. Then it is clear that of the two partitions P, P' one belongs to A and the other to B. Moreover, one has an even number of even parts and the other an odd number of even parts.

We have thus constructed a 1–1 correspondence between the partitions of S with an even number of even parts and those with an odd number of even parts. Lehmer's theorem is thus proved.

Observe that none of the transformations T, T^* is applicable to a partition with odd and unequal parts and each such partition belongs to the set of those partitions which have an even number of even parts.

The following examples illustrate the proof and have been chosen so as to include the various cases that can arise with respect to the parameters r and g.

P	transform	P'
1 4 4 7 $\underline{7\ 7}$ 8 (12) 17	T	1 $\underline{4\ 4}$ 7 8 12 (14) 17
$\underline{2\ 2}$ 3 5 (6) 7 11	T^*	2 (2) 3 $\underline{3\ 3}$ 5 7 11
1 3 $\underline{3\ 3}$ 7 (8) 9	T^*	1 3 3 3 $\underline{4\ (4)}$ 7 9
1 3 3 $\underline{3\ 3}$ 5 7	T	1 $\underline{3\ 3}$ 5 (6) 7
1 3 5 11 (14)	T^*	1 3 5 $\underline{7\ 7}$ 11

Problems

1. Find the form of n in each of the following cases

(i) $q_4(n) \equiv 0 \pmod{17}$,
(ii) $q_6(n) \equiv 0 \pmod{19}$.

2. Show that $b(n)$ is not a multiple of 8 for any $n \geqslant 0$.
3. If $p_e(n)$ and $p_o(n)$ denote the number of partitions of n with an even or odd number of parts respectively, prove that

$$p_e(n) = p_o(n) + (-1)^n q_o(n)$$

where $q_o(n)$ is the number of partitions of n into odd and unequal parts. Show that this theorem of Glaisher (1876) is equivalent to Lehmer's theorem.
4. On the lines of the binary partitions, develop the theory of m-ary partitions.

References

Andrews, George E.: Congruence properties of m-ary partitions. *J. Number Theory* **3** (1971), 104–110.
Atkin, A. O. L.: Proof of a conjecture of Ramanujan. *Glasgow Math. J.* **8** (1967), 14–32.
Chowla, S.: *J. London Math. Soc.* **9** (1934), 147–153.
Churchhouse, R. F.: Congruence properties of the binary partition function. *Proc. Cambridge Philos. Soc.* **66** (1969), 371–376.
Gupta, Hansraj: On the parity of $p(n)$. *J. Indian Math. Soc.* **10** (1946), 32–33.
——: On the coefficients of the powers of Dedekind's modular form. *J. London Math. Soc.* **39** (1964), 433–440.
(The tables in this paper give expressions for $q_r(n)$ for $r > 0$, $n \leqslant 50$.)

——: On m-ary partitions. *Proc. Cambridge Philos. Soc.* **71** (1972), 343–345.
——: A simple proof of the Churchhouse conjecture concerning binary partitions. *Indian J. Pure Appl. Math.* **3** (1972), 791–794.
Gandhi, J. M.: Congruences for $q_r(n)$ and Ramanujan's tau function. *Amer. Math. Monthly* **70** (1963), 265–274.
Hjelle, T.; Klove, T.: Congruence properties and density problems for the Fourier coefficients of modular forms. *Math. Scand.* **23** (1968), 160–166.
Lehmer, D. H.: *J. London Math. Soc.* **11** (1936), 114–118.
Klove, Torleiv.: Density problems for $p(n)$. *J. London Math. Soc.* (2), **2** (1970), 504–508.
Kolberg, O.: Note on the parity of the partition function. *Math. Scand.* **7** (1959), 377–378.
Newman, M.: *Canadian J. Math.* **9** (1957), 68–70; 549–552.
——: Periodicity modulo m and divisibility properties of the partition function. *Trans. Amer. Math. Soc.* **97** (1960), 225–238.
Rödseth, O.: Some arithmetical properties of m-ary partitions. *Proc. Cambridge Philos. Soc.* **66** (1969), 371–376.
Watson, G. N.: *J. Reine Angew. Math.* **179** (1938), 97–128.

CHAPTER 9

AN ASYMPTOTIC FORMULA FOR $p(n, k)$

1. Historical note

An event of the greatest importance in the Theory of Partitions happened in 1918 when G. H. Hardy and Srinivasa Ramanujan gave, for large n, an asymptotic expansion for $p(n)$ – the number of unrestricted partitions of n. This can be stated in the form

$$p(n) = \sum_{j=1}^{v} A_j\phi_j + O(n^{-\frac{1}{4}}), \tag{1}$$

where

$$\phi_j = \frac{B_n(u_n - j)}{u_n j}\sqrt{j}\,.\,e^{u_n/j} \tag{2}$$

$$u_n = \frac{\pi\sqrt{(24n-1)}}{6}, \quad B_n = \frac{2\sqrt{3}}{24n-1}$$

and

$$v = O(\sqrt{n}).$$

The A's in the formula are certain constants depending on n and the 24th roots of unity.

It was surprising that the first few terms in the expansion gave a value for $p(n)$ the integral part of which was $p(n)$ itself. This was found to be so at least for $n \geqslant 200$. Thus for $n = 300$, H. Gupta found that the first eight terms gave

$$\begin{array}{rr}
p(n) = 9\ 25308\ 29352\ 80987 & \cdot 1131\\
+14\ 41205 & \cdot 5722\\
+1377 & \cdot 4002\\
+26 & \cdot 9413\\
+4 & \cdot 9914\\
+0 & \cdot 3589\\
-0 & \cdot 2884\\
-0 & \cdot 0847\\
\hline
9\ 25308\ 29367\ 23602 & \cdot 0040
\end{array}$$

the integral part of which is $p(300)$ itself. Even the decimal part, that is ·0040, appears to be significant as it is the same in the case of $n = 200$ and also $n = 721$.

In 1937, D. H. Lehmer found that the Hardy–Ramanujan expansion is divergent. Soon after this, H. Rademacher showed that if

$$(u_n - j)e^{u_n/j} \text{ is replaced by } (u_n - j)e^{u_n/j} + (u_n + j)e^{-u_n/j}$$

in (2), we get a convergent series for $p(n)$, that is an exact formula for $p(n)$.

In 1938, D. H. Lehmer developed the theory of the coefficients A_j involved in the expansion.

Concerning $p(n, k)$, the number of partitions of n into exactly k summands, however, nothing like the Hardy–Ramanujan formula was available.

2. The Auluck–Chowla–Gupta conjecture

In 1942, F. C. Auluck, S. Chowla and H. Gupta noticed that for $n \geqslant n_0$ and $1 \leqslant k \leqslant n$ the values of $p(n, k)$ first increased and then decreased. This led them to conjecture that for any $n \geqslant n_0$, there is a unique $k_0 = k_0(n)$ such that

$$p(n, 1) < p(n, 2) < \cdots < p(n, k_0) \geqslant p(n, k_0 + 1) \geqslant \cdots \geqslant p(n, n).$$

In 1941, P. Erdös and J. Lehner proved that

$$\frac{P(n, k)}{p(n)} \sim \exp(-e^{-cx}/c), \qquad \text{with } c = \pi/6^{1/2},$$

where $P(n, k)$ is the number of partitions of n into *at most* k parts,

$$k = \frac{\sqrt{n}}{c} \log \sqrt{n} + x\sqrt{n},$$

and x is any fixed number independent of n. They also proved that for $k = o(n^{1/3})$,

$$k!\, p(n, k) \sim \binom{n-1}{k-1}.$$

A simple proof of this result was given by H. Gupta in 1942 and has been given in the preceding chapter.

Proceeding on the lines of Erdös and Lehner, Auluck, Chowla and Gupta proved in 1942 that

$$\frac{\sqrt{n}\, p(n, k)}{p(n)} \sim \exp(-cx - c^{-1}e^{-cx}),$$

where c, k and x are the same as in the Erdös–Lehner asymptotic formula. This showed that if their conjecture is true, then

$$k_0 \sim \frac{\sqrt{6n}}{\pi} \log \frac{\sqrt{6n}}{\pi}$$

G. Szekeres in his two important papers of 1951 and 1953, made use of the function $P(n, k)$ which is easier to handle analytically because it increases with k. He found a formula for $P(n, k)$ which is true at least for large n. After this had been done, the familiar relations

$$p(n, k) = P(n - k, k), \quad q(n, k) = P\left(n - \frac{k(k+1)}{2}, k\right)$$

were utilized to show that the Auluck–Chowla–Gupta conjecture is true at least for large n, and that it holds also for $q(n, k)$ – the number of partitions of n into k distinct parts. The value of k_0 was of course not the same in the two cases.

3. Introduction

In this chapter, we give a fairly detailed account of the method of steepest descent employed by G. Szekeres in his study of the problem. The presentation here, mainly follows his version but some explanations have been added to make the task of the reader easier.

The work of Szekeres has opened up the possibility of obtaining the infinite series for $p(n)$ without the use of elliptic modular functions. This is our only excuse for making use of Cauchy's theorem and complex variables which we have avoided so far in this book.

4. The method of steepest descent

Let

$$U = \{u_k\}, \quad k = 1, 2, 3, \ldots,$$

be a strictly increasing sequence of positive integers. Let $p_u(n)$ denote the number of partitions of a natural number n into the members of U, repetitions allowed, and $q_u(n)$ the number of partitions of n into distinct elements of U. Then we define

$$F_u(x) = \prod_u (1 - x^u)^{-1} = \sum_{n=0}^{\infty} p_u(n) x^n, \quad |x| < 1,$$

$$G_u(x) = \prod_u (1 + x^u) = \sum_{n=0}^{\infty} q_u(n) x^n, \quad |x| < 1.$$

By Cauchy's theorem

$$p_u(n) = (2\pi i)^{-1} \int F_u(z) z^{-n-1}\, dz, \tag{1}$$

$$q_u(n) = (2\pi i)^{-1} \int G_u(z) z^{-n-1}\, dz, \tag{2}$$

the integration being taken over any circle of radius <1 and centre O.

4.1. Choice of radius

We select for the radius the smallest $\rho < 1$, for which the logarithmic derivative of the integrand vanishes on the circle with radius ρ. Such a point is a saddle point of the surface formed by the absolute values of the function, provided that the second derivative does not also vanish and the integration is carried out through the saddle point or sufficiently close to it, in a direction in which the absolute value decreases rapidly as one gets away from the saddle point itself. In the neighbourhood of the saddle point the integral can be estimated very accurately by developing the logarithm of the integrand in a power series with centre at the saddle point.

In partition problems, such as those we propose to deal with here, the saddle point is at ρ itself. This simplifies the analysis very much. The saddle point condition in the two cases, is

$$n + 1 = \rho F_u'(\rho)/F_u(\rho) = \sum_u u\rho^u/(1 - \rho^u) \tag{3}$$

$$n + 1 = \rho G_u'(\rho)/G_u(\rho) = \sum_u u\rho^u/(1 + \rho^u) \tag{4}$$

or with $\rho = e^{-\alpha}$, $\alpha > 0$, (and replacing $n + 1$ by n to make the condition somewhat simpler),

$$n = \sum_u u/(e^{\alpha u} - 1) \tag{5}$$

$$n = \sum_u u/(e^{\alpha u} + 1). \tag{6}$$

Each of these equations can be solved uniquely for α and hence for ρ.

4.2. Application to $P(n, k)$

As already stated, $P(n, k)$ is the number of partitions of n into at most k parts or equivalently into parts the greatest of which does not exceed k, we have

$$F(x) = \prod_{v=1}^{k} (1 - x^v)^{-1} = \sum_{n=0}^{\infty} P(n, k)x^n,$$

so that

$$P(n, k) = (2\pi i)^{-1} \int F(z) z^{-n-1}\, dz.$$

The saddle point condition with $\rho = e^{-\alpha}$, $\alpha > 0$, is

$$n = \sum_{v=1}^{k} v/(e^{\alpha v} - 1). \tag{7}$$

Write $z = \rho e^{i\theta} = \exp(-\alpha + i\theta)$, $-\pi \leqslant \theta \leqslant \pi$. Then

$$P(n, k) = (2\pi i)^{-1} \int_{-\pi}^{\pi} F(\rho e^{i\theta})\rho^{-n} \exp(-ni\theta)\, i\, d\theta,$$

$$= (2\pi)^{-1} \exp\{n\alpha + \log F(\rho)\} \int_{-\pi}^{\pi} \frac{F(\rho e^{i\theta})}{F(\rho)} \exp(-ni\theta)\, d\theta.$$

Using condition (7) and the relation

$$\log F(z) = -\sum_{v=1}^{k} \log(1 - z^v), \tag{8}$$

we get

$$P(n, k) = (2\pi)^{-1} \exp\left\{\sum_{v=1}^{k} \left(\frac{\alpha v}{e^{\alpha v} - 1} - \log(1 - e^{-\alpha v})\right)\right\}$$

$$\cdot \int_{-\pi}^{\pi} \exp\left\{-ni\theta - \sum_{v=1}^{k} \log \frac{1 - e^{-\alpha v + iv\theta}}{1 - e^{-\alpha v}}\right\} d\theta. \tag{9}$$

To estimate the value of this expression, we need accurate information about the value of α.

4.3. An estimate for the value of α

We first assert that (7) determines α uniquely.

Proof. Notice that for each $v \leqslant k$ and $x > 0$,

$$\frac{d}{dx}\left\{\frac{v}{e^{xv} - 1}\right\} = \frac{-v^2 e^{xv}}{(e^{xv} - 1)^2} < 0,$$

so that $v/(e^{xv} - 1)$ decreases as x increases. Moreover, $v/(e^{xv} - 1)$ tends to infinity as x tends to zero and to zero as x tends to infinity. Hence

$$\sum_{v=1}^{k} v/(e^{xv} - 1) \tag{10}$$

can attain the value n only for a unique value of x, which is our α. For $x < \alpha$, (10) must be $>n$, and for $x > \alpha$ it must be $<n$. To find bounds for α, we use the relation

$$n = \sum_{v=1}^{k} v/(e^{\alpha v} - 1) = \alpha^{-1} \sum_{v=1}^{k} \alpha v/(e^{\alpha v} - 1). \tag{11}$$

Now

$$\frac{d}{dx}\{x/(e^x - 1)\} = (e^x - 1 - xe^x)/(e^x - 1)^2.$$

Write $y = e^x - 1 - xe^x$, then $y' = -xe^x < 0$ for all $x > 0$. Thus y is a decreasing function of x for all $x > 0$. Since it vanishes for $x = 0$, it must be negative for $x > 0$. Hence $x/(e^x - 1)$ is a decreasing function of x for $x > 0$. We must, therefore, have

$$\frac{1}{e^\alpha - 1} \geqslant \frac{n}{k} \quad \text{and} \quad \frac{k}{e^{\alpha k} - 1} \leqslant \frac{n}{k}.$$

The left hand inequality gives

$$\alpha \leqslant \log\left(1 + \frac{k}{n}\right) \leqslant \frac{k}{n}.$$

From the right hand inequality, we have

$$\alpha \geqslant \frac{1}{k} \log\left(1 + \frac{k^2}{n}\right) = -\frac{1}{k} \log\left(1 - \frac{k^2}{n + k^2}\right)$$

$$\geqslant k/(n + k^2).$$

Thus, for $k^2 \leqslant n$, we have

$$k/(2n) \leqslant \alpha \leqslant k/n, \tag{12}$$

and in particular for $k^2 = n$,

$$1/(2\sqrt{n}) \leqslant \alpha \leqslant 1/\sqrt{n}. \tag{13}$$

Since for a fixed n, α necessarily increases with k, it follows for $k^2 > n$, that, on the one hand

$$1/(2\sqrt{n}) \leqslant \alpha, \tag{14}$$

while on the other

$$n = \sum_{v=1}^{k} v/(e^{\alpha v} - 1) < \int_0^\infty \frac{x}{e^{\alpha x} - 1}\, dx = \pi^2/(6\alpha^2),$$

so that

$$\alpha < \pi/\sqrt{6n}. \tag{15}$$

4.4. Application of the Euler–Maclaurin summation formula

For several applications in the following sections, we need the Euler–Maclaurin summation formula in the following form:

Let $f(x)$ be any integrable function of x at least three times continuously differentiable, then

$$f(1) + f(2) + f(3) + \cdots + f(k) = \int_0^k f(x)\, dx + \{f(k) - f(0)\}/2$$

$$+ \{f'(k) - f'(0)\}/12 + \int_0^k B_3(x) f'''(x)\, dx,$$

where $B_3(x)$ is defined by the relations

$$B_3(x) = (2x^3 - 3x^2 + x)/12 \quad \text{for} \quad 0 \leqslant x \leqslant 1, \tag{16}$$

$$B_3(x+1) = B_3(x) \quad \text{for} \quad x < 0 \quad \text{or} \quad x > 1. \tag{17}$$

Proof. We have

$$\begin{aligned}\int_0^k B_3(x) f'''(x)\,dx &= \sum_{h=0}^{k-1} \int_h^{h+1} B_3(x) f'''(x)\,dx \\ &= \sum_{h=0}^{k-1} \int_0^1 B_3(x+h) f'''(x+h)\,dx \\ &= \sum_{h=0}^{k-1} \int_0^1 B_3(x) f'''(x+h)\,dx.\end{aligned}$$

Now

$$\begin{aligned}&\int_0^1 B_3(x) f'''(x+h)\,dx \\ &= [B_3(x) f''(x+h)]_0^1 - \tfrac{1}{12}\int_0^1 (6x^2 - 6x + 1) f''(x+h)\,dx \\ &= -\tfrac{1}{12}[(6x^2 - 6x + 1) f'(x+h)]_0^1 + \tfrac{1}{12}\int_0^1 (12x - 6) f'(x+h)\,dx \\ &= -\tfrac{1}{12}\{f'(h+1) - f'(h)\} + \tfrac{1}{2}[(2x-1) f(x+h)]_0^1 - \tfrac{1}{2}\int_0^1 2f(x+h)\,dx \\ &= -\tfrac{1}{12}\{f'(h+1) - f'(h)\} + \tfrac{1}{2}\{f(h+1) + f(h)\} - \int_h^{h+1} f(x)\,dx.\end{aligned}$$

Hence

$$\begin{aligned}\int_0^k B_3(x) f'''(x)\,dx &= -\tfrac{1}{12}\{f'(k) - f'(0)\} + \tfrac{1}{2}\{f(k) - f(0)\} \\ &\quad + f(0) + f(1) + f(2) + \cdots + f(k-1) - \int_0^k f(x)\,dx.\end{aligned}$$

This proves the formula.

In particular, for $f(x) = x/(e^{\alpha x} - 1)$, the Euler–Maclaurin formula gives

$$\sum_{v=1}^{k} \frac{v}{e^{\alpha v}-1} = \int_0^k \frac{x}{e^{\alpha x}-1}\,dx + \frac{1}{2}\left\{\frac{k}{e^{\alpha k}-1} - \frac{1}{\alpha}\right\}$$
$$+\frac{1}{12}\left\{\frac{(e^{\alpha k}-1)-\alpha k e^{\alpha k}}{(e^{\alpha k}-1)^2} + \frac{1}{2}\right\} + \int_0^k B_3(x) f'''(x)\,dx.$$

4.5. A preliminary estimate

We here prove the following

LEMMA. *For $x > 0$, and $f(x) = x/(e^{\alpha x}-1)$, $f'''(x)$ is negative.*

Proof. Write y for $(e^{\alpha x}-1)$ and t for $1/y$. Then

$$f(x) = xt,$$
$$y' = \alpha(y+1), \quad \frac{dt}{dy} = -y^{-2},$$
$$t' = -\alpha(y^{-1} + y^{-2}),$$
$$t'' = \alpha^2(y^{-2} + 2y^{-3})(y+1) = \alpha^2(y^{-1} + 3y^{-2} + 2y^{-3}),$$
$$t''' = -\alpha^3(y^{-2} + 6y^{-3} + 6y^{-4})(y+1).$$

Therefore

$$f'''(x) = xt''' + 3t''$$
$$= \frac{\alpha^2(y+1)}{y^4}\{3y^2 + 6y - \alpha x(y^2 + 6y + 6)\}.$$

Since $\alpha > 0$, y is positive for $x > 0$. We need, therefore, show only that the expression in the curly brackets is negative for $x > 0$. Let

$$w = 3y^2 + 6y - \alpha x(y^2 + 6y + 6).$$

Then

$$w' = \alpha\{5y^2 + 6y - \alpha x(2y^2 + 8y + 6)\},$$
$$w'' = 4\alpha^2(y+1)\{2y - \alpha x(y+2)\}.$$

We notice that w, w', w'' all vanish for $x = 0$. For $x > 0$, we now show that w'' is negative. Write

$$w^* = 2y - \alpha x(y+2).$$

Then

$$w^{*\prime} = \alpha\{y - \alpha x(y+1)\},$$
$$w^{*\prime\prime} = -\alpha^3 x(y+1).$$

Observe that w^* and $w^{*\prime}$ both vanish for $x = 0$, while $w^{*\prime\prime}$ is negative for $x > 0$.

Hence for $x > 0$, $w^{*\prime}$, w^*, w'', w', w and finally $f'''(x)$ are all decreasing functions of x and are, therefore, all negative for $x > 0$. This proves the lemma.

COROLLARY. $$\int_0^k |B_3(x) f'''(x)|\, dx < \alpha/600.$$

Proof. It is easy to see that for all x,

$$|B_3(x)| < 1/100.$$

Hence, the given integral is

$$< (1/100) \int_0^k |f'''(x)|\, dx$$
$$< (1/100)\,.\,[\,|f''(x)|\,]_0^\infty = (1/100)\,.\,[\,|\{\alpha^2 x(y^2 + 3y + 2) - 2(y^2 + y)\alpha\}/y^3\,|\,]_0^\infty.$$

The expression on the right is of an indeterminate form both for $x = 0$ and for $x = \infty$. Evaluating in the usual way, we get the desired result.

Writing β for αk (here and in what follows), the saddle point condition (7) for $P(n, k)$ takes the form

$$n = \frac{k^2}{\beta^2}\int_0^\beta \frac{x}{e^x - 1}\, dx + \frac{k}{2}\left\{\frac{1}{e^\beta - 1} - \frac{1}{\beta}\right\} + \frac{1}{12}\left\{\frac{1}{e^\beta - 1} - \frac{\beta e^\beta}{(e^\beta - 1)^2} + \frac{1}{2}\right\} + R, \tag{18}$$

where $|R| < \beta/(600\,k)$.

5. Estimation of the integral in (4.9)

We dissect the interval of integration into three parts:

$$I_1 = \int_{-\theta_0}^{\theta_0}, \quad I_2 = \int_{\theta_0}^{\pi}, \quad I_3 = \int_{-\pi}^{-\theta_0}.$$

We deal with the principal part of I_1 first when $k^2 \leqslant n$. Write

$$T = -\log\left\{\frac{1 - e^{-\alpha v + iv\theta}}{1 - e^{-\alpha v}}\right\} = -\log\left\{1 - \frac{e^{iv\theta} - 1}{e^{\alpha v} - 1}\right\}$$

Then, since $|\sin\theta| \leqslant |\theta|$, we have for $0 \leqslant \theta \leqslant \theta_0$,

$$\frac{|e^{iv\theta} - 1|}{e^{\alpha v} - 1} = \frac{2\,|\sin(v\theta/2)|}{e^{\alpha v} - 1} \leqslant \frac{v\theta}{\alpha v} = \frac{\theta}{\alpha} \leqslant \theta_0/\alpha.$$

Choose $\theta_0 = \dfrac{k^{4/7}}{n}$, then since $\dfrac{k}{2n} \leqslant \alpha$,

$$\frac{\theta_0}{\alpha} \leqslant 2k^{-3/7}.$$

Since this tends to zero as k tends to infinity, we have

$$T = \frac{e^{iv\theta}-1}{e^{\alpha v}-1} + \frac{(e^{iv\theta}-1)^2}{2(e^{\alpha v}-1)^2} + \frac{(e^{iv\theta}-1)^3}{3(e^{\alpha v}-1)^3} + \frac{(e^{iv\theta}-1)^4}{4(e^{\alpha v}-1)^4} + O(\theta^5/\alpha^5).$$

Thus, we get

$$\begin{aligned} I_1 &= \int_{-\theta_0}^{\theta_0} \exp\left\{-ni\theta - \sum_{v=1}^{k} \log \frac{1-e^{-\alpha v + iv\theta}}{1-e^{-\alpha v}}\right\} d\theta \\ &= \int_{-\theta_0}^{\theta_0} \exp\left\{-ni\theta + \sum_{v=1}^{k} T\right\} d\theta. \end{aligned} \tag{1}$$

Expanding in series, we have

$$\begin{aligned} T = {} & \frac{1}{e^{\alpha v}-1}\left\{iv\theta - \frac{v^2\theta^2}{2!} - \frac{iv^3\theta^3}{3!} + \frac{v^4\theta^4}{4!} + O(v^5\theta^5)\right\} \\ & + \frac{1}{2(e^{\alpha v}-1)^2}\left\{-v^2\theta^2 - iv^3\theta^3 + \frac{7}{12}v^4\theta^4 + O(v^5\theta^5)\right\} \\ & + \frac{1}{3(e^{\alpha v}-1)^3}\left\{-iv^3\theta^3 + \frac{3}{2}v^4\theta^4 + O(v^5\theta^5)\right\} \\ & + \frac{1}{4(e^{\alpha v}-1)^4}\{v^4\theta^4 + O(v^5\theta^5)\} + O(\theta^5/\alpha^5). \end{aligned}$$

Hence

$$\begin{aligned} S = {} & \sum_{v=1}^{k} T = \sum T \\ = {} & i\theta \sum \frac{v}{e^{\alpha v}-1} - \frac{\theta^2}{2!}\sum\left\{\frac{v^2}{e^{\alpha v}-1} + \frac{v^2}{(e^{\alpha v}-1)^2}\right\} \\ & - \frac{i\theta^3}{6}\sum\left\{\frac{v^3}{e^{\alpha v}-1} + \frac{3v^3}{(e^{\alpha v}-1)^2} + \frac{2v^3}{(e^{\alpha v}-1)^3}\right\} \\ & + \frac{\theta^4}{24}\sum\left\{\frac{v^4}{e^{\alpha v}-1} + \frac{7v^4}{(e^{\alpha v}-1)^2} + \frac{12v^4}{(e^{\alpha v}-1)^3} + \frac{6v^4}{(e^{\alpha v}-1)^4}\right\} \\ & + O\left(\theta^5 \sum_{v=1}^{k}\sum_{j=1}^{4} \frac{v^5}{(e^{\alpha v}-1)^j}\right) + O(k\theta^5/\alpha^5). \end{aligned} \tag{2}$$

We note that for all $m \geqslant 0$, absolute positive constants $c = c_m$ exist, such that

$$\frac{x^m}{(e^x - 1)^j} \leqslant c, \quad j \leqslant m. \tag{3}$$

Hence

$$\sum_{v=1}^{k} \sum_{j=1}^{4} \frac{v^5}{(e^{\alpha v} - 1)^j} \leqslant \frac{4kc}{\alpha^5},$$

so that

$$S = ni\theta - \frac{\theta^2}{2} \sum_{v=1}^{k} \frac{v^2 e^{\alpha v}}{(e^{\alpha v} - 1)^2} - \frac{i\theta^3}{6} \sum_{v=1}^{k} \frac{v^3 (e^{\alpha v} + e^{2\alpha v})}{(e^{\alpha v} - 1)^3} + O(k^{-\frac{5}{7}}).$$

Hence

$$I_1 = \int_{-\theta_0}^{\theta_0} \exp\left\{-\frac{1}{2}\frac{\theta^2}{\alpha^3} A - \frac{1}{6}\frac{i\theta^3}{\alpha^3} kB + O(\theta^4/\alpha^4)\right\} d\theta, \tag{4}$$

where

$$A = \sum_{v=1}^{k} \frac{v^2 \alpha^3 e^{\alpha v}}{(e^{\alpha v} - 1)^2}, \quad B = \sum_{v=1}^{k} \frac{v^3 \alpha^3 (e^{\alpha v} + e^{2\alpha v})}{k(e^{\alpha v} - 1)^3}. \tag{5}$$

Making use of (3), it is readily seen that for some suitable absolute positive constant $c_1, B \leqslant c_1$, so that

$$\frac{\theta^3}{\alpha^3} kB = kO(\theta_0^3/\alpha^3) = O(k^{-2/7}).$$

Consequently, we have

$$\begin{aligned} I_1 &= \int_{-\theta_0}^{\theta_0} \left\{1 - \frac{1}{6}\frac{i\theta^3}{\alpha^3} kB + O\left(\frac{\theta^4}{\alpha^4} k + \frac{\theta^6}{\alpha^6} k^2\right)\right\} \exp\left(-\frac{1}{2}\frac{\theta^2}{\alpha^3} A\right) d\theta \\ &= \int_{-\theta_0}^{\theta_0} \left\{1 + O\left(\frac{\theta^4}{\alpha^4} k + \frac{\theta^6}{\alpha^6} k^2\right)\right\} \exp\left(-\frac{1}{2}\frac{\theta^2}{\alpha^3} A\right) d\theta \\ &\quad - \int_{-\theta_0}^{\theta_0} \frac{1}{6}\frac{i\theta^3}{\alpha^3} \exp\left(-\frac{1}{2}\frac{\theta^2}{\alpha^3} A\right) d\theta \\ &= \int_{-\theta_0}^{\theta_0} \left\{1 + O\left(\frac{\theta^4}{\alpha^4} k + \frac{\theta^6}{\alpha^6} k^2\right)\right\} \exp\left(-\frac{1}{2}\frac{\theta^2}{\alpha^3} A\right) d\theta. \end{aligned}$$

(The second integral vanishes because the integrand is an odd function of θ.) Change the variable of integration to $x = A^{1/2}(2\alpha^3)^{-1/2}\theta$. Then

$$I_1 = (2\alpha^3/A)^{1/2} \int_{-x_0}^{x_0} \left\{1 + O\left(\frac{\alpha^2 k}{A^2} x^4 + \frac{\alpha^3 k^2}{A^3} x^6\right)\right\} \; e^{-x^2}\, dx,$$

where

$$x_0 = A^{1/2}(2\alpha^3)^{-1/2}\theta_0. \tag{6}$$

Now

$$A = \sum_{v=1}^{k} \frac{v^2\alpha^3 e^{\alpha v}}{(e^{\alpha v} - 1)^2}.$$

Since $k^2 \leqslant n$ and $\alpha \leqslant k/n$, we have $\beta = \alpha k \leqslant 1$. Therefore, absolute positive constants c_2 and c_3 exist, such that for $0 \leqslant t \leqslant \beta$,

$$c_2 \leqslant \frac{t^2 e^t}{(e^t - 1)^2} \leqslant c_3,$$

or

$$c_2 k \leqslant \sum_{v=1}^{k} \frac{\alpha^2 v^2 e^{\alpha v}}{(e^{\alpha v} - 1)^2} \leqslant c_3 k,$$

so that

$$c_2\beta \leqslant A \leqslant c_3\beta.$$

Therefore,

$$\alpha^2 k/A^2 = O(\alpha^2 k/\beta^2) = O(k^{-1}),$$
$$\alpha^3 k^2/A^3 = O(k^{-1}).$$

Also

$$x_0 \geqslant c_2^{1/2}\beta^{1/2}(2\alpha^3)^{-1/2}\theta_0 \geqslant (c_2/2)^{1/2}k^{1/14},$$

because

$$\theta_0 = k^{4/7}/n \quad \text{and} \quad k/(2n) \leqslant \alpha \leqslant k/n.$$

Now

$$\int_0^{x_0} e^{-x^2}\, dx = \int_0^{\infty} e^{-x^2}\, dx - \int_{x_0}^{\infty} e^{-x^2}\, dx.$$

But for $x_0 \geqslant 1$, the second integral is

$$\leqslant \int_{x_0}^{\infty} 2x\, e^{-x^2}\, dx = e^{-x_0^2} \leqslant \exp(-c_2 k^{1/7}/2).$$

Moreover

$$\int_{-\infty}^{\infty} e^{-x^2}\,dx = \sqrt{\pi}, \qquad \int_{-\infty}^{\infty} x^4 e^{-x^2}\,dx = \frac{3}{4}\sqrt{\pi}.$$

and

$$\int_{-\infty}^{\infty} x^6 e^{-x^2}\,dx = \frac{15}{8}\sqrt{\pi}.$$

Hence as the final result, we have

$$I_1 = (2\pi\,\alpha^3/A)^{1/2}(1 + O(k^{-1})). \tag{7}$$

5.1. Estimation of I_2 and I_3 when $k^2 \leqslant n$

In what follows, the c's are absolute positive constants independent of n and k, with n and k large, $k \leqslant n$, as before.

We first consider the case when $\theta_0 \leqslant \theta \leqslant k^{-2/3}$. Let

$$H(\theta) = \exp\left\{-\sum_{v=1}^{k} \log\left|\frac{1 - e^{-\alpha v + iv\theta}}{1 - e^{-\alpha v}}\right|\right\}$$

$$= \exp\left\{-\frac{1}{2}\sum_{v=1}^{k} \log\left(1 + \frac{4e^{\alpha v}\sin^2(v\theta/2)}{(e^{\alpha v} - 1)^2}\right)\right\}.$$

Then, for $v\theta \leqslant \pi/2$ and for some c_4,

$$S_1 = \sum_{v=1}^{k} \log\left\{1 + \frac{4e^{\alpha v}\sin^2(v\theta/2)}{(e^{\alpha v} - 1)^2}\right\}$$

$$\geqslant \sum_{v} \log\left\{1 + \frac{c_4 v^2\theta^2}{\alpha^2 v^2}\right\} = \sum_{v} \log\left(1 + \frac{c_4\theta^2}{\alpha^2}\right),$$

where after a sign of inequality $v \leqslant \min(\pi/(2\theta), 1/\alpha, k)$. If $\pi/(2\theta) \geqslant 1/\alpha$ i.e. $\theta \leqslant \alpha\pi/2$, then

$$S_1 \geqslant \sum_{v \leqslant k} (c_5\theta^2/\alpha^2) = c_5 k\theta^2/\alpha^2 \geqslant c_5 k\theta_0^2/\alpha^2 \geqslant c_6 k^{1/7}.$$

If $\theta > \alpha\pi/2$, then with $v \leqslant \min(\pi/(2\theta), k)$,

$$S_1 \geqslant \sum_{v} c_7 = \min(c_7\pi/(2\theta), c_7 k).$$

Since

$$\theta_0 \leqslant \theta \leqslant k^{-2/3},$$

even for $\theta > \alpha\pi/2$, we have $S_1 \geqslant c_8 k^{2/3}$. In any case, therefore

$$H(\theta) \leqslant \exp(-c_9 k^{1/7}).$$

Next, let $k^{-2/3} < \theta \leqslant \pi$. Define k^* to be $\min(k, [1/(2\alpha)])$. Take

$$m_0 = \left[\frac{\theta k^*}{2\pi} - 1\right].$$

Then for every positive integer $m \leqslant m_0$, there are at least

$$[\pi/\theta] \geqslant \pi/(2\theta)$$

numbers $v \leqslant k^*$, for which the relation

$$\left(2m - \frac{3}{2}\right)\pi \leqslant v\theta \leqslant \left(2m - \frac{1}{2}\right)\pi$$

holds. For these values of v, the point $e^{iv\theta}$ lies on the left half of the semicircle of unit radius with centre O, as shown in the following figure

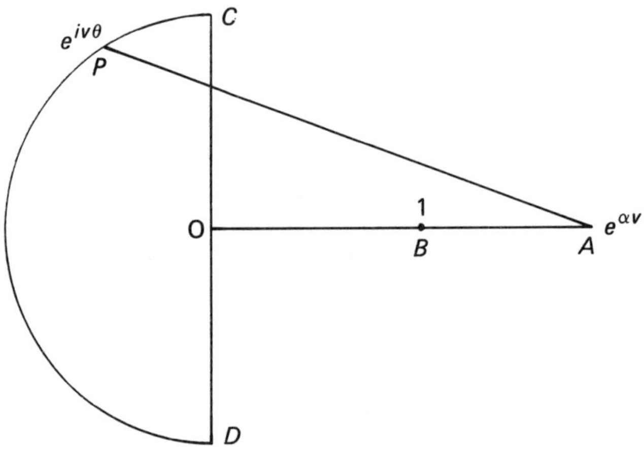

We thus have

$$\left|\frac{e^{\alpha v} - 1}{e^{\alpha v} - e^{iv\theta}}\right| = \frac{AB}{AP} \leqslant \frac{AB}{AC} = \frac{e^{\alpha v} - 1}{\sqrt{(e^{2\alpha v} + 1)}}.$$

Hence

$$\log H(\theta) \leqslant \frac{\pi}{4\theta} \sum_{m=1}^{m_0} \log \frac{(\exp\{(2m - 0{\cdot}5)\alpha\pi/\theta\} - 1)^2}{\exp\{(4m - 1)\alpha\pi/\theta\} + 1}$$

$$\leqslant \frac{1}{4} \int_1^{m_0+1} \frac{\pi}{\theta} \log \frac{(e^{2\alpha\pi x/\theta} - 1)^2}{e^{4\alpha\pi x/\theta} + 1}\, dx$$

$$= \frac{1}{8\alpha} \int_{2\pi\alpha/\theta}^{\alpha k^*} \log \frac{(e^t - 1)^2}{e^{2t} + 1}\, dt$$

$$< -\frac{1}{8}\left(k^* - \frac{2\pi}{\theta}\right).$$

This follows from the facts that, by definition, $\alpha k^* \leqslant \frac{1}{2}$, and that for $0 \leqslant t \leqslant \frac{1}{2}$, the integrand is an increasing function of t, and therefore

$$\leqslant \log\{(e^{1/2} - 1)^2/(e + 1)\} < -1.$$

Now

$$k^* - \frac{2\pi}{\theta} > k^* - 2\pi\, k^{2/3}.$$

Hence, if k is sufficiently large, then

$$k^* - \frac{2\pi}{\theta} > k/2, \quad \text{if } k^* = k,$$

and

$$> k/4, \quad \text{if } k^* = \frac{1}{2\alpha} > \frac{n}{3k} > \frac{k}{3}.$$

Thus for $k \geqslant$ some k_0,

$$H(\theta) \leqslant \exp(-k/32).$$

It follows, therefore, that

$$|I_2|, |I_3| = O(-c_{10}k^{1/7}). \tag{8}$$

5.2. Expression for $P(n, k)$ when $k^2 \leqslant n$

To get the expression for $P(n, k)$, we assume that

$$k^{1/7} \geqslant \frac{2}{c_{10}} \log n.$$

Then

$$2 \log n - c_{10}k^{1/7} \leqslant 0,$$

so that

$$\tfrac{1}{2} \log k + \log n - c_{10}k^{1/7} \leqslant 0,$$

i.e.

$$k^{1/2} n \exp(-c_{10}k^{1/7}) \leqslant 1.$$

Consequently

$$\begin{aligned}\exp(-c_{10}k^{1/7}) &= O(k^{-1/2}n^{-1}) = O(\alpha/k^{3/2})\\ &= O(\alpha^{3/2}/(k\beta^{1/2})) = O\left(\frac{\alpha^3}{A^{1/2}}\, k^{-1}\right).\end{aligned}$$

Hence

$$P(n,k)=\frac{\alpha^{3/2}}{(2\pi A)^{1/2}}\{1+O(k^{-1})\}\exp\left\{\sum_{v=1}^{k}\left(\frac{\alpha v}{e^{\alpha v}-1}-\log(1-e^{-\alpha v})\right)\right\},$$

where α is obtained from (4.7) and A from (5).

5.3. Estimation of I_1 when $\sqrt{n}<k\leqslant n$

We proceed on the same lines as in the case of $k^2\leqslant n$. Choose $\theta_0=n^{-5/7}$. Then

$$\frac{2\,|\sin(v\theta/2)|}{e^{\alpha v}-1}\leqslant\frac{v\theta}{\alpha v}\leqslant\frac{\theta_0}{\alpha}\leqslant 2n^{-3/14}.$$

The expansion of

$$T=-\log\left(1-\frac{e^{iv\theta}-1}{e^{\alpha v}-1}\right)$$

given earlier is, therefore, still valid, and proceeding exactly as before, we get

$$I_1=\int_{-\theta_0}^{\theta_0}\exp\left\{-\frac{1}{2}\frac{\theta^2}{\alpha^3}A-\frac{1}{6}\frac{\theta^3}{\alpha^4}B+\frac{1}{24}\frac{\theta^4}{\alpha^5}C+\cdots\right\}d\theta,$$

where

$$A=\sum_{v=1}^{k}\frac{v^2\alpha^3e^{\alpha v}}{(e^{\alpha v}-1)^2},$$

$$B=\sum_{v=1}^{k}\frac{v^3\alpha^4(e^{\alpha v}+e^{2\alpha v})}{(e^{\alpha v}-1)^3},$$

$$C=\sum_{v=1}^{k}\frac{v^4\alpha^5(e^{\alpha v}+4e^{2\alpha v}+e^{3\alpha v})}{(e^{\alpha v}-1)^4},$$

and so on. Notice that

$$B=O(1),\qquad C=O(1),$$

therefore

$$\frac{\theta^3}{\alpha^4}B=O(n^{-1/7}),\qquad\frac{\theta^4}{\alpha^5}C=O(n^{-5/14}).$$

Hence

$$I_1 = \int_{-\theta_0}^{\theta_0} \left\{1 + \frac{1}{24}\frac{\theta^4}{\alpha^5} C - \frac{1}{72}\frac{\theta^6}{\alpha^8} B^2 + O\left(\frac{\theta^6}{\alpha^7} + \frac{\theta^8}{\alpha^{10}} + \frac{\theta^{10}}{\alpha^{13}} + \frac{\theta^{12}}{\alpha^{16}}\right)\right\}$$

$$\cdot \exp\left(-\frac{1}{2}\frac{\theta^2}{\alpha^3} A\right) d\theta.$$

Changing the variable of integration to $x = A^{1/2}(2\alpha^3)^{-1/2}\theta$, we get as before,

$$I_1 = (2\alpha^3/A)^{1/2} \int_{-x_0}^{x_0} \left\{ 1 + \left(\frac{x^4}{6A^2} C - \frac{x^6}{9A^3} B^2\right) \alpha + O(x^6 + x^8 + x^{10} + x^{12})\alpha^2 \right\}$$

$$\cdot \exp(-x^2)\, dx$$

$$= (2\pi\alpha^3/A)^{1/2}\{1 + g(\beta)\alpha + O(\alpha^2)\}. \tag{9}$$

Here it is important to note that $g(\beta)$ is an analytic bounded function of β.

5.4. Estimation of I_2 and I_3 when $\sqrt{n} < k \leqslant n$

The method of estimation is again very similar to that used earlier, except that the two cases to be considered now are

$$n^{-5/7} \leqslant \theta \leqslant n^{-1/2} \quad \text{and} \quad n^{-1/2} < \theta \leqslant \pi.$$

The final result obtained is

$$P(n, k) = \frac{\alpha^{3/2}}{(2\pi A)^{1/2}} \{1 + g(\beta)\alpha + O(\alpha^2)\}$$

$$\cdot \exp\left\{ \sum_{v=1}^{k} \left(\frac{\alpha v}{e^{\alpha v} - 1} - \log(1 - e^{-\alpha v})\right)\right\}, \tag{10}$$

where, explicitly,

$$g(\beta) = \frac{1}{24A_0^3} (3A_0C_0 - 5B_0^2), \quad \beta = \alpha k,$$

$$A_0 = \int_0^\beta \frac{t^2 e^t}{(e^t - 1)^2}\, dt, \quad B_0 = \int_0^\beta \frac{t^3(e^t + e^{2t})}{(e^t - 1)^3}\, dt,$$

$$C_0 = \int_0^\beta \frac{t^4(e^t + 4e^{2t} + e^{3t})}{(e^t - 1)^4}\, dt.$$

6. Evaluation of $p(n, k)$

We have

$$p(n, k) = P(n - k, k).$$

Our asymptotic formulae for $P(n, k)$ will hold for $p(n, k)$ provided α is now determined from the relation

$$n = \sum_{v=1}^{k} \left(\frac{v}{e^{\alpha v} - 1} + 1 \right).$$

Let $\delta\alpha$ be the increment in the value of α when k is increased by 1. Then, the relations

$$n = \sum_{v=1}^{k+1} \left\{ \frac{v}{e^{(\alpha+\delta\alpha)v} - 1} + 1 \right\}$$

$$n = \sum_{v=1}^{k} \left\{ \frac{v}{e^{\alpha v} - 1} + 1 \right\}$$

must hold simultaneously. Hence

$$-1 = \sum_{v=1}^{k+1} \frac{v}{e^{(\alpha+\delta\alpha)v} - 1} - \sum_{v=1}^{k} \frac{v}{e^{\alpha v} - 1}.$$

Using Taylor's theorem, for the determination of $\delta\alpha$, we get

$$-1 = \frac{k+1}{e^{(\alpha+\delta\alpha)(k+1)} - 1} - \sum_{v=1}^{k} \frac{v^2 e^{\alpha v}}{(e^{\alpha v} - 1)^2} \delta\alpha + \text{terms of higher order in } \delta\alpha$$

$$= \frac{1}{\alpha} \frac{\beta}{e^{\beta} - 1} - \sum_{v=1}^{k} \frac{v^2 e^{\alpha v}}{(e^{\alpha v} - 1)^2} \delta\alpha + O\left(\frac{\delta\alpha}{\alpha^2} + \frac{(\delta\alpha)^2}{\alpha^4} \right), \quad \beta = \alpha k.$$

Hence, with A having the same meaning as before, we have

$$\frac{A\delta\alpha}{\alpha^3} - \frac{1}{\alpha} \frac{\beta}{e^{\beta} - 1} = O\left(1 + \frac{\delta\alpha}{\alpha^2} + \frac{(\delta\alpha)^2}{\alpha^4} \right).$$

Thus

$$\delta\alpha = \frac{\alpha^2}{A} \frac{\beta}{e^{\beta} - 1} + O(\alpha^3),$$

$$\begin{aligned} \delta\beta &= (\alpha + \delta\alpha)(k + 1) - \alpha k, \\ &= \alpha + k\delta\alpha + \delta\alpha = \alpha + (\beta/\alpha)\,\delta\alpha + \delta\alpha \\ &= \alpha + \frac{\beta^2}{e^{\beta} - 1} \frac{\alpha}{A} + O(\alpha^2) = O(\alpha), \end{aligned}$$

$$\begin{aligned}
\delta\{\log(1+\alpha g(\beta))\} &= \frac{\alpha g'(\beta)\,\delta\beta + g(\beta)\,\delta\alpha}{1+\alpha g(\beta)} + O(\alpha^3) \\
&= \{\alpha g'(\beta)\,\delta\beta + g(\beta)\,\delta\alpha\}(1-\alpha g(\beta) + O(\alpha^2)) \\
&= O(\alpha^2),
\end{aligned} \tag{1}$$

$$\begin{aligned}
\delta\left\{\sum_{v=1}^{k} \frac{\alpha v}{e^{\alpha v}-1}\right\} &= \sum_{v=1}^{k+1} \frac{(\alpha+\delta\alpha)v}{e^{(\alpha+\delta\alpha)v}-1} - \sum_{v=1}^{k} \frac{\alpha v}{e^{\alpha v}-1} \\
&= (\alpha+\delta\alpha)\sum_{v=1}^{k+1} \frac{v}{e^{(\alpha+\delta\alpha)v}-1} - \alpha\sum_{v=1}^{k} \frac{v}{e^{\alpha v}-1} \\
&= \delta\alpha \sum_{v=1}^{k+1} \frac{v}{e^{(\alpha+\delta\alpha)v}-1} \\
&\quad + \alpha\left\{\sum_{v=1}^{k+1} \frac{v}{e^{(\alpha+\delta\alpha)v}-1} - \sum_{v=1}^{k} \frac{v}{e^{\alpha v}-1}\right\} \\
&= \delta\alpha\left\{\sum_{v=1}^{k} \frac{v}{e^{\alpha v}-1} - 1\right\} - \alpha \\
&= -\alpha + \delta\alpha \sum_{v=1}^{k} \frac{v}{e^{\alpha v}-1} + O(\alpha^2),
\end{aligned} \tag{2}$$

$$\begin{aligned}
&\delta\left\{\sum_{v=1}^{k} \log(1-e^{-\alpha v})\right\} \\
&= \sum_{v=1}^{k+1} \log(1-e^{-(\alpha+\delta\alpha)v}) - \sum_{v=1}^{k} \log(1-e^{-\alpha v}) \\
&= \log(1-e^{-(\alpha+\delta\alpha)(k+1)}) + \sum_{v=1}^{k} \{\log(1-e^{-(\alpha+\delta\alpha)v}) - \log(1-e^{-\alpha v})\} \\
&= \log(1-e^{-\beta}) + \frac{1}{e^{\beta}-1}\,\delta\beta \\
&\quad + \sum_{v=1}^{k}\left\{\frac{v}{e^{\alpha v}-1}\,\delta\alpha - \frac{1}{2!}\frac{v^2 e^{\alpha v}}{(e^{\alpha v}-1)^2}(\delta\alpha)^2\right\} + O(\alpha^2) \\
&= \log(1-e^{-\beta}) + \frac{1}{e^{\beta}-1}\left(\alpha + \frac{\beta^2}{e^{\beta}-1}\frac{\alpha}{A}\right) \\
&\quad + \sum_{v=1}^{k} \frac{v}{e^{\alpha v}-1}\,\delta\alpha - \frac{1}{2}\,\frac{A}{\alpha^3}\,\frac{\alpha^4}{A^2}\,\frac{\beta^2}{(e^{\beta}-1)^2} + O(\alpha^2)
\end{aligned}$$

$$= \log(1 - e^{-\beta}) + \frac{\alpha}{e^{\beta} - 1} + \frac{1}{2} \frac{\beta^2}{(e^{\beta} - 1)^2} \frac{\alpha}{A} + \sum_{v=1}^{k} \frac{v}{e^{\alpha v} - 1} \delta\alpha + O(\alpha^2) \qquad (3)$$

$$\delta(\log \alpha) = \frac{\delta\alpha}{\alpha} + O(\alpha^2) = \frac{\alpha}{A} \frac{\beta}{e^{\beta} - 1} + O(\alpha^2) \qquad (4)$$

$$\delta A = \alpha \frac{\beta^2 e^{\beta}}{(e^{\beta} - 1)^2} + \frac{\beta^3 e^{\beta}}{(e^{\beta} - 1)^2} \frac{\delta\alpha}{\alpha} + O(\alpha^2)$$

$$= \alpha \left\{ \frac{\beta^2 e^{\beta}}{(e^{\beta} - 1)^2} + \frac{1}{A} \frac{\beta^4 e^{\beta}}{(e^{\beta} - 1)^3} \right\} + O(\alpha^2) \qquad (5)$$

$$\delta(\log A) = \alpha \left\{ \frac{1}{A} \frac{\beta^2 e^{\beta}}{(e^{\beta} - 1)^2} + \frac{1}{A^2} \frac{\beta^4 e^{\beta}}{(e^{\beta} - 1)^3} \right\} + O(\alpha^2). \qquad (6)$$

From (5.10), we thus obtain

$$\delta\{\log p(n, k)\} = \log p(n, k + 1) - \log p(n, k)$$

$$= -\log(1 - e^{-\beta}) - \alpha \left\{ 1 + \frac{1}{e^{\beta} - 1} + \frac{1}{2A} \frac{\beta^2 (e^{\beta} + 1)}{(e^{\beta} - 1)^2} \right.$$

$$\left. - \frac{3}{2A} \frac{\beta}{e^{\beta} - 1} + \frac{1}{2A^2} \frac{\beta^4 e^{\beta}}{(e^{\beta} - 1)^3} \right\} + O(\alpha^2). \qquad (7)$$

To bring the right side of (7) close to zero, we set

$$\beta = \tfrac{1}{2} \log n + \mu.$$

Since α is of the order $n^{-1/2}$, the right side of (7) becomes positive if μ is a large negative number and it will become negative if μ is a large positive number. We want to determine the value of μ for which the expression in (7) becomes zero. This requires an accurate evaluation of α. For this purpose, we make use of (4.18) which now takes the form

$$n = \frac{1}{\alpha^2} \int_0^{\beta} \frac{t}{e^t - 1} \, dt + \frac{1}{2\alpha} \left(\frac{\beta}{e^{\beta} - 1} + 2\beta - 1 \right)$$

$$+ \frac{1}{12} \left(\frac{1}{e^{\beta} - 1} - \frac{\beta e^{\beta}}{(e^{\beta} - 1)^2} + \frac{1}{2} \right) + O(\alpha). \qquad (8)$$

Here, if β is large, then

$$\int_0^{\beta} \frac{t}{e^t - 1} \, dt = \frac{\pi^2}{6} - \int_{\beta}^{\infty} \frac{t}{e^t - 1} \, dt$$

$$= \frac{\pi^2}{6} - (\beta + 1) e^{-\beta} + O(\beta e^{-2\beta})$$

and a simple calculation gives, when $\beta = \frac{1}{2} \log n + \mu$,

$$\alpha = \frac{\pi}{\sqrt{(6n)}} - \frac{1}{2}\left(\frac{\sqrt{6}}{\pi} e^{-\mu} - 1\right)\frac{\beta}{n} - \frac{1}{2}\left(\frac{1}{2} + \frac{\sqrt{6}}{\pi} e^{-\mu}\right)\frac{1}{n} + O(n^{-3/2}(\log n)^2),$$

and

$$A = \int_0^\infty \frac{t^2 e^t}{(e^t - 1)^2}\, dt + O(\log n/\sqrt{n}) = \frac{\pi^2}{3} + O\left(\frac{\log n}{\sqrt{n}}\right).$$

The right hand expression in (7) now becomes

$$\left(e^{-\mu} - \frac{\pi}{\sqrt{6}}\right)\frac{1}{\sqrt{n}} - \frac{\sqrt{6}}{4\pi} e^{-\mu}\frac{\beta^2}{n} + \left(\frac{5\sqrt{6}}{4\pi} e^{-\mu} - \frac{1}{2}\right)\frac{\beta}{n} + \left(\frac{1}{4} + \frac{\sqrt{6}}{2\pi} e^{-\mu} - \frac{\pi}{\sqrt{6}} e^{-\mu} + \frac{1}{2} e^{-2\mu}\right)\frac{1}{n} + O(n^{-3/2}\log^4 n).$$

This is zero only if

$$\frac{\sqrt{6}}{\pi} e^{-\mu} = 1 + w,$$

where $w = O(n^{-1/2}\log^2 n)$. Thus

$$\mu = -\log(\pi/\sqrt{6}) - w + O(n^{-1}\log^4 n),$$

so that

$$\beta = \frac{1}{2}\log n + \mu = \log\frac{\sqrt{(6n)}}{\pi} - w + O(n^{-1}\log^4 n).$$

The right hand expression in (7), therefore, vanishes only if

$$\frac{\pi}{\sqrt{(6n)}} w - \frac{1}{4n} L^2 + \frac{3}{4n} L + \frac{1}{2}\left(\frac{3}{2} - \frac{\pi^2}{6}\right)\frac{1}{n} = O(n^{-1/2}\log^4 n),$$

where (here and in what follows)

$$L = \log\left\{\frac{\sqrt{(6n)}}{\pi}\right\}.$$

This gives

$$w = \frac{1}{4}\frac{\sqrt{6}}{\pi}\frac{1}{\sqrt{n}} L^2 - \frac{3}{4}\frac{\sqrt{6}}{\pi}\frac{1}{\sqrt{n}} L - \frac{1}{2}\left(\frac{3}{2}\frac{\sqrt{6}}{\pi} - \frac{\pi}{\sqrt{6}}\right)\frac{1}{\sqrt{n}} + O(n^{-1}\log^4 n)$$

and finally

$$\alpha = \frac{\pi}{\sqrt{(6n)}} - \frac{3}{4n} + O(n^{-3/2}\log^2 n).$$

Thus, the expression in (7) is positive if $k < k_{\max}$ and it is negative if $k > k_{\max}$, where

$$k_{\max} = \frac{\beta}{\alpha} = \frac{\sqrt{(6n)}}{\pi} L - \frac{6}{\pi^2}\left(\frac{1}{4}L^2 - \frac{3}{2}L - \frac{3}{4}\right) - \frac{1}{2} + O(n^{-1/2}\log^2 n). \tag{9}$$

If $k^2 \leqslant n$, then the expression for $P(n, k)$ in section 5.2 gives similarly, and more easily,

$$\delta\{\log p(n, k)\} = -\log(1 - e^{-\beta}) + O(k^{-1}).$$

This is positive, since $\beta \leqslant 1$.

We thus obtain the following

THEOREM 1. *There exist positive integers k_0 and n_0, and for each positive integer $n \geqslant n_0$, a positive number $\chi > k_0$, with the following property:*

If $k_0 \leqslant k < \chi - \frac{1}{2}$, then $p(n, k+1) > p(n, k)$;

If $k > \chi - \frac{1}{2}$, then $p(n, k+1) < p(n, k)$.

The numbers χ satisfy the relation

$$\chi = \frac{\sqrt{(6n)}}{\pi} L - \frac{6}{\pi^2}\left(\frac{1}{4}L^2 - \frac{3}{2}L - \frac{3}{4}\right) + O(n^{-1/2}\log^2 n). \tag{10}$$

This establishes the validity of the Auluck–Chowla–Gupta conjecture at least for large values of n.

7. The case of $q(n, k)$

We have

$$q(n, k) = P\left(n - \frac{k(k+1)}{2}, k\right).$$

Hence the asymptotic formulae for $P(n, k)$ hold also for $q(n, k)$ provided α is determined from the relation

$$n = \sum_{v=1}^{k}\left(\frac{v}{e^{\alpha v} - 1} + v\right) = \sum_{v=1}^{k}\frac{ve^{\alpha v}}{e^{\alpha v} - 1}.$$

In place of (6.8), we now get

$$n = \frac{1}{\alpha^2}\int_0^{\beta}\frac{te^t}{e^t - 1}\,dt + \frac{1}{2\alpha}\left(\frac{\beta e^{\beta}}{e^{\beta} - 1} - 1\right) + O(1), \quad \beta = \alpha k. \tag{1}$$

We further have now

$$k + 1 = \sum_{v=1}^{k} \frac{v}{e^{\alpha v} - 1} - \sum_{v=1}^{k+1} \frac{v}{e^{(\alpha+\delta\alpha)v} - 1}$$

$$= \sum_{v=1}^{k} \frac{v^2 e^{\alpha v}}{(e^{\alpha v} - 1)^2}\, \delta\alpha - \frac{k+1}{e^{\alpha(k+1)} - 1} + O\left(\frac{\beta}{1+\beta}\, \frac{(\delta\alpha)^2}{\alpha^4} + \frac{\delta\alpha}{\alpha^2}\right),$$

or with

$$A = \sum_{v=1}^{k} \frac{\alpha^3 v^2 e^{\alpha v}}{(e^{\alpha v} - 1)^2},$$

$$A\alpha^{-3}\, \delta\alpha - \alpha^{-1}\left(\frac{\beta}{e^{\beta} - 1} + \beta\right) = O\left(\frac{\beta}{1+\beta}\, \frac{(\delta\alpha)^2}{\alpha^4} + \frac{\delta\alpha}{\alpha^2} + \frac{1+\beta}{\beta}\right).$$

The factor $\beta/(1 + \beta)$ serves the purpose of making the constant in the O-notation, valid for all positive values of β. Since

$$A = O\left(\frac{\beta}{1+\beta}\right)$$

we get the following results

$$\delta\alpha = \frac{\alpha^2}{A}\, \frac{\beta e^{\beta}}{e^{\beta} - 1} + O\left(\frac{(1+\beta)^2}{\beta^2}\, \alpha^3\right) = O((\beta^{-1} + \beta)\alpha^2) \tag{2}$$

$$\delta\left\{\sum_{v=1}^{k} \frac{\alpha v}{e^{\alpha v} - 1} - \sum_{v=1}^{k} \log(1 - e^{-\alpha v})\right\}$$

$$= -\log(e^{\beta} - 1) - \alpha\left(\frac{e^{\beta}}{e^{\beta} - 1} + \frac{1}{2A}\, \frac{\beta^2 e^{2\beta}}{(e^{\beta} - 1)^2}\right) + O\left(\left(\frac{1+\beta}{\beta}\right)^2 \alpha^2\right) \tag{3}$$

$$\log(\alpha + \delta\alpha) - \log\alpha = \frac{\alpha}{A}\, \frac{\beta e^{\beta}}{e^{\beta} - 1} + O((\beta^{-2} + \beta^2)\alpha^2) \tag{4}$$

$$\delta A = \alpha\left(\frac{\beta^2 e^{\beta}}{(e^{\beta} - 1)^2} + \frac{1}{A}\, \frac{\beta^4 e^{2\beta}}{(e^{\beta} - 1)^3}\right) + O((\beta^{-1} + 1)\alpha^2) \tag{5}$$

$$\delta(\log A) = \alpha\left\{\frac{1}{A}\, \frac{\beta^2 e^{\beta}}{(e^{\beta} - 1)^2} + \frac{1}{A^2}\, \frac{\beta^4 e^{2\beta}}{(e^{\beta} - 1)^3}\right\} + O\left(\left(\frac{1+\beta}{\beta}\right)^2 \alpha^2\right) \tag{6}$$

$$\delta\left\{\log\left(1 + \frac{1+\beta}{\beta}\, g(\beta)\alpha\right)\right\} = O((\beta^{-2} + \beta^2)\alpha^2). \tag{7}$$

Hence

$$\delta\{\log q(n, k)\} = -\log(e^{\beta} - 1) - \alpha\left\{\frac{e^{\beta}}{e^{\beta} - 1} - \frac{3}{2A}\, \frac{\beta e^{\beta}}{e^{\beta} - 1} + \frac{1}{2A}\, \frac{\beta^2 (e^{\beta} + e^{2\beta})}{(e^{\beta} - 1)^2} + \frac{1}{2A^2}\, \frac{\beta^4 e^{2\beta}}{(e^{\beta} - 1)^3}\right\} + O((\beta^{-2} + \beta^2)\alpha^2). \tag{8}$$

The right hand expression in (8) is expected to vanish when β is near log 2. We, therefore, set

$$\beta = \log 2 + \mu n^{-1/2}.$$

Here μ may be assumed to be bounded because the expression is easily seen to be negative if μ is large positive, and positive if μ is large negative. The corresponding value of α can be obtained from (1). We now use the well-known formula

$$\int_0^{\log 2} \frac{t}{e^t - 1}\,dt = \frac{\pi^2}{12} - \frac{1}{2}(\log 2)^2. \tag{9}$$

Substituting this result in (1), we obtain

$$\alpha = \frac{\pi}{\sqrt{(12n)}} + \left(\frac{1}{2}\log 2 - \frac{1}{4} + \frac{\sqrt{12}}{\pi}\mu \log 2\right)\frac{1}{n} + O(n^{-3/2}).$$

Also

$$\begin{aligned} A &= \int_0^{\beta} \frac{t^2 e^t}{(e^t - 1)^2}\,dt + O(\alpha) \\ &= 2\int_0^{\log 2} \frac{t}{e^t - 1}\,dt - (\log 2)^2 + O(n^{-1/2}) \\ &= \frac{\pi^2}{6} - 2(\log 2)^2 + O(n^{-1/2}). \end{aligned}$$

This enables us to write (8) as

$$\delta\{\log q(n,k)\} = \left\{-2\mu - \frac{\pi}{\sqrt{3}} + \frac{3\sqrt{3}}{\pi}\frac{(1-\log 2)\log 2}{1-w} - \frac{12\sqrt{3}}{\pi^3}\frac{(\log 2)^4}{(1-w)^2}\right\}\frac{1}{\sqrt{n}} + O(n^{-1}), \tag{10}$$

where

$$w = \frac{12}{\pi^2}(\log 2)^2.$$

Now (10) can vanish only if

$$\mu = -\frac{\pi}{\sqrt{12}} + \frac{3\sqrt{3}}{2\pi}\frac{(1-\log 2)\log 2}{1-w} - \frac{6\sqrt{3}}{\pi^3}\frac{(\log 2)^4}{(1-w)^2} + O(n^{-1/2}).$$

Hence

$$k_0^* = k_{\max} = \frac{\beta}{\alpha} = \left(\frac{\sqrt{12}}{\pi}\log 2\right)\sqrt{n} - \left(1 + \frac{w}{4(1-w)} - \frac{w}{\log 2}\right) + O(n^{-1/2}).$$

We thus obtain

THEOREM 2. *There exist positive integers k_0^* and n_0^*, and for each $n \geqslant n_0^*$, a positive number $\chi^* > k_0^*$ with the following property. If*

$$k_0^* \leqslant k < \chi^* - \tfrac{1}{2}, \quad \textit{then } q(n, k+1) > q(n, k).$$

If

$$k > \chi^* - \tfrac{1}{2}, \quad \textit{then } q(n, k+1) < q(n, k).$$

The number χ^* satisfies the relation

$$\chi^* = \frac{\sqrt{12n}}{\pi} \log 2 - \left(\frac{1}{2} + \frac{1}{4}\frac{w}{1-w} - \frac{w}{\log 2}\right) + O(n^{-1/2}),$$

where

$$w = \frac{12}{\pi^2} (\log 2)^2 .$$

Numerically

$$\chi^* = 0.7643041 \sqrt{n} - 0.0084280 + O(n^{-1/2})$$

8. Explicit form of the asymptotic formula for $P(n, k)$

We have seen that

$$P(n, k) \sim \left(\frac{\alpha^3}{2\pi A_0}\right)^{1/2} \exp\left\{\sum_{v=1}^{k} \left(\frac{\alpha v}{e^{\alpha v} - 1} - \log(1 - e^{-\alpha v})\right)\right\}$$

uniformly when k (and hence n) tend to infinity. Here

$$A_0 = \int_0^\beta \frac{t^2 e^t}{(e^t - 1)^2}\, dt = 2 \int_0^\beta \frac{t}{e^t - 1}\, dt - \frac{\beta^2}{e^\beta - 1},$$

$$\alpha = \beta/k,$$

and β is determined from

$$n = \frac{k^2}{\beta^2} \int_0^\beta \frac{t}{e^t - 1}\, dt + \frac{k}{2}\left(\frac{1}{e^\beta - 1} - \frac{1}{\beta}\right) + \frac{1}{12}\left(\frac{1}{e^\beta - 1} - \frac{\beta e^\beta}{(e^\beta - 1)^2} + \frac{1}{2}\right).$$

Throughout this section and elsewhere, the symbol $\sim$ means that the quotient of the two sides of the asymptotic equation tends to 1 as k tends to infinity.

From the Euler–Maclaurin formula we have

$$\sum_{v=1}^{k} \frac{\alpha v}{e^{\alpha v} - 1} \sim \frac{k}{\beta} \int_0^\beta \frac{t}{e^t - 1}\, dt + \frac{1}{2}\left(\frac{\beta}{e^\beta - 1} - 1\right)$$

and

$$\sum_{v=1}^{k} \log \frac{v}{1-e^{-\alpha v}} = \int_0^k \log \frac{x}{1-e^{-\alpha x}}\, dx + \frac{1}{2} \log \frac{\beta}{1-e^{-\beta}} + o(1)$$

$$= k \log k - k + \frac{k}{\beta} \int_0^\beta \frac{t}{e^t - 1}\, dt$$

$$-\left(k + \frac{1}{2}\right) \log(1 - e^{-\beta}) + \frac{1}{2} \log \beta + o(1).$$

By Stirling's formula

$$\sum_{v=1}^{k} \log v = \left(k + \frac{1}{2}\right) \log k - k + \frac{1}{2} \log(2\pi) + o(1).$$

Hence

$$\sum_{v=1}^{k} \left\{ \frac{\alpha v}{e^{\alpha v} - 1} - \log(1 - e^{-\alpha v}) \right\}$$

$$= 2\frac{k}{\beta} \int_0^\beta \frac{t}{e^t - 1}\, dt + \frac{1}{2}\left(\frac{\beta}{e^\beta - 1} - 1\right) - \left(k + \frac{1}{2}\right) \log(1 - e^{-\beta})$$

$$+ \frac{1}{2} \log \frac{\beta}{2\pi k} + o(1).$$

Thus

$$P(n, k) \sim \frac{\beta^2}{2\pi k^2} \left\{ 2\int_0^\beta \frac{t}{e^t - 1}\, dt - \frac{\beta^2}{e^\beta - 1} \right\}^{-1/2}$$

$$\cdot \exp\left\{ \frac{2k}{\beta} \int_0^\beta \frac{t}{e^t - 1}\, dt + \frac{1}{2}\left(\frac{\beta}{e^\beta - 1} - 1\right) - \left(k + \frac{1}{2}\right) \log(1 - e^{-\beta}) \right\}, \tag{1}$$

where β is determined from the relation given earlier. The formula describes the asymptotic behaviour of $P(n, k)$ for the full range of k, provided k tends to infinity with n. To examine how the formula works for small k, suppose that k^3/n is bounded. Then

$$\beta \sim k^2/n = O(k^{-1}).$$

Since $k\beta^2$ is small, we have

$$n = \frac{k^2}{\beta} - \frac{1}{4} k^2 + O(k).$$

that is

$$\beta = \frac{k^2}{n}\left(1 - \frac{k^2}{4n} + O\left(\frac{k}{n}\right)\right),$$

and a simple calculation gives

$$P(n, k) \sim \frac{1}{2\pi n}(e^2 n/k^2)^k e^{k^3/(4n)}.$$

If k^3/n is small, we get

$$P(n, k) \sim \frac{1}{2\pi n}(e^2 n/k^2)^k.$$

But for $k = o(n^{1/3})$, Erdös and Lehner have proved that for large n,

$$P(n, k) \sim \frac{1}{k!}\binom{n-1}{k-1} \sim \frac{n^{k-1}}{k!(k-1)!}.$$

Hence the asymptotic formula

$$P(n, k) \sim \frac{k^{2k-1}}{e^{2k}(k!)^2}\beta^2\left\{2\int_0^\beta \frac{t}{e^t - 1}\,dt - \frac{\beta^2}{e^\beta - 1}\right\}^{-1/2}$$

$$\cdot \exp\left\{\frac{2k}{\beta}\int_0^\beta \frac{t}{e^t - 1}\,dt + \frac{1}{2}\left(\frac{\beta}{e^\beta - 1} - 1\right) - \left(k + \frac{1}{2}\right)\log(1 - e^{-\beta})\right\}$$

where β is determined as before, is valid for all $k \leqslant n$, uniformly when n tends to infinity. In particular, for $k = n$, β is given by the relation

$$\frac{n^2}{\beta^2}\frac{\pi^2}{6} - \frac{n}{2\beta} = n + O(1),$$

so that

$$\frac{n}{\beta} = \frac{\sqrt{(6n)}}{\pi} + \frac{3}{2\pi^2} + O(n^{-1/2}).$$

Hence from (1), we readily obtain

$$p(n) = P(n, n) \sim \frac{1}{4n\sqrt{3}}\exp(2\pi\sqrt{(n/6)}),$$

which is the asymptotic term in the Hardy–Ramanujan formula.

9. Tables

The following tables by G. Szekeres give the values of χ and χ^* for certain values of n.

n	χ
6	2·22
24	6·53
54	11·59
96	17·22
150	23·29
216	29·73
294	36·47
384	43·48
486	50·71
600	58·16

n	χ^*
4	1·52
9	2·28
16	3·05
25	3·81
36	4·58
49	5·34
64	6·11
81	6·87
100	7·63
121	8·40
144	9·16
169	9·93
196	10·69
225	11·46
256	12·22
289	12·98
324	13·75
361	14·51
400	15·28

References

G. H. Hardy; S. Ramanujan: *Proc. London Math. Soc.* (2) **17** (1918), 75–115.
D. H. Lehmer: *J. London Math. Soc.* **12** (1937), 171–176.
H. Rademacher: *Proc. London Math. Soc.* **43** (1937), 241–254.
——: *Ann. of Math.* **44** (1943), 416–422.
D. H. Lehmer: *Trans. Amer. Math. Soc.* **43** (1938), 271–295.
F. C. Auluck; S. Chowla; H. Gupta: *J. Indian Math. Soc.* **6** (1942), 105–112.
P. Erdös; J. Lehner: *Duke Math. J.* **8** (1941), 335–345.
H. Gupta: *Proc. Indian Acad. Sci.* **16** (1942), 101–102.
G. Szekeres: *Quart. J. Math. Oxford Ser.* (2) **2** (1951), 85–108; **4** (1953), 96–111.

CHAPTER 10

PARTITION OF j-PARTITE NUMBERS

1. Introduction

In an earlier chapter, we have considered some problems concerning partitions of positive integers. We now take up the more general question of partitioning a j-partite number

$$N_j = (n_1, n_2, n_3, \ldots, n_j) = (n_i),$$

where each n_i is a positive integer. We shall speak of such N_j as positive. The j-partite number

$$N_j = (0, 0, 0, \ldots, 0)$$

shall be called null and denoted by 0. j-partite numbers in which some of the elements are zero, are also used sometimes. These are said to be degenerate. We shall, however, make very little use of them.

A j-partite number is just a row matrix or a vector and the addition of j-partite numbers has the same meaning as the addition of matrices or vectors. A partition of N_j into exactly k parts is a way of expressing N_j as a sum of k positive j-partite numbers, the order in which the summands occur being irrelevant. Thus the bipartite number (5, 4) has the following partitions into exactly two parts: (1, 1), (4, 3); (2, 1), (3, 3); (1, 2), (4, 2); (2, 2), (3, 2); (1, 3), (4, 1); (2, 3), (3, 1). In this sense, the earlier work can be considered to be concerned with partitions of unipartite numbers.

2. Notations

All small letters, other than x, but including α and β, denote positive integers unless expressly stated otherwise. i runs from 1 to j. $n^{(m)}$ stands for the product $n(n-1)(n-2)\ldots(n-m+1)$. N_j, M_j, N_{1j}, N_{2j}, etc., denote j-partite numbers (n_i), (m_i), (n_{1i}), (n_{2i}), etc. We write $N_j > M_j$, or $N_j - M_j > 0$ or $M_j < N_j$ when $n_i > m_i$ for each $i \leqslant j$. tM_j stands for the j-partite number (tm_i). When M_j occurs t times as a part in a partition, we express this by writing $M_j(t)$. Further we say, $t \mid N_j$ when $t \mid n_i$ for each $i \leqslant j$. $p(N_j, k)$ denotes the number of partitions of N_j into exactly k summands and $q(N_j, k)$ the number of partitions of N_j into exactly k distinct parts. We write $P_k = P(N_j, k)$ for $k!p(N_j, k)$ and $Q_k = Q(N_j, k)$ for $k!q(N_j, k)$. We take $p(N_j, 0) = 0 = q(N_j, 0)$ when $N_j \neq 0$, and $p(N_j, 0) = 1 = q(N_j, 0)$ when $N_j = 0$.

If $f(N_j)$ be any number-theoretic function defined for all N_j, we define the operator σ_r by the relation

$$\sigma_r f(N_j) = \sum_{0<rM_j\leqslant N_j} f(N_j - rM_j).$$

We leave it to the reader to show that the operators σ are commutative, that is,

$$\sigma_r\sigma_s f(N_j) = \sigma_s\sigma_r f(N_j).$$

Finally, the number of those partitions of N_j into exactly k parts, which have exactly t parts equal, $t \geqslant 1$, and all others distinct from each other is denoted by $q^*(N_j, k; t)$ or simply by $q^*(t)$. We speak of such partitions as the t^*-partitions of N_j, it being understood that the partitions are into exactly k parts.

3. Recurrence formula for $q(N_j, k)$

LEMMA 1. *The partitions of $N_j - tM_j$, where M_j runs through all numbers for which $tM_j \leqslant N_j$, into $(k-t)$ distinct summands, provide for $t > 1$, all the t^*- and $(t+1)^*$-partitions of N_j into exactly k parts. For $t = 1$, these provide the partitions of N_j into k distinct parts, k times over, besides all the 2^*-partitions of N_j.*

Proof. For a given M_j and $t > 1$, $tM_j \leqslant N_j$, let

$$N_{1j}, N_{2j}, N_{3j}, \ldots, N_{k-t,j}, \quad t \leqslant k,$$

be any partition of $N_j - tM_j$ into $(k-t)$ distinct summands. Then

$$M_j(t), N_{1j}, N_{2j}, \ldots, N_{k-t,j}, \quad t \leqslant k,$$

is a t^*-partition of N_j when M_j is distinct from each of the summands

$$N_{1j}, N_{2j}, \ldots, N_{k-t,j};$$

otherwise, it is a $(t+1)^*$-partition of N_j.

When $t = 1$, let $N_{1j}, N_{2j}, \ldots, N_{kj}$ be any partition of N_j into exactly k distinct summands. Then from the partitions of $N_j - M_j$ into $(k-1)$ distinct parts, as M_j runs through all values $<N_j$, the partition

$$N_{1j}, N_{2j}, \ldots, N_{kj}$$

of N_j will be obtained every time that M_j passes through one of the values N_{sj}, $s = 1, 2, 3, \ldots, k$. The partitions of $N_j - M_j$ into $(k-1)$ distinct parts will give also all the 2^*-partitions of N_j.

Thus, we have

$$kq(N_j, k) + q^*(2) = \sigma_1 q(N_j, k-1)$$

$$q^*(2) + q^*(3) = \sigma_2 q(N_j, k-2)$$

$$\vdots$$

$$q^*(k-1) + q^*(k) = \sigma_{k-1} q(N_j, 1)$$

and finally

$$q^*(k) = \sigma_k q(N_j, 0).$$

Hence

$$kq(N_j, k) = \sum_{s=1}^{k} (-1)^{s-1} \sigma_s q(N_j, k-s),$$

or what is the same thing

$$Q_k = \sum_{s=1}^{k} (-1)^{s-1}(k-1)^{(s-1)}\sigma_s Q_{k-s}, \tag{1}$$

where

$$Q_0 = q(N_j, 0).$$

In particular

$$Q_1 = \sigma_1 Q_0,$$

$$Q_2 = (\sigma_1^2 - \sigma_2)Q_0,$$

$$Q_3 = (\sigma_1^3 - 3\sigma_1\sigma_2 + 2\sigma_3)Q_0.$$

4. The recurrence formula for $p(N_j, k)$

Consider a fixed $M_j < N_j$. Then M_j occurs as a part in a partition of N_j into k summands, at least once in $p(N_j - M_j, k-1)$ partitions, at least twice in $p(N_j - 2M_j, k-2)$ partitions, at least thrice in $p(N_j - 3M_j, k-3)$ partitions and so on; because if we take M_j as a part t times, we are left to partition $(N_j - tM_j)$ into $(k-t)$ parts.

Hence M_j occurs as a part exactly once in

$$p(N_j - M_j, k-1) - p(N_j - 2M_j, k-2)$$

partitions; exactly twice in

$$p(N_j - 2M_j, k-2) - p(N_j - 3M_j, k-3)$$

partitions and so on. Thus if all the $p(N_j, k)$ partitions of N_j into exactly k parts were to be written out at length, the number of times M_j will have to be written is exactly

$$p(N_j - M_j, k-1) + p(N_j - 2M_j, k-2) + \cdots + p(N_j - rM_j, k-r),$$

where r is the greatest integer $\leqslant k$ for which

$$N_j - rM_j \geqslant 0.$$

Letting M_j vary now, the number of all the parts in the $p(N_j, k)$ partitions is given by

$$\sigma_1 p(N_j, k-1) + \sigma_2 p(N_j, k-2) + \cdots + \sigma_{k-1} p(N_j, 1) + \sigma_k p(N_j, 0).$$

Since the number of all the parts is also $kp(N_j, k)$ we must have

$$kp(N_j, k) = \sum_{s=1}^{k} \sigma_s p(N_j, k-s)$$

or what is the same thing

$$P_k = \sum_{s=1}^{k} (k-1)^{(s-1)} \sigma_s P_{k-s}, \tag{1}$$

where $P_0 = p(N_j, 0)$.

The rather surprising similarity between the recursion formulae for $p(N_j, k)$ and $q(N_j, k)$ shows that the expression for $p(N_j, k)$ will be the same as that for $q(N_j, k)$ with the minus sign in the latter replaced by plus signs. Thus

$$P_4 = (\sigma_1^4 + 6\sigma_1^2\sigma_2 + 3\sigma_2^2 + 8\sigma_1\sigma_3 + 6\sigma_4)P_0,$$

while

$$Q_4 = (\sigma_1^4 - 6\sigma_1^2\sigma_2 + 3\sigma_2^2 + 8\sigma_1\sigma_3 - 6\sigma_4)Q_0.$$

5. The degree and the weight of an operator

The degree of the operator

$$\sigma_{\alpha_1}^{\beta_1} \sigma_{\alpha_2}^{\beta_2} \cdots \sigma_{\alpha_t}^{\beta_t}$$

is defined as the sum

$$\sum_{r=1}^{t} \beta_r$$

and its weight as the sum

$$\sum_{r=1}^{t} \alpha_r \beta_r.$$

If

$$P_k = \phi_k(\sigma) \cdot P_0,$$

and

$$Q_k = \psi_k(\sigma) \cdot Q_0,$$

where $\phi_k(\sigma)$ and $\psi_k(\sigma)$ are polynomials in the operators $\sigma_1, \sigma_2, \ldots, \sigma_k$, then it can be readily shown that the weight of each term in these polynomials is k. Moreover, in the expression for Q_k, the sign of any term for which the operator is of degree $k - s$ is given by $(-1)^s$. It would, therefore, be quite useful, sometimes, to group together terms of the same degree in the operators.

6. More on the product of σ-operators

Two questions still beset us, first to find an expression for $T_k P_0$, where

$$T_k = \sigma_{\alpha_1}^{\beta_1} \sigma_{\alpha_2}^{\beta_2} \sigma_{\alpha_3}^{\beta_3} \ldots \sigma_{\alpha_t}^{\beta_t}$$

is an operator with α's all distinct and

$$\sum_{r=1}^{t} \alpha_r \beta_r = k;$$

second, to find the numerical coefficient of T_k in $\phi_k(\sigma)$. We proceed to answer these questions.

Let X_s denote the expression

$$\frac{x^s}{1 - x^s}, \quad |x| < 1,$$

and $\{f(x)\}_{n_i}$ the coefficient of x^{n_i} in the formal expansion of $f(x)$ in ascending powers of x. Then we have the

THEOREM 1.

$$\sigma_{\alpha_1}^{\beta_1} \sigma_{\alpha_2}^{\beta_2} \ldots \sigma_{\alpha_t}^{\beta_t} P_0 = \prod_i \{X_{\alpha_1}^{\beta_1} X_{\alpha_2}^{\beta_2} \ldots X_{\alpha_t}^{\beta_t}\}_{n_i}$$

Proof. We have

$$\sigma_r P_0 = \sum_{0 < rM_j \leqslant N_j} P(N_j - rM_j, 0)$$

$$= \sum_{N_j - rM_j = 0} 1.$$

In other words $\sigma_r P_0$ is the number of solutions of the equation

$$rM_j = N_j, \tag{1}$$

that is of the equations

$$rm_i = n_i, \quad i = 1, 2, 3, \ldots, j. \tag{2}$$

This is 1 if $r \mid N_j$ and 0 otherwise. Hence

$$\sigma_r P_0 = \prod_i \{X_r\}_{n_i}.$$

Again

$$\sigma_r \sigma_s P_0 = \sum_{\substack{rN_{1j}+sN_{2j} \leqslant N_j \\ 0<N_{1j},N_{2j}}} P(N_j - rN_{1j} - sN_{2j}, 0)$$

$$= \sum_{\substack{rN_{1j}+sN_{2j}=N_j \\ 0<N_{1j},N_{2j}}} 1.$$

In other words, $\sigma_r \sigma_s P_0$ is the number of positive solutions of the equation

$$rN_{1j} + sN_{2j} = N_j \tag{3}$$

that is, of the equations

$$rn_{1i} + sn_{2i} = n_i, \quad i = 1, 2, 3, \ldots, j. \tag{4}$$

If $h(i)$ denotes the number of solutions of (4) then the number of solutions of (3) is given by the product

$$h(1)h(2)h(3) \ldots h(j). \tag{5}$$

We notice that $h(i)$ is the coefficient of x^{n_i} in $X_r X_s$. This is true even when $r = s$.

In the general case,

$$\sigma_{\alpha_1} \sigma_{\alpha_2} \sigma_{\alpha_3} \ldots \sigma_{\alpha_t} P_0$$

is the number of non-null solutions of the equation

$$\alpha_1 N_{1j} + \alpha_2 N_{2j} + \alpha_3 N_{3j} + \cdots + \alpha_t N_{tj} = N_j \tag{6}$$

that is, of the equations

$$\alpha_1 n_{1i} + \alpha_2 n_{2i} + \alpha_3 n_{3i} + \cdots + \alpha_t n_{ti} = n_i, \quad i = 1, 2, 3, \ldots, j. \tag{7}$$

The number of solutions $h(i)$ of (7) is the coefficient of x^{n_i} in

$$X_{\alpha_1} X_{\alpha_2} X_{\alpha_3} \ldots X_{\alpha_t}.$$

Hence

$$\sigma_{\alpha_1} \sigma_{\alpha_2} \sigma_{\alpha_3} \ldots \sigma_{\alpha_t} P_0 = \prod_i \{X_{\alpha_1} X_{\alpha_2} X_{\alpha_3} \ldots X_{\alpha_t}\}_{n_i}. \tag{8}$$

The result holds even when the α's are not all distinct. The theorem is a direct consequence of this result.

The answer to the second question is contained in the

THEOREM 2. *The coefficient of T_k in $\phi_k(\sigma)$, where*

$$T_k = \sigma_{\alpha_1}^{\beta_1} \sigma_{\alpha_2}^{\beta_2} \sigma_{\alpha_3}^{\beta_3} \ldots \sigma_{\alpha_t}^{\beta_t}, \quad \sum_{r=1}^{t} \alpha_r \beta_r = k,$$

is

$$k!\Big/\prod_{r=1}^{t}\{\beta_r!\alpha_r^{\beta_r}\}.$$

Proof. Let $C(T_u)$ denote the coefficient of the term corresponding to the operator T_u in $\phi_u(\sigma)$. Assume that the theorem holds for all operators of weight $\leqslant(k-1)$. Then, since

$$P_k = \sum_{s=1}^{k} (k-1)^{(s-1)}\sigma_s P_{k-s},$$

we have

$$C(T_k) = \sum_{r=1}^{t} (k-1)^{(\alpha_r-1)} C(T_k/\sigma_{\alpha_r})$$

$$= \sum_{r=1}^{t} (k-1)^{(\alpha_r-1)} \cdot \frac{(k-\alpha_r)!}{\prod_{h=1}^{t}\{\beta_h!\alpha_h^{\beta_h}\}/\alpha_r\beta_r}$$

$$= \frac{(k-1)!}{\prod_{h=1}^{t}\{\beta_h!\alpha_h^{\beta_h}\}} \cdot \sum_{r=1}^{t}\alpha_r\beta_r$$

$$= \frac{k!}{\prod_{h=1}^{t}\{\beta_h!\alpha_h^{\beta_h}\}}.$$

The theorem follows by induction on k.

Remark

It is noteworthy that the coefficient of every term in $\phi_k(\sigma)$ is independent of j.

In fact, if λ be a partition of k into a_1 parts each equal to 1, a_2 parts each equal to 2, a_3 parts each equal to 3, and so on, a's being non-negative integers, then we may write $C(\lambda)$ for

$$k!/\{a_1!a_2!a_3!\dots a_k!\ 1^{a_1}2^{a_2}3^{a_3}\dots k^{a_k}\},$$

$J(n, \lambda)$ for the coefficient of x^n in

$$X_1^{a_1}X_2^{a_2}X_3^{a_3}\dots X_k^{a_k},$$

and $A(\lambda)$ for the sum

$$a_1 + a_2 + a_3 + \cdots + a_k.$$

We then have

$$Q(N_j, k) = \sum_{\lambda} \{(-1)^{k-A(\lambda)} C(\lambda) \prod_i J(n_i, \lambda)\}$$

and

$$P(N_j, k) = \sum_{\lambda} \{C(\lambda) \prod_i J(n_i, \lambda)\},$$

where λ runs over all the partitions of k. In what follows, we shall use $A(\lambda)$, $C(\lambda)$ and $J(n, \lambda)$ in this sense.

For an example, take the case $k = 6$. The eleven partitions of 6 are given by

(i) $a_1 = 6, a_2 = a_3 = a_4 = a_5 = a_6 = 0,$

(ii) $a_1 = 4, a_2 = 1, a_3 = a_4 = a_5 = a_6 = 0,$

(iii) $a_1 = 3, a_2 = 0, a_3 = 1, a_4 = a_5 = a_6 = 0,$

(iv) $a_1 = 2, a_2 = 2, a_3 = a_4 = a_5 = a_6 = 0,$

(v) $a_1 = 2, a_2 = 0, a_3 = 0, a_4 = 1, a_5 = a_6 = 0,$

(vi) $a_1 = a_2 = a_3 = 1, a_4 = a_5 = a_6 = 0,$

(vii) $a_1 = 0, a_2 = 3, a_3 = a_4 = a_5 = a_6 = 0,$

(viii) $a_1 = 0, a_2 = 1, a_3 = 0, a_4 = 1, a_5 = a_6 = 0,$

(ix) $a_1 = a_2 = 0, a_3 = 2, a_4 = a_5 = a_6 = 0,$

(x) $a_1 = 1, a_2 = a_3 = a_4 = 0, a_5 = 1, a_6 = 0,$

(xi) $a_1 = a_2 = a_3 = a_4 = a_5 = 0, a_6 = 1,$

and we have

$$\begin{aligned} Q_6 = {} & \frac{6!}{6!} \prod_i (X_1^6)_{n_i} - \frac{6!}{4!2} \prod_i (X_1^4 X_2)_{n_i} \\ & + \left\{ \frac{6!}{3!3} \prod_i (X_1^3 X_3)_{n_i} + \frac{6!}{2!2!2^2} \prod_i (X_1^2 X_2^2)_{n_i} \right\} \\ & - \left\{ \frac{6!}{2!4} \prod_i (X_1^2 X_4)_{n_i} + \frac{6!}{1.2.3} \prod_i (X_1 X_2 X_3)_{n_i} + \frac{6!}{3!2^3} \prod_i (X_2^3)_{n_i} \right\} \\ & + \left\{ \frac{6!}{2.4} \prod_i (X_2 X_4)_{n_i} + \frac{6!}{2!3^2} \prod_i (X_3^2)_{n_i} + \frac{6!}{1.5} \prod_i (X_1 X_5)_{n_i} \right\} \\ & - \frac{6!}{6} \prod_i (X_6)_{n_i}. \end{aligned}$$

In P_6 all the above terms have positive signs.

When $k = 2, 3$ we have the explicit formulae

$$p(N_j, 2) = \left[\tfrac{1}{2}\left(1 + \prod_i (n_i - 1)\right)\right]$$

and

$$p(N_j, 3) = \left[\left\{\prod_i \binom{n_i - 1}{2} + 3\prod_i \left[\frac{n_i - 1}{2}\right] + 2\right\}\Big/6\right]$$

where $[x]$ indicates as usual the greatest integer in x.

7. P_k in the form of a determinant

Eliminating $P_1, P_2, P_3, \ldots, P_{k-1}$ from the relations

$$P_t = \sum_{s=1}^{t} (t-1)^{(s-1)}\sigma_s P_{t-s}, \quad t = k, k-1, k-2, \ldots, 3, 2, 1,$$

we get

$$P_k = \begin{vmatrix} \sigma_1 & \sigma_2 & \sigma_3 & \ldots & \sigma_{k-1} & \sigma_k \\ -(k-1) & \sigma_1 & \sigma_2 & \ldots & \sigma_{k-2} & \sigma_{k-1} \\ 0 & -(k-2) & \sigma_1 & \ldots & \sigma_{k-3} & \sigma_{k-2} \\ \cdot & \cdot & \cdot & \ldots & \cdot & \cdot \\ 0 & 0 & 0 & \ldots & \sigma_1 & \sigma_2 \\ 0 & 0 & 0 & \ldots & -1 & \sigma_1 \end{vmatrix} P_0$$

A similar formula can be given for Q_k.

If we replace each σ by x in the determinant for P_k, the value of the determinant comes out to be

$$x(x+1)(x+2)\ldots(x+k-1). \tag{1}$$

Hence the sum of the coefficients of terms for which the operator is of degree $(k-s)$, $s \geqslant 0$, is the coefficient of x^{k-s} in (1).

The reader will readily recognize this to be $G(k-1, s)$ that is, the sum of the products of the first $(k-1)$ integers taken s at a time. Thus the G-function has made its appearance once more! Thus, we have the important relation

$$\sum_{A(\lambda)=k-s} C(\lambda) = G(k-1, s), \quad 0 \leqslant s \leqslant k-1. \tag{2}$$

From (1), it also follows that the sum of the coefficients of all the terms in P_k is $k!$, while the sum of the coefficients of the terms in Q_k is zero. The number of terms in each expression is of course $p(n)$ – the number of unrestricted partitions of n.

As an example of (2), we have for $k = 6, s = 3$,

$$\sum_{A(\lambda)=3} C(\lambda) = \frac{6!}{2!4} + \frac{6!}{1.2.3} + \frac{6!}{3!2^3} = 225 = G(5, 3).$$

8. $J(n, \lambda)$ – a polynomial

It would be reasonable to ask if anything can be said about $J(n, \lambda)$, the coefficient of x^n in

$$X_1^{a_1} X_2^{a_2} X_3^{a_3} \ldots X_k^{a_k}, \tag{1}$$

where the a's are nonnegative integers, such that

$$\sum_{r=1}^{k} r a_r = k.$$

Let

$$B(r) = \sum_{t=1}^{[k/r]} a_{rt}, \quad 1 \leqslant r \leqslant k.$$

Rewrite the expression (1) in the form

$$X_{\alpha_1}^{\beta_1} X_{\alpha_2}^{\beta_2} X_{\alpha_3}^{\beta_3} \ldots X_{\alpha_k}^{\beta_k}, \tag{2}$$

so that

$$B(\alpha_1) \geqslant B(\alpha_2) \geqslant B(\alpha_3) \geqslant \cdots \geqslant B(\alpha_k).$$

Evidently, α_1 can always be taken equal to 1, and this we shall do. Then, the answer to the above question can be given in the form of the

THEOREM 3. *The coefficient of x^n in (2), is a polynomial of degree at most $B(1) - 1$, in which the coefficient of n^c, where*

$$B(\alpha_{r+1}) \leqslant c < B(\alpha_r), \tag{3}$$

depends only upon the residue of n modulo the l.c.m. of the numbers $\alpha_1, \alpha_2, \alpha_3, \ldots, \alpha_r$.

Proof. By a process akin to that of breaking up into partial fractions, the expression

$$1 \Big/ \prod_{r=1}^{k} (1 - x^{\alpha_r})^{\beta_r}$$

can be written in the form

$$\sum_{r=1}^{k} \sum_{g=1}^{B(\alpha_r)} \frac{f_r(x, g)}{(1 - x^{\alpha_r})^g}, \tag{4}$$

where $f_r(x, g)$ is a polynomial in x of degree at most $(\alpha_r - 1)$ with rational coefficients, not necessarily integral. Let

$$n - k = t_r\alpha_r + s_r, \quad 0 \leqslant s_r \leqslant \alpha_r - 1. \tag{5}$$

Then the coefficient of x^{n-k} in

$$\frac{f_r(x, g)}{(1 - x^{\alpha_r})^g}$$

is given by

$$G\binom{g + t_r - 1}{g - 1}, \tag{6}$$

where G is the coefficient of x^{s_r} in $f_r(x, g)$. Replacing t_r by $\dfrac{n - k - s_r}{\alpha_r}$ in (6), we get a polynomial in n of degree $g - 1$ in which the coefficients of powers of n depend upon the residue of $n - k$, and therefore of n modulo α_r. However n^c can occur in (6) only if $c < g$, therefore, if

$$B(\alpha_{r+1}) \leqslant c < B(\alpha_r) \leqslant B(\alpha_{r-1}) \leqslant \cdots \leqslant B(\alpha_1) \tag{7}$$

then the coefficient of n^c depends upon the residues of n modulo $\alpha_1, \alpha_2, \alpha_3, \ldots, \alpha_r$, and, therefore, on the residue of n modulo their l.c.m.

We observe that for $\alpha_1 = 1$ in (4), we have

$$f_1(x, B(1)) = \lim_{x \to 1} \frac{(1 - x)^{B(1)}}{\prod_{r=1}^{k} (1 - x^{\alpha_r})^{\beta_r}}$$

$$= \frac{1}{\prod_{r=1}^{k} \alpha_r^{\beta_r}}. \tag{8}$$

Hence the polynomial occurring as the coefficient of x^n in (2) is of degree exactly $B(1) - 1$ in n, provided

$$B(1) > B(\alpha_2). \tag{9}$$

We illustrate our argument by an example. Take the expression

$$X_1^2 X_3 X_5 X_6^3.$$

Here

$$B(1) = \sum_{t=1}^{28} a_t = 7,$$

$$B(2) = \sum_{t=1}^{14} a_{2t} = 3,$$

$$B(3) = \sum_{t=1}^{9} a_{3t} = 4,$$

$$B(4) = \sum_{t=1}^{7} a_{4t} = 0,$$

$$B(5) = \sum_{t=1}^{5} a_{5t} = 1,$$

$$B(6) = \sum_{t=1}^{4} a_{6t} = 3,$$

$$B(7) = B(8) = \cdots = B(28) = 0.$$

This gives

$$\alpha_1 = 1, \quad \alpha_2 = 3, \quad \alpha_3 = 2, \quad \alpha_4 = 6, \quad \alpha_5 = 5, \quad \alpha_6 = 4,$$

and

$$\alpha_r = r, \quad 7 \leqslant r \leqslant 28;$$

also

$$B(\alpha_1) = 7, \quad B(\alpha_2) = 4, \quad B(\alpha_3) = B(\alpha_4) = 3, \quad B(\alpha_5) = 1, \quad B(\alpha_6) = 0,$$

and

$$B(\alpha_r) = 0, \quad 7 \leqslant r \leqslant 28.$$

What the theorem now states is that the coefficient of x^n in

$$X_1^2 X_3 X_2^0 X_6^3 X_5 X_4^0 X_7^0 \ldots X_{28}^0$$

is a polynomial of degree six in n, in which the coefficients of n^6, n^5, n^4 are independent of n, the coefficient of n^3 depends upon the residue of n modulo 3 (the l.c.m. of 1, 3), the coefficients of n^2 and n depend upon the residue of n modulo 6 (the l.c.m. of 1, 3, 2, 6), and the constant term depends upon the residue of n modulo 30 (the l.c.m. of 1, 3, 2, 6, 5).

Exercises

1. Consider the coefficient of x^n in $X_2^3 X_3 X_5^4 X_6$ and state what theorem 3 has to say in this case.
2. State the form which theorem 3 takes in the case of the expression $X_1 X_2 X_3 \ldots X_t$.
3. Express $1/(1 - x^2)(1 - x^3)^2$ in the form (4), and find the coefficient of x^n when (i) $n \equiv 2 \pmod 6$, (ii) $n \equiv 5 \pmod 6$.

9. Some important lemmas

The following set of five lemmas shall be useful to us in finding the bounds for $P(N_j, k)$.

LEMMA 2. *If $A(\lambda) = k - s$, then*

$$J(n, \lambda) \leqslant \binom{n-s-1}{k-s-1}, \quad 0 \leqslant s \leqslant k-1.$$

Proof. The coefficient of x^{n-k} in

$$(1 - x^t)^{-a_t}$$

is less than or equal to the coefficient of x^{n-k} in

$$(1 - x)^{-a_t}.$$

Then $J(n, \lambda)$ is the coefficient of x^{n-k} in

$$\{(1-x)^{a_1}(1-x^2)^{a_2} \ldots (1-x^k)^{a_k}\}^{-1}$$

and so is at most the coefficient of x^{n-k} in

$$\{(1-x)^{a_1}(1-x)^{a_2} \ldots (1-x)^{a_k}\}^{-1}.$$

This equals the coefficient of x^{n-k} in

$$\{(1-x)^{k-s}\}^{-1}$$

which is

$$\binom{n-s-1}{k-s-1}.$$

LEMMA 3. *If $a_1 > 0$ and $A(\lambda) = k - s$, then*

$$1^{a_1}2^{a_2}3^{a_3} \ldots k^{a_k} J(n, \lambda) \geqslant \binom{n-s-1}{k-s-1}.$$

Proof. Since $\binom{m}{r} \geqslant \binom{m-1}{r}$ for all positive m and r, we have

$$\binom{m}{r} \geqslant \frac{1}{t}\left\{\binom{m}{r} + \binom{m-1}{r} + \binom{m-2}{r} + \cdots + \binom{m-t+1}{r}\right\}. \quad (1)$$

Again, for $a_1 > 0$, the coefficient of x^m in

$$(1-x)^{-a_1}(1-x^t)^{-1}$$

is

$$\binom{m+a_1-1}{a_1-1}+\binom{m-t+a_1-1}{a_1-1}+\binom{m-2t+a_1-1}{a_1-1}+\cdots$$
$$+\binom{m-qt+a_1-1}{a_1-1}, \tag{2}$$

where $q=\left[\dfrac{m}{t}\right]$. Making use of (1) we readily see that (2)

$$\geqslant \frac{1}{t}\left\{\binom{m+a_1-1}{a_1-1}+\binom{m+a_1-2}{a_1-1}+\binom{m+a_1-3}{a_1-1}+\cdots+\binom{a_1-1}{a_1-1}\right\}$$
$$=\frac{1}{t}\binom{m+a_1}{a_1}$$
$$=\frac{1}{t}\cdot G, \tag{3}$$

where G is the coefficient of x^m in

$$(1-x)^{-a_1-1}.$$

Repeatedly applying (3), it follows that $J(n,\lambda)$ which is the coefficient of x^{n-k} in

$$\{(1-x)^{a_1}(1-x^2)^{a_2}\ldots(1-x^k)^{a_k}\}^{-1}$$

is $\geqslant$ coefficient of x^{n-k} in

$$\frac{1}{2^{a_2}3^{a_3}\ldots k^{a_k}}(1-x)^{-A(\lambda)},$$

whence

$$J(n,\lambda)\geqslant\{1^{a_1}2^{a_2}3^{a_3}\ldots k^{a_k}\}^{-1}\binom{n-s-1}{k-s-1}. \tag{4}$$

This is the result in the lemma.

LEMMA 4. *If $A(\lambda)>\frac{1}{2}k$, then the partition λ has at least one part equal to* 1.

Proof. We have

$$a_1+2a_2+3a_3+\cdots+ka_k=k, \tag{5}$$

and

$$a_1+a_2+a_3+\cdots+a_k>\frac{k}{2}. \tag{6}$$

Subtracting (6) twice from (5), we get

$$-a_1+a_3+2a_4+\cdots+(k-2)a_k<0. \tag{7}$$

Since the a's are nonnegative integers, (7) can hold only if $a_1 > 0$ and this is the lemma.

LEMMA 5. *If λ_s is the partition of k with $(k-2s)$ parts each equal to* 1 *and* s *parts each equal to* 2, *then*

$$C(\lambda_s) \geqslant \frac{k^{2s}}{s!\,e^{2s}2^s}, \quad s < \frac{k}{2}.$$

Proof. We have

$$C(\lambda_s) = \frac{k!}{(k-2s)!\,s!\,2^s}.$$

Now

$$\begin{aligned}\log \frac{k!}{(k-2s)!} &= \sum_{t=k-2s+1}^{k} \log t \\ &\geqslant \int_{k-2s}^{k} \log x\, dx \\ &= k \log\left(\frac{k}{e}\right) - (k-2s)\log\left(\frac{k-2s}{e}\right) \\ &\geqslant k \log\left(\frac{k}{e}\right) - (k-2s)\log\left(\frac{k}{e}\right) \\ &= 2s \log\left(\frac{k}{e}\right),\end{aligned}$$

so that

$$\frac{k!}{(k-2s)!} \geqslant \left(\frac{k}{e}\right)^{2s}, \tag{8}$$

and the lemma follows.

LEMMA 6. *For* $0 \leqslant s \leqslant k-1$, *we have*

$$G(k-1, s) \leqslant \frac{k^{2s}}{s!\,2^s}.$$

Proof. Since

$$G(k-1, 0) = 1$$

and

$$G(k-1, 1) = k(k-1)/2,$$

the lemma holds for $s = 0$ and $s = 1$. Suppose the lemma holds for every $s \leqslant t-1$, $t \geqslant 2$. Then, since

$$G(k-1,t)-G(k-2,t)=(k-1)G(k-2,t-1),$$

we must have

$$\begin{aligned}
G(k-1,t) &= \sum_{u=t+1}^{k} (u-1)G(u-2,t-1)\\
&\leqslant \sum_{u=2}^{k} (u-1)G(u-2,t-1)\\
&\leqslant \sum_{u=2}^{k} \frac{(u-1)^{2t-1}}{(t-1)!\,2^{t-1}}\\
&\leqslant \frac{\int_1^{k+1} (u-1)^{2t-1}\,du}{(t-1)!\,2^{t-1}}\\
&= \frac{k^{2t}/2t}{(t-1)!\,2^{t-1}}\\
&= \frac{k^{2t}}{t!\,2^t}.
\end{aligned}$$

The lemma follows by induction.

10. Bounds for $P(N_j, k)$

Since $P(N_j, k) = 0$, if in $N_j = (n_1, n_2, n_3, \ldots, n_j)$, any of the n_i's is less than k, we can assume without loss of generality that

$$n_1 \geqslant n_2 \geqslant n_3 \geqslant \cdots \geqslant n_j \geqslant k. \tag{1}$$

For the lower bound of $P(N_j, k)$, we have the

THEOREM 4.

$$\frac{P(N_j, k)}{\prod_i \binom{n_i - 1}{k-1}} \geqslant \exp_{\left[\frac{k+1}{2}\right]}\left(\frac{k^{j+1}}{2^{2j+1}e^2 n_1 n_2 n_3 \ldots n_j}\right),$$

where $\exp_r(x)$ *denotes the sum of the first* r *terms in the expansion of* e^x.

Proof. We have

$$\begin{aligned}
P(N_j, k) &= \sum_{\lambda} \left\{C(\lambda) \prod_i J(n_i, \lambda)\right\}\\
&\leqslant \sum_{\lambda_s} \left\{C(\lambda_s) \prod_i J(n_i, \lambda_s)\right\},
\end{aligned} \tag{2}$$

where λ_s is the partition of k into $(k-2s)$ parts each equal to 1, and s parts each equal to 2, $0 \leqslant s \leqslant \left[\dfrac{k-1}{2}\right]$.

From Lemmas 3 and 5, we then have

$$P(N_j, k) \geqslant \sum_{0 \leqslant s \leqslant [(k-1)/2]} \frac{k^{2s}}{s!2^s e^{2s}} \prod_i \left\{ \frac{1}{2^s} \binom{n_i - s - 1}{k - s - 1} \right\}$$

$$= \sum_{0 \leqslant s \leqslant [(k-1)/2]} \frac{k^{2s}}{s!2^{(j+1)s} e^{2s}} \prod_i \binom{n_i - s - 1}{k - s - 1}. \tag{3}$$

Hence

$$\frac{P(N_j, k)}{\prod_i \binom{n_i - 1}{k - 1}} \geqslant 1 + \frac{k^2}{2^{j+1}e^2} \prod_i \binom{k-1}{n_i - 1} + \frac{1}{2!}\left(\frac{k^2}{2^{j+1}e^2}\right)^2 \prod_i \frac{(k-1)(k-2)}{(n_i-1)(n_i-2)}$$

$$+ \cdots \text{ to } \left[\frac{k+1}{2}\right] \text{ terms.} \tag{4}$$

Since for $s \leqslant \left[\dfrac{k-1}{2}\right]$,

$$\frac{k-s}{n-s} \geqslant \frac{k}{2n}, \tag{5}$$

we get

$$\frac{P(N_j, k)}{\prod_i \binom{n_i - 1}{k - 1}} \geqslant 1 + \frac{k^2}{2^{j+1}e^2} \prod_i \frac{k}{2n_i} + \frac{1}{2!}\left(\frac{k^2}{2^{j+1}e^2}\right)^2 \prod_i \left(\frac{k}{2n_i}\right)^2$$

$$+ \cdots \text{ to } \left[\frac{k+1}{2}\right] \text{ terms}$$

$$= 1 + \frac{k^{j+2}}{2^{2j+1}e^2} \prod_i \frac{1}{n_i} + \frac{1}{2!}\left(\frac{k^{j+2}}{2^{2j+1}e^2}\right)^2 \prod_i \left(\frac{1}{n_i}\right)^2$$

$$+ \cdots \text{ to } \left[\frac{k+1}{2}\right] \text{ terms}$$

$$= \exp_{\left[\frac{k+1}{2}\right]} \left\{ \frac{k^{j+2}}{2^{2j+1}e^2 n_1 n_2 n_3 \ldots n_j} \right\}.$$

For the upper bound, we prove the

THEOREM 5.

$$\frac{P(N_j, k)}{\prod_i \binom{n_i - 1}{k - 1}} \leqslant \exp_k \left(\frac{k^{j+2}}{2n_1 n_2 n_3 \ldots n_j} \right) < \exp \left(\frac{k^{j+2}}{2n_1 n_2 n_3 \ldots n_j} \right).$$

Proof. We have

$$P(N_j, k) = \sum_{s=0}^{k-1} \sum_{A(\lambda)=k-s} \left\{ C(\lambda) \prod_i J(n_i - \lambda) \right\}. \tag{6}$$

Since

$$\sum_{A(\lambda)=k-s} C(\lambda) = G(k-1, s),$$

using Lemma 2, we obtain,

$$P(N_j, k) \leqslant \sum_{s=0}^{k-1} \left\{ G(k-1, s) \prod_i \binom{n_i - s - 1}{k - s - 1} \right\}.$$

Lemma 6 now gives

$$P(N_j, k) \leqslant \sum_{s=0}^{k-1} \left\{ \frac{k^{2s}}{s!2^s} \prod_i \binom{n_i - s - 1}{k - s - 1} \right\}. \tag{7}$$

Hence

$$\frac{P(N_j, k)}{\prod_i \binom{n_i - 1}{k - 1}} \leqslant 1 + \frac{k^2}{2} \prod_i \binom{k-1}{n_i - 1} + \frac{1}{2!}\left(\frac{k^2}{2}\right)^2 \prod_i \frac{(k-1)(k-2)}{(n_i - 1)(n_i - 2)}$$

$$+ \cdots \text{ to } k \text{ terms.} \tag{8}$$

Since for $s < k \leqslant n$,

$$\frac{k-s}{n-s} \leqslant \frac{k}{n}. \tag{9}$$

The theorem follows immediately.

As a direct consequence of the theorems 4 and 5, we can state that

For some μ, $0 < \mu < 1$,

$$\frac{P(N_j, k)}{\prod_i \binom{n_i - 1}{k - 1}} = \exp\left(\frac{\mu k^{j+2}}{2 n_1 n_2 n_3 \ldots n_j} \right).$$

If $k^{j+2} = o(n_1 n_2 n_3 \ldots n_j)$, we get

$$P(N_j, k) = \{1 + o(1)\} \prod_i \binom{n_i - 1}{k - 1}.$$

In particular, if

$$n_1 \geqslant n_2 \geqslant n_3 \geqslant \cdots \geqslant n_j \geqslant k,$$

and

$$k = o(n_1^{1/3}),$$

we have

$$P(N_j, k) \sim \prod_i \binom{n_i - 1}{k - 1}.$$

Remarks

Nothing is known about the behaviour of μ in general. In the particular case when $n_1 = n_2 = n_3 = \cdots = n_{j=k}$, however,

$$P(N_j, k) = k!,$$

and we get for large k,

$$\mu \sim \frac{2 \log k}{k}.$$

11. Ordering of *j*-partite numbers

In partitioning of positive integers into exactly k summands, we arrange the parts in any partition according to their magnitude, placing either the greatest or the smallest part at the beginning. Can something similar be done for the parts in a partition of N_j? To be able to do this, we must establish some sort of an order among j-partite numbers. This we do as follows.

Let $N_j = (n_i)$ be any positive j-partite number, then we define the modulus of N_j as the positive real number

$$|N_j| = \sqrt{\sum n_i^2}. \tag{1}$$

Considering N_j as a vector in j-dimensional space and taking the axes to be rectangular, we define the angle θ_i which N_j makes with the X_i-axis by the relation

$$\cos \theta_i = n_i / |N_j|. \tag{2}$$

Now let N_j and M_j be any two vectors in space. If r is the least positive integer for which

$$\frac{n_r}{|N_j|} \neq \frac{m_r}{|M_j|}, \quad 1 \leqslant r \leqslant j, \tag{3}$$

then we say that N_j and M_j are in ascending order if

$$\frac{n_r}{|N_j|} > \frac{m_r}{|M_j|} \tag{4}$$

and we write

$$N_j, M_j \uparrow. \tag{5}$$

If there is no r satisfying (3) then the two vectors are collinear and we say N_j and M_j are in ascending order if

$$|N_j| \leqslant |M_j|. \tag{6}$$

With this convention the parts in the partition of any j-partite number N_j can be arranged in an ascending order.

Exercises

1. List all the tri-partite numbers less than (6, 5, 4), and arrange them in ascending order.
2. Write down the partitions of (7, 5) into three parts arranging the parts in an ascending order in each case.

12. Graph of a partition

Let

$$N_j = N_{1j} + N_{2j} + N_{3j} + \cdots + N_{kj} \tag{1}$$

be any partition of N_j into exactly k parts such that

$$N_{1j}, N_{2j}, N_{3j}, \ldots, N_{kj}\uparrow. \tag{2}$$

Starting from the origin O, place the vectors

$$N_{1j}, N_{2j}, N_{3j}, \ldots, N_{kj}$$

end to end in this order. We shall thus get a sort of a stairway which ascends from the origin to the point N_j in j-dimensional space. This we call the graph of the partition (1). Observe that if we move along the stairway from the origin to the point N_j, at each turning we turn through an acute angle only.

Exercise

Draw the graphs of the partitions of (7, 4) into exactly three parts.

13. Ordered partitions or decompositions

When the order in which the parts in a partition of N_j occur is relevant, the partition is spoken of as an ordered partition or a decomposition of N_j. We prefer to use the second word because it is free from any ambiguity.

Let

$$N_j = N_{1j} + N_{2j} + N_{3j} + \cdots + N_{kj} \tag{1}$$

be a decomposition of N_j into exactly k parts. Then a graphical representation of the decomposition is obtained by placing the vectors $N_{1j}, N_{2j}, N_{3j}, \ldots, N_{kj}$ end to end, in this order between the origin and N_j as before. The co-ordinates of the end points are given by

$$O, N_{1j}, N_{1j} + N_{2j}, \ldots, N_{1j} + N_{2j} + \cdots + N_{k-1,j}, N_j. \tag{2}$$

The projections of the end points strictly between O and N_j, on the axis OX_i, are distinct lattice points of OX_i strictly between O and the projection of N_j on OX_i.

Every decomposition of N_j into exactly k summands determines and is uniquely determined by these sets of $(k-1)$ lattice points on the axes $OX_1, OX_2, OX_3, \ldots, OX_j$. The total number of decompositions of N_j into k parts is, therefore, the same as the number of such sets of lattice points.

The number of ways in which the set of $k-1$ lattice points on OX_i can be chosen is given by

$$\binom{n_i - 1}{k - 1}. \tag{3}$$

Hence, the total number of decompositions of N_j into k parts is

$$\prod_i \binom{n_i - 1}{k - 1}. \tag{4}$$

Since any partition of N_j into exactly k parts can lead to at most $k!$ decompositions and that too only when all the parts are distinct, we immediately have

$$p(N_j, k) \geqslant \frac{1}{k!} \prod_i \binom{n_i - 1}{k - 1}, \tag{5}$$

and

$$q(N_j, k) \leqslant \frac{1}{k!} \prod_i \binom{n_i - 1}{k - 1}. \tag{6}$$

14. A conjecture

Dealing with partitions of integers $n > 0$, Auluck, Chowla and Gupta had conjectured that there exists a unique integer $s = s(n)$ such that

$$\begin{aligned} p(n, 1) < p(n, 2) < \cdots < p(n, s) \geqslant p(n, s+1) \\ \geqslant p(n, s+2) \geqslant \cdots \geqslant p(n, n). \end{aligned}$$

The conjecture appears to hold even for partitions of N_j and can be stated in the form:

There exists a unique integer $s = s(N_j)$ such that

$$\begin{aligned} p(N_j, 1) < p(N_j, 2) < \cdots < p(N_j, s) \geqslant p(N_j, s+1) \\ \geqslant p(N_j, s+2) \geqslant \cdots \geqslant p(N_j, n_j) \end{aligned}$$

where

$$n_1 \geqslant n_2 \geqslant n_3 \geqslant \cdots \geqslant n_j.$$

Assuming that $p(N_j, k)$ behaves very much like the function

$$\frac{1}{k!}\prod_i \binom{n_i - 1}{k - 1}, \tag{1}$$

the relation

$$p(N_j, s) \geqslant p(N_j, s + 1) \tag{2}$$

gives

$$s = \min(r, n_j), \tag{3}$$

where r is the least positive integer for which

$$\prod_i (n_i - r) \geqslant r^j(r + 1). \tag{4}$$

Roughly, such an r is given by

$$r = (n_1 n_2 n_3 \ldots n_j)^{1/(j+1)}. \tag{5}$$

15. Partitions into degenerate parts

We end this chapter with the

THEOREM 6. *The number of partitions of any j-partite number $N_j = (n_i)$, degenerate or otherwise, into k parts, degenerate or non-degenerate, is the same as the number of partitions of the positive j-partite number*

$$N_j \oplus k = (n_1 + k, n_2 + k, n_3 + k, \ldots, n_j + k)$$

into exactly k non-degenerate parts.

The proof is simple and is left to the reader.

The results obtained in this chapter had been obtained earlier by E. M. Wright using generating functions. [E. M. Wright: Partitions of multipartite numbers, *Proc. Amer. Math. Soc.* 7 (1956), 880–890.]

CHAPTER 11

SOME DIOPHANTINE EQUATIONS

1. Introduction

In this chapter, we consider a few Diophantine equations with some of their applications. In the last part of the chapter we are occupied with the theory of the Tarry–Escott problem.

We first consider the equation

$$x_1x_2x_3 \ldots x_n = y_1y_2y_3 \ldots y_m \tag{1}$$

and express its solutions in positive integers, in terms of certain sets of natural numbers which we call parameters. These are represented by symbols such as u_{ij}.

We next consider equations of the form

$$x_1 + x_2 + x_3 + \cdots + x_n = m. \tag{2}$$

In the latter case, our main interest is in the number of solutions more than in the actual solutions themselves. In the former case, we are concerned with the number of parameters involved in the solutions. The general equation

$$a_1x_1 + a_2x_2 + \cdots + a_nx_n = m \tag{3}$$

is also considered.

We use the word set in the sense of a multiset as well. That is to say that our sets do not necessarily have distinct elements.

2. Solution of (1.1) by matrices

Evidently, solutions of (1.1) can be found by means of an $n \times m$ matrix with natural number entries:

$$A = \begin{bmatrix} a_{11} & a_{12} & a_{13} & \ldots & a_{1m} \\ a_{21} & a_{22} & a_{23} & \ldots & a_{2m} \\ a_{31} & a_{32} & a_{33} & \ldots & a_{3m} \\ \cdot & \cdot & \cdot & \ldots & \cdot \\ a_{n1} & a_{n2} & a_{n3} & \ldots & a_{nm} \end{bmatrix}$$

Let x_i denote the product of the entries in the ith row and y_j of those in the jth column of A. Then of course $x_1, x_2, x_3, \ldots, x_n; y_1, y_2, y_3, \ldots, y_m$ satisfy (1.1).

More interesting is the fact that every solution of (1.1) can be found by this scheme; that is, given a solution

$$x_1', x_2', x_3', \ldots, x_n'; \quad y_1', y_2', y_3', \ldots, y_m'$$

of (1.1), we can find a matrix A having row products $x_1', \ldots, x_n'$ and column products $y_1', y_2', \ldots, y_m'$. In fact, several matrices of natural numbers may correspond to a given solution of (1.1). Thus, for example, the matrices

$$\begin{bmatrix} 4 & 3 \\ 2 & 2 \\ 1 & 3 \end{bmatrix} \quad \text{and} \quad \begin{bmatrix} 2 & 6 \\ 4 & 1 \\ 1 & 3 \end{bmatrix}$$

both correspond to the solution $(x_1, x_2, x_3; y_1, y_2) = (12, 4, 3; 8, 18)$ of $x_1x_2x_3 = y_1y_2$.

This situation is not very desirable. We, therefore, describe here a method which enables us to select just one matrix U corresponding to any given solution of (1.1). Naturally, the elements of such a matrix will have to satisfy certain conditions which we will state.

The procedure is best explained if we consider a particular case of equation (1.1), say with $n = 4$, $m = 3$. Thus, we have

$$x_1x_2x_3x_4 = y_1y_2y_3. \tag{1}$$

For any given solution of (1), let

$$(x_1, y_1) = u_{11},$$
$$(x_1/u_{11}, y_2) = u_{12},$$
$$(x_1/u_{11}u_{12}, y_3) = u_{13},$$

so that

$$(x_1/u_{11}, y_1/u_{11}) = 1,$$
$$(x_1/u_{11}u_{12}, y_2/u_{12}) = 1,$$
$$(x_1/u_{11}u_{12}u_{13}, y_3/u_{13}) = 1.$$

Hence

$$(x_1/u_{11}u_{12}u_{13}, y_1/u_{11}) = (x_1/u_{11}u_{12}u_{13}, y_2/u_{12})$$
$$= (x_1/u_{11}u_{12}u_{13}, y_3/u_{13}) = 1. \tag{2}$$

Now, since (1) can be written

$$\frac{x_1}{u_{11}u_{12}u_{13}} x_2x_3x_4 = \frac{y_1}{u_{11}} \cdot \frac{y_2}{u_{12}} \cdot \frac{y_3}{u_{13}},$$

we have

$$\frac{x_1}{u_{11}u_{12}u_{13}} \,\Big|\, \frac{y_1}{u_{11}} \cdot \frac{y_2}{u_{12}} \cdot \frac{y_3}{u_{13}}.$$

Therefore, in view of (2), we must have

$$\frac{x_1}{u_{11}u_{12}u_{13}} = 1,$$

that is

$$x_1 = u_{11}u_{12}u_{13}. \tag{3}$$

Equation (1) now reduces to

$$x_2x_3x_4 = w_1w_2w_3 \tag{4}$$

where

$$w_1 = y_1/u_{11}, \quad w_2 = y_2/u_{12}, \quad w_3 = y_3/u_{13}.$$

As the second step in the reduction, let

$$(x_2, w_1) = u_{21}, \quad (x_2/u_{21}, w_2) = u_{22}, \quad (x_2/u_{21}u_{22}, w_3) = u_{23}.$$

Then proceeding as before, we get

$$x_2 = u_{21}u_{22}u_{23} \tag{5}$$

and (4) reduces to

$$x_3x_4 = z_1z_2z_3, \tag{6}$$

with

$$z_1 = w_1/u_{21}, \quad z_2 = w_2/u_{22}, \quad z_3 = w_3/u_{23}.$$

At the next step, we obtain

$$x_3 = u_{31}u_{32}u_{33}, \tag{7}$$

with

$$(x_3, z_1) = u_{31}, \quad (x_3/u_{31}, z_2) = u_{32}, \quad (x_3/u_{31}u_{32}, z_3) = u_{33}$$

and the equation (6) reduces to

$$x_4 = t_1t_2t_3, \tag{8}$$

where

$$t_1 = z_1/u_{31}, \quad t_2 = z_2/u_{32}, \quad t_3 = z_3/u_{33}.$$

Finally, we get

$$x_4 = u_{41}u_{42}u_{43}, \tag{9}$$

with

$$(x_4, t_1) = u_{41}, \quad (x_4/u_{41}, t_2) = u_{42}, \quad (x_4/u_{41}u_{42}, t_3) = u_{43}$$

and the equation

$$1 = r_1 r_2 r_3, \tag{10}$$

where

$$r_1 = t_1/u_{41}, \quad r_2 = t_2/u_{42}, \quad r_3 = t_3/u_{43}.$$

Since the only solution of (10) in positive integers is

$$r_1 = r_2 = r_3 = 1,$$

we readily get

$$t_1 = u_{41}, \qquad t_2 = u_{42}, \qquad t_3 = u_{43},$$
$$z_1 = u_{31}u_{41}, \qquad z_2 = u_{32}u_{42}, \qquad z_3 = u_{33}u_{43},$$
$$w_1 = u_{21}u_{31}u_{41}, \quad w_2 = u_{22}u_{32}u_{42}, \quad w_3 = u_{23}u_{33}u_{43},$$

and finally

$$y_1 = u_{11}u_{21}u_{31}u_{41}, \quad y_2 = u_{12}u_{22}u_{32}u_{42}, \quad y_3 = u_{13}u_{23}u_{33}u_{43}.$$

This completes our solution of (1) in terms of the well-defined natural numbers $u_{11}, u_{12}, \ldots, u_{43}$.

The matrix corresponding to the given solution of (1), is thus provided by

$$U = \begin{bmatrix} u_{11} & u_{12} & u_{13} \\ u_{21} & u_{22} & u_{23} \\ u_{31} & u_{32} & u_{33} \\ u_{41} & u_{42} & u_{43} \end{bmatrix}$$

It must be clearly understood that the elements of U are no longer arbitrary. For any given solution of (1) the u's are fixed natural numbers.

The same procedure works completely in the case of (1.1) and the reader will have no difficulty in proving for himself the

THEOREM 1. *Corresponding to any given solution of (1.1), there is a unique matrix* $U = (u_{ij})$, $1 \leqslant i \leqslant n$, $1 \leqslant j \leqslant m$, *such that*

$$x_i = \prod_{j=1}^{m} u_{ij}, \quad y_j = \prod_{i=1}^{n} u_{ij},$$

and the u's are defined recursively by the relation

$$(x_k/(u_{k1}u_{k2} \ldots u_{k,h-1}), \quad y_h/(u_{1h}u_{2h} \ldots u_{k-1,h})) = u_{kh}.$$

The set of natural numbers u_{ij} constitutes what we have called the parameters.

3. The g.c.d. conditions satisfied by the parameters

THEOREM 2. *For positive integers k, h, s and t such that*

$$k+t \leqslant n, \qquad h+s \leqslant m,$$

we have

$$(u_{k,h+s}, u_{k+t,h}) = 1.$$

Proof. From theorem 1, we have

$$(x_k/(u_{k1}u_{k2}\ldots u_{kh}), \quad y_h/(u_{1h}u_{2h}\ldots u_{kh})) = 1. \tag{1}$$

Also

$$x_k = u_{k1}u_{k2}\ldots u_{km}, \quad y_h = u_{1h}u_{2h}\ldots u_{nh}. \tag{2}$$

Hence

$$(u_{k,h+1}u_{k,h+2}\ldots u_{km}, u_{k+1,h}u_{k+2,h}\ldots u_{nh}) = 1$$

This implies the result stated in the theorem.

COROLLARY. *For positive integers i, j, h and k with $i, h \leqslant n; j, k \leqslant m$, if $(i-h)(j-k)$ is negative then*

$$(u_{ij}, u_{hk}) = 1.$$

The g.c.d. conditions of theorem 2 will be useful to us in reducing the number of parameters in specific cases.

In practice, the following device for finding the parameters prime to a given parameter is very useful. Write down the matrix of parameters. Separate the given parameter u_{ij} from the rest by lines parallel to the rows and columns, thus:

$$\begin{array}{cccc|c|ccc}
 & u_{11} & \cdots & & u_{1j} & \cdots & u_{1m} & \\
B_2 & \cdot & \cdots & & \cdot & \cdots & \cdot & B_1 \\
\hline
 & u_{i1} & \cdots & & u_{ij} & \cdots & u_{im} & \\
\hline
B_3 & \cdot & \cdots & & \cdot & \cdots & \cdot & B_4 \\
 & u_{n1} & \cdots & & u_{nj} & \cdots & u_{nm} &
\end{array}$$

Then u_{ij} is prime to each of the parameters in blocks B_1 and B_3.

Exercise

One solution of the equation

$$x_1x_2x_3x_4 = y_1y_2y_3$$

is given by

$$x_1 = 16, \quad x_2 = 189, \quad x_3 = 56, \quad x_4 = 36;$$
$$y_1 = 98, \quad y_2 = 108, \quad y_3 = 576.$$

Compute the parameters u_{ij} and verify theorem 2.

4. Reduction in the number of parameters

The following example will show how the number of parameters can be reduced in specific cases.

Example. Consider the equation

$$x^2 = y_1 y_2 y_3 y_4. \tag{1}$$

Replacing x^2 by $x_1 x_2$, (1) becomes

$$x_1 x_2 = y_1 y_2 y_3 y_4. \tag{2}$$

Solutions of (2) are provided by the matrix of parameters

$$\begin{bmatrix} u_1 & u_2 & u_3 & u_4 \\ u_5 & u_6 & u_7 & u_8 \end{bmatrix}$$

with the necessary g.c.d. conditions. To obtain solutions of (1), all that we need to do is to let

$$x_1 = x_2.$$

This gives

$$u_1 u_2 u_3 u_4 = u_5 u_6 u_7 u_8. \tag{3}$$

Solutions of (3) are given by the matrix

$$\begin{bmatrix} v_1 & v_2 & v_3 & v_4 \\ v_5 & v_6 & v_7 & v_8 \\ v_9 & v_{10} & v_{11} & v_{12} \\ v_{13} & v_{14} & v_{15} & v_{16} \end{bmatrix}$$

Thus, we have

$$x = v_1 v_2 \ldots v_{16},$$
$$y_1 = u_1 u_5 = v_1^2 v_2 v_3 v_4 v_5 v_9 v_{13},$$
$$y_2 = u_2 u_6 = v_5 v_6^2 v_7 v_8 v_2 v_{10} v_{14},$$
$$y_3 = u_3 u_7 = v_9 v_{10} v_{11}^2 v_{12} v_3 v_7 v_{15},$$
$$y_4 = u_4 u_8 = v_{13} v_{14} v_{15} v_{16}^2 v_4 v_8 v_{12}.$$

This solution appears to involve 16 parameters. The g.c.d. conditions, however, show that

$$(u_5, u_2u_3u_4) = 1, \quad (u_6, u_3u_4) = 1, \quad (u_7, u_4) = 1.$$

Hence

$$v_5 = v_9 = v_{13} = v_{10} = v_{14} = v_{15} = 1,$$

so that

$$x = v_1v_2v_3v_4v_6v_7v_8v_{11}v_{12}v_{16},$$
$$y_1 = v_1^2v_2v_3v_4, \quad y_2 = v_6^2v_7v_8v_2,$$
$$y_3 = v_{11}^2v_{12}v_3v_7, \quad y_4 = v_{16}^2v_4v_8v_{12}.$$

The number of parameters is, thus, reduced to 10.

Let us now consider a slightly more complicated set of equations

$$x_1x_2^2 = y_1y_2 = z_1z_2. \tag{4}$$

First solving the equation

$$y_1y_2 = z_1z_2, \tag{5}$$

we have

$$y_1 = u_1u_2, \quad y_2 = u_3u_4, \quad z_1 = u_1u_3, \quad z_2 = u_2u_4,$$

with $(u_2, u_3) = 1$.

Replacing x_2^2 by x_2x_3, we get the equation

$$x_1x_2x_3 = u_1u_2u_3u_4. \tag{6}$$

Solutions of (6) are provided by the matrix

$$\begin{bmatrix} v_1 & v_2 & v_3 & v_4 \\ v_5 & v_6 & v_7 & v_8 \\ v_9 & v_{10} & v_{11} & v_{12} \end{bmatrix}$$

To get solutions of (4), we now take $x_2 = x_3$, that is

$$v_5v_6v_7v_8 = v_9v_{10}v_{11}v_{12}. \tag{7}$$

Solving, we obtain

$$v_5 = w_1w_2w_3w_4, \qquad v_9 = w_1w_5w_9w_{13},$$
$$v_6 = w_5w_6w_7w_8, \qquad v_{10} = w_2w_6w_{10}w_{14},$$
$$v_7 = w_9w_{10}w_{11}w_{12}, \qquad v_{11} = w_3w_7w_{11}w_{15},$$
$$v_8 = w_{13}w_{14}w_{15}w_{16}, \qquad v_{12} = w_4w_8w_{12}w_{16}.$$

Now, the g.c.d. conditions require that we have

$$(u_2, u_3) = 1,$$
$$(v_9, v_6v_7v_8) = (v_{10}, v_7v_8) = (v_{11}, v_8) = 1.$$

Hence

$$w_7 = 1; \quad w_5 = w_9 = w_{13} = w_{10} = w_{14} = w_{15} = 1.$$

We thus get

$$x_1 = v_1v_2v_3v_4, \quad x_2 = w_1w_2w_3w_4w_6w_8w_{11}w_{12}w_{16},$$
$$y_1 = v_1v_2w_1^2w_2^2w_3w_4w_6^2w_8,$$
$$y_2 = v_3v_4w_3w_4w_8w_{11}^2w_{12}^2w_{16}^2,$$
$$z_1 = v_1v_3w_1^2w_2w_3^2w_4w_{11}^2w_{12},$$
$$z_2 = v_2v_4w_2w_4w_6^2w_8^2w_{12}w_{16}.$$

The number of parameters has, thus, been reduced from 20 to 13.

5. Applications

We can apply the method depicted in the foregoing sections to obtain general solutions to some well-known problems. We consider two of these here.

The equation

$$x^2 + y^2 = z^2. \tag{1}$$

We can write (1) as

$$x^2 = z^2 - y^2 = (z + y)(z - y),$$

or what is the same thing

$$x^2 = y_1y_2, \tag{2}$$

where

$$y_1 = z + y, \quad y_2 = z - y,$$

so that y_1, y_2 are of the same parity. Let $(y_1, y_2) = u_1$. Then $u_1 \mid x$ and (2) can be written as

$$(x/u_1)^2 = (y_1/u_1) . (y_2/u_1). \tag{3}$$

Since $(y_1/u_1, y_2/u_1) = 1$, (3) yields

$$(y_1/u_1) = u_2^2, \quad (y_2/u_1) = u_3^2, \quad (x/u_1) = u_2u_3 \tag{4}$$

with $(u_2, u_3) = 1$ and $u_2 > u_3$. Since y_1, y_2 are of the same parity, u_1 is odd if they are both odd and even if they are both even. Hence if y_1, y_2 are both odd then so also are u_2 and u_3. In either case, we have

$$z = \tfrac{1}{2}(y_1 + y_2) = \tfrac{1}{2}u_1(u_2^2 + u_3^2),$$
$$y = \tfrac{1}{2}(y_1 - y_2) = \tfrac{1}{2}u_1(u_2^2 - u_3^2), \quad x = u_1u_2u_3.$$

with $u_2 > u_3$, $(u_2, u_3) = 1$.

For any given solution of (1), u_1, u_2, u_3 are uniquely determined. Take for example, $x = 45, y = 24, z = 51$. Then

$$y_1 = y + z = 75, \quad y_2 = z - y = 27.$$

Hence $u_1 = (75, 27) = 3$.

$$u_2^2 = y_1/u_1 = 25,$$
$$u_3^2 = y_2/u_1 = 9,$$

so that $u_2 = 5, u_3 = 3$. Notice that $u_2 > u_3$ and $(u_2, u_3) = 1$.

We next consider the problem of finding three square numbers in arithmetical progression.

Let x^2, y^2, z^2 be any three squares in A.P. Without loss of generality, we can assume that $0 < x^2 < y^2 < z^2$.

Then we have

$$y^2 - x^2 = z^2 - y^2. \tag{5}$$

Evidently $(x, y) \mid (y, z)$ and $(y, z) \mid (x, y)$. Hence $(x, y) = (y, z) = u$ say. We can, therefore, take $x = ux_1, y = uy_1, z = uz_1$. Equation (5) now reduces to the equation

$$y_1^2 - x_1^2 = z_1^2 - y_1^2 \tag{6}$$

with

$$(x_1, y_1) = 1 = (y_1, z_1). \tag{7}$$

If y_1 is even, then in view of (7), both x_1 and z_1 will have to be odd. This will make the left side of (6) congruent to -1 (mod 4) and the right side congruent to 1 (mod 4). This shows that y_1 is odd. Similarly it can be shown that x_1 and z_1 are also odd. Now (6) can be written as

$$u_1u_2 = v_1v_2 \tag{8}$$

with

$$u_1 = (y_1 + x_1)/2, \quad u_2 = (y_1 - x_1)/2,$$
$$v_1 = (z_1 + y_1)/2, \quad v_2 = (z_1 - y_1)/2.$$

Solving (8), we have

$$u_1 = w_1w_2, \quad u_2 = w_3w_4, \quad v_1 = w_1w_3, \quad v_2 = w_2w_4, \tag{9}$$

with $(w_2, w_3) = 1$. Also

$$u_1 + u_2 = y_1 = v_1 - v_2, \tag{10}$$

or

$$u_1 + v_2 = v_1 - u_2,$$

i.e.

$$w_2(w_1 + w_4) = w_3(w_1 - w_4). \tag{11}$$

Since $(w_2, w_3) = 1$, (11) gives

$$w_1 - w_4 = tw_2, \quad w_1 + w_4 = tw_3. \tag{12}$$

Hence

$$u_1 = w_1(w_1 - w_4)/t, \quad u_2 = w_4(w_1 + w_4)/t,$$
$$v_1 = w_1(w_1 + w_4)/t, \quad v_2 = w_4(w_1 - w_4)/t.$$

In a straightforward manner, we now get

$$x_1 = (w_1^2 - 2w_1w_4 - w_4^2)/t, \qquad x = u(w_1^2 - 2w_1w_4 - w_4^2)/t,$$
$$y_1 = (w_1^2 + w_4^2)/t, \qquad y = u(w_1^2 + w_4^2)/t,$$
$$z_1 = (w_1^2 + 2w_1w_4 - w_4^2)/t, \qquad z = u(w_1^2 + 2w_1w_4 - w_4^2)/t,$$

with $t = (w_1 - w_4, w_1 + w_4)$, $w_1 > w_4$.

Example. $w_1 = 9$, $w_4 = 5$ gives

$$x_1^2 = 289, \quad y_1^2 = 2809, \quad z_1^2 = 5329.$$

These are in A.P.

Exercises

1. Find a parametric solution of the equation:

$$x^2 + y^2 = u^2 + v^2$$

and give a numerical example.

2. If t_x denotes the triangular number $x(x + 1)/2$, then find expressions for x, y, z so that t_x, t_y, t_z are in arithmetical progression.

6. Number of solutions of (1.2)

THEOREM 3. *The number of solutions $(x_1, x_2, \ldots, x_m)$ of the equation (1.2), namely*

$$x_1 + x_2 + \cdots + x_m = n,$$

in positive integers is

$$\binom{n-1}{m-1}.$$

Proof. Divide the circumference of a circle into n equal parts each of unit length. Name the points of division 1, 2, 3, . . . , n in the counter-clockwise direction. Mark point 1 and any other $(m - 1)$ of the remaining $(n - 1)$ points. Proceeding in the counter-clockwise direction again along the circumference, name the point 1 as A_1 and the remaining $(m - 1)$ marked points $A_2, A_3, \ldots, A_m$ in order. Taking the lengths of the arcs $A_1A_2, A_2A_3, \ldots, A_{m-1}A_m, A_mA_1$ as the values of $x_1, x_2, \ldots, x_{m-1}, x_m$ in order, we get a solution of (1.2) in positive integers. Conversely, given any such solution of (1.2), the points $A_1(=1)$, $A_2, \ldots, A_m$ can be uniquely located on the circumference of the circle. Hence the theorem.

THEOREM 4. *The number of solutions of (1.2) in nonnegative integers is*

$$\binom{n+m-1}{m-1}.$$

The proof is left to the reader.

Exercises

1. Prove theorems (3) and (4) by induction on m.
2. Show that each of the theorems (3) and (4) is deducible from the other.
3. In how many ways can ten copies of a book be distributed among four booksellers (i) when there is no restriction as to the number of copies that any one of them receives; (2) when each one of them must get at least one copy of the book.
4. Among five-figured numbers that can be formed from the digits 1, 2, 3, . . . , 9 find the number of those that have 13 for the sum of the digits. [Bhaskara]
5. In Bhaskara's problem, replace 5 by m and 13 by n and investigate the number of solutions.

7. An application

THEOREM 5. *Any solution of the Diophantine equation*

$$x^n = y_1y_2y_3 \ldots y_m \tag{1}$$

can be expressed in terms of

$$\binom{n+m-1}{n}$$

parameters, which are uniquely determinable.

Proof. The equation

$$u_1 + u_2 + \cdots + u_m = n$$

has $\binom{n+m-1}{n}$ solutions in nonnegative integers. Let these solutions be listed and numbered in some order. If the jth solution is

$$u_{j1}, u_{j2}, \ldots, u_{jm}$$

then a solution of the equation in the theorem is provided by

$$x = \Pi\, v_j, \quad y_k = \Pi\, v_j^{u_{jk}}, \quad k = 1, 2, \ldots, m,$$

where j runs over the integers from 1 to $\binom{n+m-1}{n}$ and the v's are any positive integers.

Given a solution of (1), how the parameters v can be uniquely determined will be clear from the following

Example. Consider the equation

$$x^4 = y_1 y_2 y_3. \tag{2}$$

The solutions of the auxiliary equation

$$u_1 + u_2 + u_3 = 4$$

are these:

j	u_{j1}	u_{j2}	u_{j3}	j	u_{j1}	u_{j2}	u_{j3}
1	4	0	0	9	1	1	2
2	3	1	0	10	1	0	3
3	3	0	1	11	0	4	0
4	2	2	0	12	0	3	1
5	2	1	1	13	0	2	2
6	2	0	2	14	0	1	3
7	1	3	0	15	0	0	4
8	1	2	1				

Thus, we have

$$x = v_1 v_2 v_3 v_4 v_5 v_6 v_7 v_8 v_9 v_{10} v_{11} v_{12} v_{13} v_{14} v_{15};$$
$$y_1 = v_1^4 v_2^3 v_3^3 v_4^2 v_5^2 v_6^2 v_7^1 v_8^1 v_9^1 v_{10}^1,$$
$$y_2 = v_2^1 v_4^2 v_5^1 v_7^3 v_8^2 v_9^1 v_{11}^4 v_{12}^3 v_{13}^2 v_{14}^1,$$
$$y_3 = v_3^1 v_5^1 v_6^2 v_8^1 v_9^2 v_{10}^3 v_{12}^1 v_{13}^2 v_{14}^3 v_{15}^4.$$

Let

$$g_1 = (x, y_1) \qquad = v_1 v_2 v_3 v_4 v_5 v_6 v_7 v_8 v_9 v_{10}$$
$$g_2 = (x, y_1/g_1) \qquad = v_1 v_2 v_3 v_4 v_5 v_6$$
$$g_3 = (x, y_1/g_1 g_2) \qquad = v_1 v_2 v_3$$
$$g_4 = (x, y_1/g_1 g_2 g_3) \qquad = v_1$$

$$\begin{aligned}
g_5 &= (x, y_2) &&= v_2 v_4 v_5 v_7 v_8 v_9 v_{11} v_{12} v_{13} v_{14}\\
g_6 &= (x, y_2/g_5) &&= v_4 v_7 v_8 v_{11} v_{12} v_{13}\\
g_7 &= (x, y_2/g_5 g_6) &&= v_7 v_{11} v_{12}\\
g_8 &= (x, y_2/g_5 g_6 g_7) &&= v_{11}\\
g_9 &= (x, y_3) &&= v_3 v_5 v_6 v_8 v_9 v_{10} v_{12} v_{13} v_{14} v_{15}\\
g_{10} &= (x, y_3/g_9) &&= v_6 v_9 v_{10} v_{13} v_{14} v_{15}\\
g_{11} &= (x, y_3/g_9 g_{10}) &&= v_{10} v_{14} v_{15}\\
g_{12} &= (x, y_3/g_9 g_{10} g_{11}) &&= v_{15}.
\end{aligned}$$

As the next step in the sieving process, let

$$\left.\begin{aligned}
w_1 &= x/g_1 &&= v_{11} v_{12} v_{13} v_{14} v_{15}\\
w_2 &= g_1/g_2 &&= v_7 v_8 v_9 v_{10}\\
w_3 &= g_2/g_3 &&= v_4 v_5 v_6\\
w_4 &= g_3/g_4 &&= v_2 v_3\\
w_5 &= g_4 &&= v_1
\end{aligned}\right\} \quad \text{(i)}$$

$$\left.\begin{aligned}
w_6 &= x/g_5 &&= v_1 v_3 v_6 v_{10} v_{15}\\
w_7 &= g_5/g_6 &&= v_2 v_5 v_9 v_{14}\\
w_8 &= g_6/g_7 &&= v_4 v_8 v_{13}\\
w_9 &= g_7/g_8 &&= v_7 v_{12}\\
w_{10} &= g_8 &&= v_{11}
\end{aligned}\right\} \quad \text{(ii)}$$

$$\left.\begin{aligned}
w_{11} &= x/g_9 &&= v_1 v_2 v_4 v_7 v_{11}\\
w_{12} &= g_9/g_{10} &&= v_3 v_5 v_8 v_{12}\\
w_{13} &= g_{10}/g_{11} &&= v_6 v_9 v_{13}\\
w_{14} &= g_{11}/g_{12} &&= v_{10} v_{14}\\
w_{15} &= g_{12} &&= v_{15}.
\end{aligned}\right\} \quad \text{(iii)}$$

The symmetry in the above relations is brought about by the manner in which we have numbered the solutions of the auxiliary equation. Notice that each of the 15 parameters appears once in each of the three sets of relations presented above.

The w's are uniquely defined and we have

$$v_1 = w_5.$$

Now, v_2 occurs in the expressions for w_4, w_7, and w_{11}/v_1. Naturally, therefore, we take

$$v_2 = (w_4, w_7, w_{11}/v_1).$$

As a parameter gets determined, we substitute its value in the set of 15 relations given above and proceed to derive the value of the next undetermined parameter.

We thus get

$$
\begin{aligned}
v_3 &= w_4/v_2,\\
v_4 &= (w_3, w_8, w_{11}/v_1v_2),\\
v_5 &= (w_3/v_4, w_7/v_2, w_{12}/v_3),\\
v_6 &= w_3/v_4v_5,\\
v_7 &= (w_2, w_9, w_{11}/v_1v_2v_4),\\
v_8 &= (w_2/v_7, w_8/v_4, w_{12}/v_3v_5),\\
v_9 &= (w_2/v_7v_8, w_7/v_2v_5, w_{13}/v_6),\\
v_{10} &= w_2/v_7v_8v_9,\\
v_{11} &= w_{10},\\
v_{12} &= w_9/v_7,\\
v_{13} &= w_8/v_4v_8,\\
v_{14} &= w_{14}/v_{10},\\
v_{15} &= w_{15}.
\end{aligned}
$$

We are thus through.

A similar problem was considered earlier in section 4. But the method presented in this section is more general. The earlier method will be very difficult to apply when n is large.

8. An application of theorem 5

For any given a, let us find the general solution of the Diophantine equation

$$ax^2 + y^2 = z^2. \tag{1}$$

Writing u for $z + y$ and v for $z - y$, the equation takes the form

$$ax^2 = uv, \tag{2}$$

with u and v having the same parity. Let $(u, a) = g_1$. Then we can take

$$u = g_1g_2, \quad a = g_1g_3.$$

Since g_3 must divide v, we take $v = g_3g_4$ and (2) reduces to

$$x^2 = g_2g_4. \tag{3}$$

The most general solution of (3) is given by

$$x = h_1h_2h_3, \quad g_2 = h_1^2h_2, \quad g_4 = h_2h_3^2. \tag{4}$$

The solution of (1) is now easily completed.

For any given solution of (1), all the parameters can be uniquely determined.

Thus, for $a = 12$, one solution of (1) is

$$x = 10, \quad y = 71, \quad z = 79,$$

and we have

$$g_1 = 6, \quad g_2 = 25, \quad g_3 = 2, \quad g_4 = 4.$$

Hence

$$h_1 = 5, \quad h_2 = 1, \quad h_3 = 2.$$

All the parameters are thus uniquely determined.

Exercise

Find a general solution of the Diophantine equations

1. $x^2 = y_1 y_2 y_3 y_4 y_5$.
2. $x^3 = y_1 y_2$.

In each case show how the parameters involved can be uniquely determined.

9. An inequality for $p(n, m)$

Let $p(n, m)$ denote the number of partitions of n into exactly m summands, and $q(n, m)$ the number of partitions of n into exactly m distinct parts. Then, since any partition of n into exactly m summands will give rise to at most $m!$ solutions of (1.2) in positive integers, it follows that

$$m!\,p(n, m) \geqslant \binom{n-1}{m-1}. \tag{1}$$

Again, each partition of n into exactly m distinct parts gives rise to $m!$ solutions of (1.2) with x's all distinct positive integers, therefore

$$m!\,q(n, m) \leqslant \binom{n-1}{m-1}. \tag{2}$$

Replacing n by $n + \frac{1}{2}m(m-1)$ in (2), and recalling that

$$q(n + \tfrac{1}{2}m(m-1), m) = p(n, m), \tag{3}$$

we get the important inequality

$$\binom{n-1}{m-1} \leqslant m!\,p(n, m) \leqslant \binom{n-1+c_m}{m-1},$$

with

$$c_m = \tfrac{1}{2}m(m-1).$$

10. Number of meanings of non-associative products

Consider the product

$$a_m * a_{m-1} * \cdots * a_3 * a_2 * a_1, \tag{1}$$

where the binary operation denoted by $*$ is non-associative. Use the operator P to indicate that the two elements on its immediate right are to be replaced by their $*$-product. Then what we have to do is to insert $(m - 1)$ operators P in the m gaps – one of which is to the left of a_m and the remaining $(m - 1)$ between consecutive a's in the places where $*$'s are at present, in such a way that the expression in (1) will be meaningful.

Let x_j denote the number of P's inserted between a_j and a_{j+1}, for $j = 1, 2, 3, \ldots, m - 1$; and x_m the number of those to the left of a_m. Then for the expression in (1) to be meaningful, it is necessary and sufficient that we have

$$x_1 = 0, \quad x_1 + x_2 \leqslant 1, \quad x_1 + x_2 + x_3 \leqslant 2, \ldots,$$
$$x_1 + x_2 + \cdots + x_{m-1} \leqslant m - 2,$$

and

$$x_1 + x_2 + \cdots + x_m = m - 1.$$

If we write the a's and the P's in order round a circle starting with a_1 and proceeding in the counter-clockwise direction, then for the expression (1) to be meaningful, it is necessary and sufficient that counting from a_1 the number of a's should exceed the number of P's at every stage.

The situation is covered by the more general

THEOREM 6. *If m similar white balls and n similar black balls are placed round a circle in any order and $n < m$, then there are exactly $(m - n)$ white balls from which one can start counting in the counter-clockwise direction, so that the number of white balls will exceed the number of black balls at every stage.*

In our problem, the a's are the white balls and P's the black balls and $n = m - 1$.

Proof. Against each white ball write 1 and against each black ball write -1, as in the following figure with $m = 7, n = 5$.

Now start with one of the -1's and proceeding along the circle in the clockwise direction, cancel it out with the first 1 you come across. Repeat the process with each of the other -1's until all the -1's and an equal number of 1's get cancelled. Since the number of -1's is less than the number of 1's, at the end of the process exactly $(m - n)$ ones will be left behind. The white balls corresponding to these 1's are those from any of which one can start counting in the counter-clockwise direction so that the count of the white balls will always be more than that of the black balls.

Thus, in any cycle of m solutions of (1.2), obtained by taking

$$x_1 = c_j, \quad x_2 = c_{j+1}, \ldots, x_{m-j} = c_{m-1}, x_{m-j+1} = c_m, \ldots, x_m = c_{j-1},$$

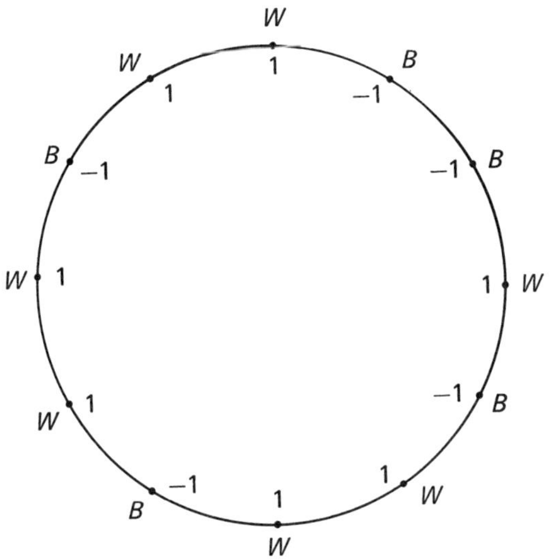

and $j = 1, 2, 3, \ldots, m$ in turn, there are exactly $(m - n)$ which satisfy the relations

$$x_1 + x_2 + x_3 + \cdots + x_t \leqslant t - 1, \quad t \leqslant m. \tag{2}$$

Hence the number of solutions of (1.2) which satisfy conditions (2) must be

$$\frac{m-n}{m}\binom{n+m-1}{m-1}.$$

In particular, the number of meanings of non-associative products of m letters is

$$\frac{1}{m}\binom{2m-2}{m-1}.$$

These are the well-known Catalan numbers about which quite a lot has been written in recent times.

11. A generalization of (1.2)

More generally, one can investigate the number of nonnegative integral solutions of the equation

$$a_1x_1 + a_2x_2 + \cdots + a_mx_m = n, \tag{1}$$

where the a's are given positive integers.

THEOREM 7. *The number of nonnegative integral solutions of* (1) *is the coefficient of* y^n *in*

$$1/(1 - y^{a_1})(1 - y^{a_2}) \ldots (1 - y^{a_m}). \tag{2}$$

Proof. We have

$$1/(1-y^a) = y^{0a} + y^{1a} + y^{2a} + \cdots. \tag{3}$$

Therefore, a typical term in the power series for (2) has the form

$$y^{x_1 a_1} y^{x_2 a_2} \ldots y^{x_m a_m} = y^n$$

and every term like this corresponds to a solution of (1). Conversely, every solution of (1) corresponds to such a term in the power-series for (2). Hence the coefficient of y^n in the expansion of (2) must be equal to the number of solutions of (1). In particular, when

$$a_1 = a_2 = \cdots = a_m = 1,$$

the number of solutions of (1), which is nothing but (1.2) now, is the coefficient of y^n in $(1-y)^{-m}$. This, being

$$\binom{n+m-1}{m-1},$$

is seen to be in agreement with the result in theorem (4). Note that (1) may have no solution at all. This happens when y^n does not appear in the power series for (2).

Example. The equation

$$9x_1 + 6x_2 = 13$$

has no solution in integers.

If, however, at least one of the a's in (1) is 1, then (1) has at least one solution in integers.

Quite often, one is interested in finding for a large n, the approximate number of solutions in nonnegative integers of an equation of the form

$$a_1x_1 + a_2x_2 + \cdots + a_{m-1}x_{m-1} + x_m = n. \tag{4}$$

This is the same as the number of positive integral solutions of

$$a_1x_1 + a_2x_2 + \cdots + a_{m-1}x_{m-1} + x_m = n + a_1 + a_2 + \cdots + a_{m-1} + 1.$$

It is also the same as the number of positive integral solutions of the inequality

$$a_1x_1 + a_2x_2 + \cdots + a_{m-1}x_{m-1} < n + a_1 + a_2 + \cdots + a_{m-1} + 1.$$

In this connection, of fundamental importance is the

THEOREM 8. *For a fixed $m > 1$ and a large n, the number of positive integral solutions of (4) is asymptotically equal to*

$$\frac{1}{(m-1)!} \cdot \frac{n^{m-1}}{a_1 a_2 \ldots a_{m-1}}.$$

Proof. The hyper-plane

$$a_1x_1 + a_2x_2 + \cdots + a_{m-1}x_{m-1} = n$$

in $(m-1)$-dimensional Euclidean space cuts a set of rectangular axes at the points $A_1, A_2, \ldots, A_{m-1}$ say. If O is the origin of the coordinate system, the number of solutions of (4) in positive integers, is the same as the number of lattice points strictly within the polyhedron $OA_1A_2 \ldots A_{m-1}$. Since the average density of lattice points in the space is 1 per unit volume, the number of lattice points within the polyhedron is asymptotic to the volume of $OA_1A_2 \ldots A_{m-1}$

$$= \frac{1}{(m-1)!} \prod_{j=1}^{m-1} OA_j = \frac{1}{(m-1)!} \prod_{j=1}^{m-1} n/a_j$$

and this is the theorem.

The error is of the order of the number of lattice points on the surface of the polyhedron. Hence the number of solutions of (4) in positive integers

$$= \frac{n^{m-1}}{(m-1)!\, a_1 a_2 \ldots a_{m-1}} + O\left(\frac{n^{m-2}(a_1 + a_2 + \cdots + a_{m-1})}{a_1 a_2 \ldots a_{m-1}}\right).$$

In particular, when

$$a_1 = a_2 = \cdots = a_{m-1} = 1,$$

the number of positive integral solutions of the equation

$$x_1 + x_2 + \cdots + x_m = n$$

is

$$\sim \frac{n^{m-1}}{(m-1)!}$$

which is in agreement with the result of theorem 3.

Again, $p(n, m)$ – the number of partitions of n into exactly m summands – is the number of solutions of the equation

$$x_1 + 2x_2 + 3x_3 + \cdots + mx_m = n; \quad x_m \geqslant 1.$$

Hence

$$(m-1)!\, p(n, m) = \frac{n^{m-1}}{m!} + O(mn^{m-2}/(m-2)!).$$

For $m = o(\sqrt[3]{n})$ and large n, we thus have

$$p(n, m) \sim n^{m-1}/(m-1)!\, m!$$

This is a famous result of P. Erdös and J. Lehner.

12. The Tarry–Escott problem

The two sets of integers

$$A = \{1, 4, 5, 8\}, \quad B = \{2, 2, 7, 7\},$$

have a queer property. For example, we have

$$1 + 4 + 5 + 8 = 2 + 2 + 7 + 7$$
$$1^2 + 4^2 + 5^2 + 8^2 = 2^2 + 2^2 + 7^2 + 7^2$$
$$1^3 + 4^3 + 5^3 + 8^3 = 2^3 + 2^3 + 7^3 + 7^3$$

but

$$1^4 + 4^4 + 5^4 + 8^4 \neq 2^4 + 2^4 + 7^4 + 7^4.$$

Denoting by $\sigma_k(S)$ the sum of the kth powers of the members of a set S, for the two sets A and B above, we have

$$\sigma_k(A) = \sigma_k(B), \tag{1}$$

for $k = 1, 2, 3$ but not for $k = 4$. More briefly, we write this in the form

$$A \overset{3}{=} B. \tag{2}$$

In its simplest form the Tarry–Escott problem is to find two distinct sets A_1, A_2 of m integers each – distinct in the sense that members of one set are not merely a permutation of those of the other – such that

$$A_1 \overset{r}{=} A_2; \tag{3}$$

that is

$$\sigma_k(A_1) = \sigma_k(A_2)$$

for $k = 1, 2, 3, \ldots, r$ but not for $k = r + 1$. We shall show that such sets exist. Our main interest then will be to find sets with as few members as we can. The least value of m for which sets A_1, A_2 satisfying (3) exist for a given r will be denoted by $M(r)$. When the condition 'but not for $k = r + 1$' is not insisted upon, the least m for which sets A_1, A_2 with

$$\sigma_k(A_1) = \sigma_k(A_2), \quad 1 \leqslant k \leqslant r,$$

exist, will be denoted by $N(r)$. In this case, we write

$$A_1 \underset{r}{=} A_2. \tag{4}$$

13. Some notations

In what follows, capital letters A, B, C etc. denote sets (or multisets) of integers; small letters denote positive integers unless stated otherwise; $C + D$ is written for the set of integers obtained by adding each member of C to each member of D. When D has only one member, say t, we write $C + t$ for the set of integers obtained by adding t to each member of C. $C - D$ is the set of integers obtained by subtracting each member of D from each member of C. Thus if C has h members and D has j members, both $C + D$ and $C - D$ have hj members each. Finally, the set consisting of all the members of the sets $A_1, A_2, \ldots, A_t$ is denoted by $(A_1, A_2, \ldots, A_t)$

or, for convenience of writing, simply by

$$A_1, A_2, A_3, \ldots, A_t.$$

Thus, if $A_1 = \{1, 1, 2\}, A_2 = \{1, 2, 3, 3\}$ and $A_3 = \{1, 2, 3\}$ then

$$A_1, A_2, A_3 = \{1, 1, 2, 1, 2, 3, 3, 1, 2, 3\} = \{1, 1, 1, 1, 2, 2, 2, 3, 3, 3\}.$$

14. The generating polynomial

Let $A = \{a_1, a_2, \ldots, a_m\}, \quad B = \{b_1, b_2, \ldots, b_m\}$

be two sets of integers, such that

$$A \stackrel{r}{=} B, \tag{1}$$

then the polynomial

$$g(x) = x^{a_1} - x^{b_1} + x^{a_2} - x^{b_2} + \cdots + x^{a_m} - x^{b_m} \tag{2}$$

is called the generating polynomial of (1).

THEOREM 9. *Every polynomial of the form*

$$f_r(x) = xh_r(x) \prod_{j=0}^{r} (x^{d_j} - 1), \quad r \geqslant 1;$$

where $h_r(x)$ is a polynomial in x with integral coefficients and $h_r(1) \neq 0$, and d's are positive integers, generates sets A, B satisfying the relation

$$A \stackrel{r}{=} B,$$

and conversely.

Proof. Since 1 is an $(r + 1)$-ple root of the equation

$$f_r(x) = 0$$

but not an $(r + 2)$-ple root, we must have

$$f_r(1) = f_r^{(1)}(1) = f_r^{(2)}(1) = \cdots = f_r^{(r)}(1) = 0$$

but not

$$f_r^{(r+1)}(1) \neq 0.$$

(Here $f_r^{(j)}(x)$ denotes the jth differential coefficient of $f_r(x)$.) Since $f_r(x)$ is a polynomial in x with integral coefficients, we can assume that

$$f_r(x) = c_0 x^n + c_1 x^{n-1} + \cdots + c_{n-1} x, \tag{3}$$

where c's are integers, not necessarily positive. Replacing $c_{n-j} x^j$ by

$$x^j + x^j + \cdots + x^j \quad \text{or} \quad -x^j - x^j - \cdots - x^j \ |c_{n-j}| \text{ terms}$$

according as c_{n-j} is positive or negative, we can write (3) as

$$f_r(x) = \sum x^{a_i} - \sum x^{b_j}. \tag{4}$$

Since $f_r(1) = 0$, the two sums in (4) must each have the same number, m say, of terms, where

$$2m = \sum_{j=0}^{n-1} |c_j|. \tag{5}$$

Thus, we have

$$f_r(x) = \sum_{i=1}^{m} (x^{a_i} - x^{b_i}). \tag{6}$$

On differentiating with respect to x, (6) gives

$$F_1(x) = x f_r^{(1)}(x) = \sum_{i=1}^{m} (a_i x^{a_i} - b_i x^{b_i}). \tag{7}$$

For $r \geqslant 1, f_r^{(1)}(1) = 0$, hence

$$\sum_{i=1}^{m} (a_i - b_i) = 0,$$

that is

$$a_1 + a_2 + \cdots + a_m = b_1 + b_2 + \cdots + b_m.$$

Moreover, 1 is a root of $F_1(x) = 0$ repeated r times. Now differentiating (7) with respect to x, we get

$$F_2(x) = x F_1^{(1)}(x) = \sum_{i=1}^{m} (a_i^2 x^{a_i} - b_i^2 x^{b_i}).$$

If $r \geqslant 2, F_2(1) = 0$ and we have

$$a_1^2 + a_2^2 + \cdots + a_m^2 = b_1^2 + b_2^2 + \cdots + b_m^2.$$

In general, setting

$$F_k(x) = x F_{k-1}^{(1)}(x)$$

and noting that for $r \geqslant k, F_k(1) = 0$, we get for each $k \leqslant r$,

$$a_1^k + a_2^k + \cdots + a_m^k = b_1^k + b_2^k + \cdots + b_m^k.$$

Since 1 is an unrepeated root of $F_r(x) = 0$, we do not have

$$a_1^{r+1} + a_2^{r+1} + \cdots + a_m^{r+1} = b_1^{r+1} + b_2^{r+1} + \cdots + b_m^{r+1}.$$

This proves the first part of the theorem.

Conversely, for a certain $r \geqslant 1$, let

$$A \overset{r}{=} B, \tag{8}$$

so that

$$g(x) = \sum_{i=1}^{m} (x^{a_i} - x^{b_i}) \tag{9}$$

is the generating polynomial for (8). Then

$$G_1(x) = xg^{(1)}(x) = \sum_{i=1}^{m} (a_i x^{a_i} - b_i x^{b_i}). \tag{10}$$

Since

$$a_1 + a_2 + \cdots + a_m = b_1 + b_2 + \cdots + b_m,$$

we must have $G_1(x) = 0$, and therefore $g^{(1)}(x) = 0$, for $x = 1$. Hence 1 is at least a 2-ple root of $g(x) = 0$. Now

$$G_2(x) = xG_1^{(1)}(x) = \sum_{i=1}^{m} (a_i^2 x^{a_i} - b_i^2 x^{b_i})$$

If $r \geqslant 2$, (8) has

$$a_1^2 + a_2^2 + \cdots + a_m^2 = b_1^2 + b_2^2 + \cdots + b_m^2.$$

Hence we infer that 1 is at least a 3-ple root of $g(x) = 0$. Proceeding in this manner, the reader can prove the second part of the theorem for himself.

As a consequence of theorem 9, we have the following important result:

If for some r,

$$A \overset{r}{=} B$$

then

$$A + t, \quad B \overset{r+1}{=\!=} A, \quad B + t. \tag{11}$$

In fact if $g(x)$ is the generating polynomial of (8), then the generating polynomial of the relation in (11) is

$$(x^t - 1)g(x).$$

We have thus shown that sets A, B each with the same number of integral elements exist such that

$$A \overset{r}{=} B,$$

for each $r \geqslant 1$.

Exercises

1. If $A \overset{r}{=} B$ and $C \overset{r}{=} D$, then $A, C \underset{r}{=} B, D$.
2. If $A \overset{r}{=} B$ and $C \overset{s}{=} D, s > r$; then $A, C \overset{r}{=} B, D$.
3. Construct sets A, B, C, D such that

$$A \overset{r}{=} B, \quad C \overset{r}{=} D \quad \text{and} \quad A, C \overset{r+2}{=\!=} B, D.$$

4. If $A \overset{r}{=} B$, then for every t, $A + t \overset{r}{=} B + t$. Show also that

$$A + C \overset{r}{=} B + C.$$

5. Find the generating polynomial for

$$\{1, 4, 5, 8\} \overset{3}{=} \{2, 2, 7, 7\}$$

in the simplest form.

6. Starting with

$$\{1, 4\} \overset{1}{=} \{2, 3\}$$

and taking $t = 3, 5, 7, 8, 11, 13, 9, 19, 17, 6, 4, 15$ in succession, use (11) to show that $M(13) \leqslant 32$. Further taking

$$C = \{0, 6, 8, 14\}$$

use the result of exercise 4 above to prove that $M(13) \leqslant 30$.

7. Use (11) and exercise 4 to improve upon any of the following known results:

$$M(14) \leqslant 30, \quad M(15) \leqslant 30, \quad M(16) \leqslant 38, \quad M(17) \leqslant 48,$$
$$M(18) \leqslant 58, \quad M(19) \leqslant 58, \quad M(20) \leqslant 65, \quad M(21) \leqslant 80.$$

15. Bounds for $M(r)$

THEOREM 10. *For any* $r \geqslant 1$,

$$M(r) \geqslant r + 1.$$

Proof. Let

$$A = \{a_1, a_2, \ldots, a_m\}, \quad B = \{b_1, b_2, \ldots, b_m\},$$

be two sets of integers, such that

$$A \overset{r}{=} B. \tag{1}$$

We can assume that A and B are disjoint and a's and b's are positive integers, for if not, we can choose a positive integer t such that the sets $A + t, B + t$ consist of positive elements alone. The relation (1) holds good for $A + t$ and $B + t$ in place of A and B. We can further assume that

$$a_1 \leqslant a_2 \leqslant \cdots \leqslant a_m, \quad b_1 \leqslant b_2 \leqslant \cdots \leqslant b_m.$$

Let $a_1, a_2, \ldots, a_m$ be the roots of the equation

$$x^m - c_1 x^{m-1} + c_2 x^{m-2} - \cdots + (-1)^m c_m = 0 \tag{2}$$

and $b_1, b_2, \ldots, b_m$ those of the equation

$$y^m - d_1 y^{m-1} + d_2 y^{m-2} - \cdots + (-1)^m d_m = 0. \tag{3}$$

Then, writing A_k for the sum of the kth powers of the elements of A and B_k for the sum of the kth powers of the elements of B, we have

$$A_k - c_1 A_{k-1} + c_2 A_{k-2} - \cdots + (-1)^k k c_k = 0, \quad k = 1, 2, 3, \ldots, \tag{4}$$

where we take $c_k = 0$ for $k > m$. Similarly,

$$B_k - d_1 B_{k-1} + d_2 B_{k-2} - \cdots + (-1)^k k d_k = 0, \quad k \geqslant 1, \tag{5}$$

where we take $d_k = 0$ for $k > m$.

Relations (4) give

$$k! c_k = \begin{vmatrix} A_1 & 1 & 0 & 0 & . & 0 \\ A_2 & A_1 & 2 & 0 & . & 0 \\ A_3 & A_2 & A_1 & 3 & . & 0 \\ . & . & . & . & . & . \\ A_{k-1} & A_{k-2} & A_{k-3} & A_{k-4} & . & k-1 \\ A_k & A_{k-1} & A_{k-2} & A_{k-3} & . & A_1 \end{vmatrix} \tag{6}$$

A similar expression with B's in place of A's gives $k!\, d_k$. Now, for $k = 1, 2, 3, \ldots, r$ we have $A_k = B_k$. Hence for $1 \leqslant k \leqslant r$ we get $c_k = d_k$. If $m \leqslant r$, the two equations (2) and (3) become identical and they then have the same set of roots. This means that sets A and B have the same elements. The theorem follows immediately. The theorem also implies that

$$N(r) \geqslant r + 1.$$

THEOREM 11. *For any $r \geqslant 1, M(r) \leqslant 2^r$.*

Proof. We have already shown that if

$$A \stackrel{r}{=} B$$

then

$$A + t, \quad B \stackrel{r+1}{=\!=} B + t, \quad A.$$

If A and B have m elements each, then

$$C = A + t, B; \quad D = B + t, A$$

have $2m$ elements each. Hence

$$M(r + 1) \leqslant 2M(r).$$

Since $M(1) = 2$, the theorem follows.

The following examples show that at least for $r \leqslant 9$

$$M(r) = r + 1.$$

$$1, 3 \stackrel{1}{=} 2, 2$$

$$1, 4, 4 \stackrel{2}{=} 2, 2, 5$$

$$1, 5, 8, 12 \stackrel{3}{=} 2, 3, 10, 11$$

$$1, 5, 9, 17, 18 \stackrel{4}{=} 2, 3, 11, 15, 19$$

$$1, 4, 6, 12, 14, 17 \stackrel{5}{=} 2, 2, 9, 9, 16, 16$$

$$1, 19, 28, 59, 65, 90, 102 \stackrel{6}{=} 2, 14, 39, 45, 76, 85, 103$$

$$1, 5, 10, 24, 28, 42, 47, 51 \stackrel{7}{=} 2, 3, 12, 21, 31, 40, 49, 50$$

$$1, 25, 31, 84, 87, 134, 158, 182, 198 \stackrel{8}{=} 2, 18, 42, 66, 113, 116, 169, 175, 199$$

$$\begin{matrix} 1, 3084, 3302, 11894, 23315, 24187, \\ 35608, 44200, 44418, 47501 \end{matrix} \stackrel{9}{=} \begin{matrix} 13, 2866, 3520, 11870, 23739, 23763, \\ 35632, 43982, 44636, 47489 \end{matrix}$$

Exercises

1. Prove that the system of equations

$$a_1^k + a_2^k + \cdots + a_{n-1}^k = b_1^k + b_2^k + \cdots + b_{n-1}^k, \qquad k = 2, 3, \ldots, n,$$

has no non-trivial solutions in positive integers.

2. If A, B and C are three distinct sets of positive integers such that

$$A \stackrel{r}{=} B \stackrel{r}{=} C$$

then show that for t_1, t_2, t_3 not all equal,

$$A + t_1, B + t_2, C + t_3 \stackrel{r+1}{=\!=} A + t_2, B + t_3, C + t_1 \stackrel{r+1}{=\!=} A + t_3, B + t_1, C + t_2.$$

3. Extend the result of exercise 2 to a system of q sets and show that for such a system

$$M_q(r + 1) \leqslant qM_q(r),$$

where $M_q(r)$ is the least number of elements in each of the q sets for which such sets exist.

16. An upper bound for $N(r)$

THEOREM 12. $N(r) \leqslant \frac{1}{2}r(r + 1) + 1$.

Proof. Write h for $\frac{1}{2}r(r + 1) + 1$. Let n be any arbitrarily large fixed positive integer and let

$$n = a_1 + a_2 + \cdots + a_h, \quad a_i \leqslant a_j, \quad i < j \leqslant h, \tag{1}$$

be any partition of n into exactly h summands. There are $p(n, h)$ such partitions and corresponding to each one of them there is a unique sequence

$$s_2, s_3, \ldots, s_r \tag{2}$$

with

$$s_k = a_1^k + a_2^k + \cdots + a_h^k. \tag{3}$$

Since

$$s_k < (a_1 + a_2 + \cdots + a_h)^k = n^k \tag{4}$$

the number of distinct sequences (2) cannot exceed

$$n^2 \,.\, n^3 \,.\, \cdots n^r = n^{h-2}. \tag{5}$$

Now

$$p(n, h) \geqslant \binom{n-1}{h-1} / h!. \tag{6}$$

We can, therefore, select an n for which

$$p(n, h) > n^{h-2}.$$

For any such n, there will be at least two partitions of n into h summands for which the sequence (2) is the same. These provide the elements of two sets A, B satisfying the relation

$$A \underset{r}{=} B.$$

Since each of the sets has h elements, the theorem is proved.

Exercise

Modify the proof of theorem 12 so that it will cover the case of q sets with equal sums of the first r powers.

Among the papers which deal with problems discussed in this chapter are the following:

E. T. Bell: Reciprocal arrays and Diophantine analysis; and M. Ward: A type of multiplicative Diophantine system. *American J. Math.* **55** (1933), 50–67.

E. M. Wright: On Tarry's problem. *Quarterly J. Math.* **7** (1936), 43–45; also *J. London Math. Soc.* **9** (1934), 267–272.

S. Chowla: *Proc. Indian Acad. Sci.* **1** (1935), 528–530.

S. Sastry: *Proc. Indian Acad. Sci.* **1** (1935), 928–929.

P. Erdös and J. Lehner: The distribution of the number of summands in the partitions of a positive integer. *Duke Math. J.* **8** (1941), 335–345.

H. Gupta: A system of equations having no nontrivial solutions. *J. of Research, National Bureau of Standards* **71B** (1967), 181–182.

CHAPTER 12

PERMUTATIONS OF THE FIRST *n* NATURAL NUMBERS

1. Notation

We write S_n for the set of the first n natural numbers, $(n;r)$ for $\binom{n}{r}$, $p(n)$ for the number of unrestricted partitions of n, and $p(n, r)$ for the number of partitions of n into r non-negative integers. The parts in a partition are invariably written in the non-decreasing order from left to right. Results in the theory of partitions are freely used and so are the Tables of Partitions.*

2. Ordering of permutations

From any permutation

$$a_1 a_2 a_3 \ldots a_{n-1} \tag{1}$$

of S_{n-1}, one can derive n permutations of S_n by first writing n on the immediate right of a_{n-1} and then letting it step over

$$a_{n-1}, a_{n-2}, \ldots, a_1$$

in turn. Thus the seven permutations of S_7 that can be derived from the permutation 352146 of S_6 are

3521467
3521476
3521746
3527146
3572146
3752146
7352146

Starting from the unique permutation

1

of S_1, we get

12
21

*Tables of Partitions, University Press, Cambridge 1958.

the two permutations of S_2. Taking each of these *in turn*, we get

123
132
312

213
231
321

the six permutations of S_3. Proceeding in this manner, taking each permutation *in turn*, we can derive from the $(n-1)!$ permutations of S_{n-1} all the $n!$ permutations of S_n. It is noteworthy that the order in which the permutations of S_n appear is, for $n > 2$, distinct from the lexicographic order in which the permutations are usually arranged.

3. Summits of a permutation and their indices

If in the permutation

$$a_1 a_2 a_3 \dots a_n \tag{1}$$

of S_n, for some $k \leqslant n-1$, we have

$$a_k > a_{k+1}$$

then we speak of a_k as a summit of (1) and of k as the index of this summit. If a_k is not a summit, then its index is taken to be zero.

The sum of the indices of the elements of (1) is called the index of (1). We shall denote the index of (1) by i.

If the indices of the summits of (1) are $j_1, j_2, \dots, j_s$, then we say that the index set of (1) is

$$J(s) = \{j_1, j_2, \dots, j_s\} \tag{2}$$

Here we invariably assume that

$$1 \leqslant j_1 < j_2 < \cdots < j_s \leqslant n-1.$$

Evidently, we have

$$j_1 + j_2 + \cdots + j_s = i. \tag{3}$$

The only permutation of S_n for which the index set is empty is

$$1\ 2\ 3 \dots n$$

and its index is zero. On the other hand, the only permutation of which the index is $(n; 2)$, is

$$n\ n-1\ n-2 \dots 3\ 2\ 1.$$

Directly connected with the index set (3) of (1) is the interval set

$$R(s) = \{r_1, r_2, \ldots, r_s\}$$

defined by the relations

$$r_k = j_k - j_{k-1}, \quad 1 \leqslant k \leqslant s,$$

with $j_0 = 0$. The elements of the interval set are all positive integers, not necessarily distinct. The elements of the index set are, however, distinct positive integers.

Relation (3) implies that

$$sr_1 + (s-1)r_2 + \cdots + 2r_{s-1} + r_s = i.$$

4. The right hand and left hand degrees of the elements of a permutation

From among the a's on the right side of a_k in (3.1), $1 \leqslant k \leqslant n$, let f_k denote the number of those that are $<a_k$. Similarly let g_k denote the number of a's which are $>a_k$ and are to its left in (3.1). Then f_k and g_k are said to be the right hand and left hand degrees respectively of a_k in (3.1). We write

$$F = F[a_1, a_2, \ldots, a_n] = [f_1, f_2, \ldots, f_n]$$

and

$$G = G[a_1, a_2, \ldots, a_n] = [g_1, g_2, \ldots, g_n]$$

to mean that the right hand degree $F(a_k)$ of a_k in (3.1) is f_k and its left hand degree $G(a_k)$ in (3.1) is g_k.

We also write

$$B = B[1, 2, 3, \ldots, n] = [b_1, b_2, b_3, \ldots, b_n]$$

to mean that the right hand degree of k in (3.1) is b_k.

In what follows, we use f_k, g_k, b_k and F, G, B in this sense throughout and call F, G, B the degree vectors of (3.1). Where nothing is specifically mentioned, degree will mean the right hand degree of an element.

THEOREM 1. *For each $k \leqslant n$, we have*

$$a_k + g_k = k + f_k.$$

Proof. The number of a's which are $<a_k$ and appear on its left in (3.1) is $k - 1 - g_k$. Hence

$$f_k + k - 1 - g_k = a_k - 1.$$

From this the theorem follows immediately.

COROLLARY. *For each k,*

$$1 + f_k \leqslant a_k \leqslant k + f_k \leqslant n.$$

Proof. Evidently

$$0 \leqslant f_k \leqslant n - k \quad \text{and} \quad a_k > f_k.$$

Hence the inequalities follow directly.

THEOREM 2. *For any permutation of S_n,*

$$\sum_{k=1}^{n} f_k = \sum_{k=1}^{n} g_k = \sum_{k=1}^{n} b_k.$$

Proof. From theorem 1, we have

$$\sum_{k=1}^{n} a_k + \sum_{k=1}^{n} g_k = \sum_{k=1}^{n} k + \sum_{k=1}^{n} f_k,$$

whence the left hand equality follows at once. The right hand equality follows from the fact that the b's are just the f's in some order.

The theorem shows that the sum of the elements of each of the three degree vectors F, G, B is the same. We will denote this common sum by d and call it the degree of the permutation (3.1).

THEOREM 3. *a_k is a summit of (3.1), if and only if*

$$f_k > f_{k+1}.$$

Proof. If a_k is a summit, then we assert that

$$f_k \geqslant 1 + f_{k+1}.$$

Since a_k is a summit,

$$a_{k+1} < a_k.$$

Hence, the number of a's which are $<a_k$ and are on its right in (3.1) is certainly at least one more than the number of a's which are $<a_{k+1}$ and appear on the right side of a_{k+1} in (3.1). On the other hand, if

$$a_k < a_{k+1},$$

then it is clear that

$$f_k \leqslant f_{k+1}.$$

This proves the theorem.

Theorem 3 will be of immense use to us later.

DEFINITION. *A permutation will be said to be of the type (s, i, d) if it has s summits, index i and is of degree d.*

Example. The permutation 4367125 of S_7 is of type (2, 5, 11).

5. Determination of a permutation from its degree vector B

Let b_k be the number of a's which are $<k$ and are on its right side in (3.1). Then the permutation is uniquely determined from its degree vector B, in the following manner.

Mark n positions in a line:

$$\circ \quad \circ \quad \circ \quad \circ \quad \ldots \quad \circ \quad \circ \quad \circ.$$

Let n move along the line from the right, leave b_n positions empty and occupy the next position. Next, let $(n-1)$ move along the line from the right like n, leave b_{n-1} *unoccupied* positions empty and occupy the next *empty* position. Continue the operation with $(n-2), (n-3), \ldots, 3, 2, 1$ in turn and you get the desired permutation. The rule is simple: When k has its turn, it moves along the line from the right, leaves b_k *unoccupied* positions empty and occupies the next *empty* position.

Example. Let $B = [0, 0, 2, 0, 2, 2, 3, 1]$, then the required permutation is constructed as follows:

○	○	○	○	○	○	○	○
○	○	○	○	○	○	8	○
○	○	○	7	○	○	8	○
○	○	○	7	6	○	8	○
○	○	5	7	6	○	8	○
○	○	5	7	6	○	8	4
3	○	5	7	6	○	8	4
3	○	5	7	6	2	8	4
3	1	5	7	6	2	8	4

The desired permutation appears at the last step.

We have a nice criterion for determining whether or not a given figure will be a summit. It can be stated in the form:

If the given figure has an empty position on its immediate right when it appears for the first time in the above scheme, then it is a summit, otherwise not.

In our example, 8, 7, 6 and 3 alone are summits.

It will be observed that for the above construction to work, it is necessary and sufficient that for each $k \leqslant n$, we have

$$0 \leqslant b_k \leqslant k-1.$$

6. The determination of a permutation from its F vector

The method is very similar to the one in the preceding section. We first note that an F vector is legitimate if and only if

$$0 \leqslant f_k \leqslant n-k.$$

The following example describes the scheme in this case.

Let $F = [2, 3, 0, 1, 2, 0, 0]$. Evidently $n = 7$. We mark 7 positions and number them from left to right. Now a_1 moves along the line from left to right, leaves f_1 positions empty and occupies the next position. Next a_2 moves along the line from left to right like a_1, leaves f_2 *unoccupied* positions empty and occupies the next *empty* position. The operation is continued with the remaining a's. The value of any a is given by the number of the position occupied by it. In our example, we have

1	2	3	4	5	6	7
○	○	○	○	○	○	○
○	○	a_1	○	○	○	○
○	○		○	a_2	○	○
a_3	○		○		○	○
	○		a_4		○	○
	○				○	a_5
	a_6				○	
					a_7	

The required permutation is thus seen to be 3 5 1 4 7 2 6. For any legitimate F vector the permutation is again uniquely determined.

7. The position of a given permutation in our ordering

Given a permutation of S_n, the following algorithm determines the number of permutations which precede it from among the $n!$ permutations of S_n.

Elements:	1	2	3	4	...	n
b's	b_1	b_2	b_3	b_4	...	b_n
	m_1	m_2	m_3	m_4	...	m_n

We start with $m_1 = b_1 = 0$, and then calculate the values of the other m's from the formula

$$m_k = km_{k-1} + b_k, \quad k = 2, 3, \ldots, n. \tag{1}$$

In the algorithm, m_n gives the number of permutations which precede the given permutation (that is the one corresponding to the B vector) in our ordering.

From (1), it will be readily seen that

$$m_n = b_n + nb_{n-1} + n(n-1)b_{n-2} + \cdots + n!\,b_1$$

$$= n! \sum_{k=1}^{n} b_k/k! \tag{2}$$

The justification for the algorithm is provided by the fact that the given permutation arises from a permutation of S_{n-1} obtainable from the given permutation by the deletion of n. If this permutation had m_{n-1} permutations preceding it in its own set, then the given permutation is preceded by $nm_{n-1} + b_n$ permutations in the derived set.

Example. Let us determine the position of the permutation

3521746

among the permutations of S_7.

The algorithm takes the form:

	1	2	3	4	5	6	7
b's	0	1	2	0	3	0	2
m's	0	1	5	20	103	618	4328

The given permutation is, therefore, the 4329th in the set of 7! permutations of S_7.

Given the position of a permutation, the reverse process will give its B vector, from which the permutation itself can be determined. From the above algorithm, it will also be clear that if the permutations of S_n are arranged in our way, then their B vectors are in lexicographic order but not so their F vectors.

Thus for $n = 4$, we have

Position	Permutation	B vector	F vector
1	1234	0000	0000
2	1243	0001	0010
3	1423	0002	0200
4	4123	0003	3000
5	1324	0010	0100
6	1342	0011	0110
7	1432	0012	0210
8	4132	0013	3010
9	3124	0020	2000
10	3142	0021	2010
11	3412	0022	2200
12	4312	0023	3200
13	2134	0100	1000
14	2143	0101	1010
15	2413	0102	1200
16	4213	0103	3100
17	2314	0110	1100
18	2341	0111	1110
19	2431	0112	1210
20	4231	0113	3110
21	3214	0120	2100
22	3241	0121	2110
23	3421	0122	2210
24	4321	0123	3210

8. The graphs of a permutation, its conjugate and its inverse

The graph of the permutation in (3.1) is obtained by plotting in the usual way on graph paper, the points

$$(1, a_1), (2, a_2), (3, a_3), \ldots, (n, a_n) \tag{1}$$

and joining them by straight lines in the order in which the abscissae increase. The images of the points (1), in the lines

$$2y = n + 1 \quad \text{and} \quad y = x$$

provide respectively the graphs of the *conjugate* and the *inverse* of the permutation in (3.1). In the following diagrams are shown, for example, the graphs of the permutation 352164 and those of its conjugate and inverse.

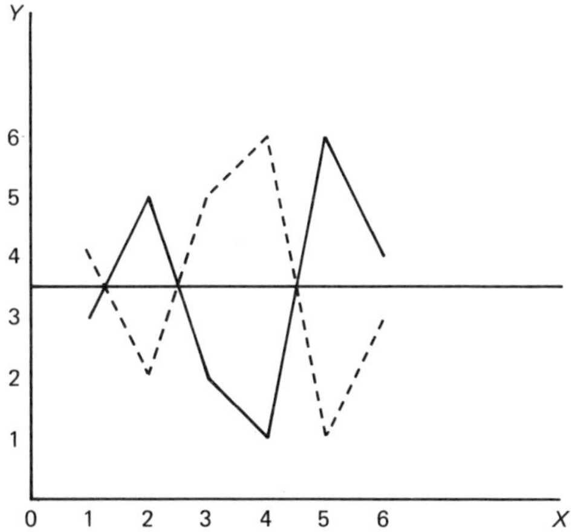

Graph of 352164 ———

Graph of the conjugate - - - - -

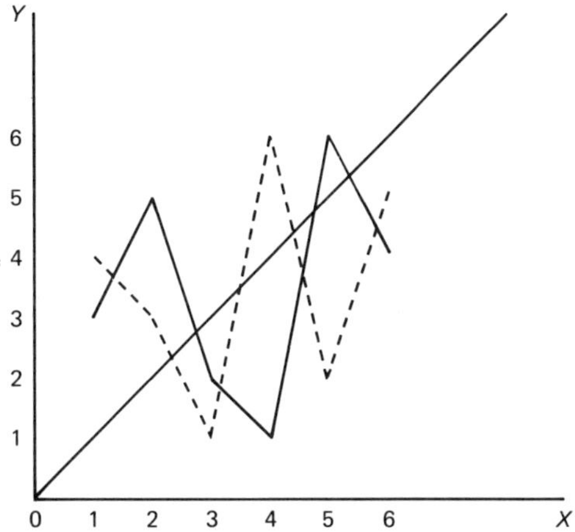

Graph of 352164 ———

Graph of the inverse - - - - -

9. Properties of the conjugate of a permutation

Let

$$h_1 h_2 h_3 \ldots h_n \tag{1}$$

be the conjugate of (3.1). Denote by $I(a_k)$ the index and by $F(a_k)$ the degree of a_k in (3.1). Similarly, let $I(h_k)$ and $F(h_k)$ denote respectively the index and the degree of h_k in (1). Then, we have

THEOREM 4.

(i) $h_k + a_k = n + 1$;

(ii) $I(h_k) + I(a_k) = k, \quad 1 \leqslant k \leqslant n - 1,$
$\qquad = 0, \quad k = n,$

(iii) $F(h_k) + F(a_k) = n - k, \quad k \leqslant n.$

Proof. (i) is a direct consequence of the definition. The definition also implies that for any two distinct positive integers k_1 and k_2 each $\leqslant n$,

$$a_{k_1} > a_{k_2} \quad \text{or} \quad < a_{k_2}$$

according as

$$h_{k_1} < h_{k_2} \quad \text{or} \quad > h_{k_2}.$$

(ii) and (iii) follow readily.

Furthermore, we have

$$\sum_{k=1}^{n} I(h_k) = \sum_{k=1}^{n} k - \sum_{k=1}^{n} I(a_k),$$

and

$$\sum_{k=1}^{n} F(h_k) = \sum_{k=1}^{n-1} (n-k) - \sum_{k=1}^{n} F(a_k).$$

Therefore, if (3.1) is of the type (s, i, d), then (1) is of the type

$$(n - 1 - s, (n; 2) - i, (n; 2) - d). \tag{2}$$

Example. The permutation 4367125 is of the type (2, 5, 11). Its conjugate 4521763 is of the type (4, 16, 10).

10. The type of a derived permutation

Let the permutation

$$a_1 a_2 a_3 \ldots a_{n-1} \tag{1}$$

of S_{n-1} be of the type (s, i, d).

First consider the permutation derived from (1) by writing n to the immediate right of a_{n-1}. Since in this operation no new summits are introduced nor is the index or the degree of any element altered, while the index and the degree of n are each zero, the derived permutation is still of the type (s, i, d).

Let us next insert n between a_{n-k-1} and a_{n-k} and assume that there are exactly t summits of (1) to its right, $k \geqslant 1$, $t \geqslant 0$. Then, two cases arise:

(i) When (1) has a summit at a_{n-k-1}.

In this case, the derived permutation still has s summits, its index i^* is given by

$$i^* = i + t + 1$$

and its degree d^* is given by

$$d^* = d + k.$$

This is because with the insertion of n, the summit has moved from a_{n-k-1} to n

and the index of each of the t summits to its right has increased by 1. Moreover, the degree of n is k while the degrees of all other elements have remained unaltered. The derived permutation is, therefore, of the type $(s, i + t + 1, d + k)$.

(ii) When a_{n-k-1} is not a summit of (1).

The derived permutation has now a new summit at n with index $(n - k)$ and, reasoning as before, we have

$$i^* = i + t + (n - k),$$
$$d^* = d + k.$$

The derived permutation is clearly of the type $(s + 1, i + t + n - k, d + k)$.

It will be seen that of the n permutations derivable from (1):

(i) Exactly $(s + 1)$ have s summits each, the remaining $(n - s - 1)$ have $(s + 1)$ summits each. (2)

(ii) The indices of the derived permutations are all distinct and run from i to $(n + i - 1)$ in some order. (3)

(iii) The degrees of the derived permutations are distinct and run from d to $(n + d - 1)$. (4)

Example. Let us consider the eight permutations that can be derived from the permutation 3521647 of the type (3, 10, 7). The work of writing the indices and the degrees of the derived permutations can be presented conveniently in the form:

Permutation	with s summits	index	with $(s + 1)$ summits	index	degree
35216478	×	10			7
35216487			×	17	8
35216847	×	11			9
35218647			×	16	10
35281647	×	12			11
35821647	×	13			12
38521647			×	15	13
83521647			×	14	14

It will be observed that the degrees can be entered in the table in a straight forward manner. For the indices, the derived permutations are divided into two subsets – one consisting of those with s summits, and the other of those with $(s + 1)$ summits. In the case of the first subset, the indices increase from top down, in the case of the second from bottom up.

11. Permutations with s summits

Let $t(n, s)$ denote the number of those permutations of S_n which have exactly s summits. Then, we have

THEOREM 5. *For* $n \geqslant 2$, *and* $1 \leqslant s \leqslant n-1$.

$$t(n, s) = (s+1)\,t(n-1, s) + (n-s)\,t(n-1, s-1),$$

with $t(m, 0) = 1$ *for each* $m > 0$ *and* $t(m, k) = 0$ *if* $0 < m < k$.

The theorem is a direct consequence of (10.2).

THEOREM 6. *For all non-negative integers n and s,*

$$t(n, s) = \sum_{k=0}^{s} (-1)^k (n+1; k)(s-k+1)^n.$$

Proof. The expression on the right can be shown to satisfy the recurrence relation in theorem 5.

Note. Let p be any prime. Then theorem 6 gives

$$t(p-1, s) \equiv 1 \pmod{p}.$$

Since

$$\sum_{s=0}^{p-2} t(p-1, s) = (p-1)!$$

we get

$$(p-1)! \equiv p-1 \pmod{p}.$$

This is Wilson's Theorem.

The numbers $t(n, s)$ are the Eulerian numbers and have been studied in detail by several writers. It will not, however, be out of place here to give a short table of

Values of $t(n, s)$

$s \downarrow$ \ $n \rightarrow$	1	2	3	4	5	6	7	8	9
0	1	1	1	1	1	1	1	1	1
1		1	4	11	26	57	120	247	502
2			1	11	66	302	1191	4293	14608
3				1	26	302	2416	15619	88234
4					1	57	1191	15619	156190
5						1	120	4293	88234
6							1	247	14608
7								1	502
8									1

The symmetry in the column entries is explained by (9.2).

12. Permutations with a given index

Let $u(n, i)$ denote the number of those permutations of S_n which have given index i. Then we have

THEOREM 7. *For* $n \geqslant 1$, *and* $0 \leqslant i \leqslant (n; 2)$,

$$u(n, i) = \sum_k u(n-1, k),$$

where $u(0, 0) = 1$ *and* k *runs over nonnegative integers from* $(i - n + 1)$ *to* i.

The theorem is a direct consequence of (10.3).

We can also give a generating function for $u(n, i)$. Let

$$U(x, n) = \sum_{i \geqslant 0} u(n, i) x^i.$$

Then theorem 7 implies that

$$\begin{aligned} U(x, n) &= (1 + x + x^2 + \cdots + x^{n-1}) \,.\, U(x, n-1), \\ &= (1 + x)(1 + x + x^2) \cdots (1 + x + x^2 + \cdots + x^{n-1}), \\ &= (1 - x)(1 - x^2)(1 - x^3) \cdots (1 - x^n)/(1 - x)^n. \end{aligned} \tag{1}$$

13. Permutations with a given degree

Let $v(n, d)$ denote the number of permutations of S_n which are of degree d.

Then (10.4) clearly shows that the number-theoretic functions $u(n, k)$ and $v(n, k)$ have the same recurrence relation and the same initial value. They are therefore identical, though they arise independently. We give a short table of values for the two.

k ↓	$n \to 1$	2	3	4	5	6
0	1	1	1	1	1	1
1		1	2	3	4	5
2			2	5	9	14
3			1	6	15	29
4				5	20	49
5				3	22	71
6				1	20	90
7					15	101
8					9	101
9					4	90

One can use (9.2) to supply the missing entries in the last two columns.

The problem of this section can also be dealt with as follows. Evidently $v(n, d)$ is the number of solutions of the Diophantine equation

$$b_1 + b_2 + \cdots + b_n = d, \tag{1}$$

where

$$0 \leqslant b_k \leqslant k - 1. \tag{2}$$

Letting b_n take the values 0, 1, 2, . . . , $n - 1$ in succession in the equation

$$b_1 + b_2 + \cdots + b_{n-1} = d - b_n \tag{3}$$

we readily get the recurrence formula for $v(n, d)$.

When however d is fixed and n is arbitrary, the following procedure is better suited to our needs. Since b_1 is necessarily zero, (1) provides a decomposition of d into at most $n - 1$ parts. If the largest of these parts is c and it occurs m times in the decomposition, we select m of the numbers

$$b_{c+1}, b_{c+2}, \ldots, b_n$$

(which are the only ones eligible for the value c) and take each of them equal to c. This requires that we have

$$c \leqslant n - 1 \quad \text{and} \quad m \leqslant n - c. \tag{4}$$

The m b's selected go out of our list and we repeat the process with the next largest part in the decomposition and with the b's now left over. This is continued till all the parts in the decomposition have been dealt with. If at any stage conditions such as those in (4) are not satisfied, the decomposition makes no contribution to the number $v(n, d)$. Since we have to take the largest part at each stage, it will be enough first to list all the partitions of d into at most $(n - 1)$ parts and the largest part not exceeding $(n - 1)$.

Example. Let us find the number of permutations of S_n which are of degree 5.

We first note that for $n < 4$, no permutation of S_n can have degree 5. We can, therefore assume that $n \geqslant 4$.

The partitions of 5 are

(i) 5; (ii) 1 + 4; (iii) 2 + 3; (iv) 1 + 2 + 2;
(v) 1 + 1 + 3; (vi) 1 + 1 + 1 + 2; (vii) 1 + 1 + 1 + 1 + 1.

(i) makes a contribution only if $n \geqslant 6$. The eligible b's are

$$b_6, b_7, \ldots, b_n.$$

Only one of these has to be selected. The required contribution is

$$(n - 5; 1).$$

(ii) This makes a contribution only if $n \geqslant 5$. We select one b from the following and take it to be 4. The b's are

$$b_5, b_6, \ldots, b_n.$$

After the selection, $n - 5$ of these b's will be left and b_2, b_3, b_4 will also be eligible to take the value 1. The required contribution will thus be

$$(n - 4; 1)(n - 2; 1).$$

Proceeding in this manner, the contributions from the other partitions of 5, will be found to be

(iii) $(n-3;1)(n-3;1)$ for $n \geqslant 4$;

(iv) $(n-2;2)(n-3;1)$ for $n \geqslant 4$;

(v) $(n-3;1)(n-2;2)$ for $n \geqslant 4$;

(vi) $(n-2;1)(n-2;3)$ for $n \geqslant 4$;

(vii) $(n-1;5)$ for $n \geqslant 6$.

Hence $v(4,5)=3$, $v(5,5)=3+4+6+6+3=22$; and for $n \geqslant 6$, $v(n,5)$ is the sum of the contributions noted under (i)–(vii) above.

Let

$$d = m_1c_1 + m_2c_2 + \cdots + m_rc_r \tag{5}$$

be any partition of d with

$$1 \leqslant c_1 < c_2 < \cdots < c_r \leqslant n-1,$$

and

$$m_r \leqslant n - c_r, \quad m_{r-1} + m_r \leqslant n - c_{r-1}, \ldots, m_1 + m_2 + \cdots + m_r \leqslant n - c_1.$$

Then the contribution which (5) makes to $v(n, d)$ is given by

$$(n - c_r; m_r)(n - c_{r-1} - m_r; m_{r-1})(n - c_{r-2} - m_{r-1} - m_r; m_{r-2}) \ldots$$
$$\ldots (n - c_1 - m_2 - m_3 - \cdots - m_r; m_1) \tag{6}$$

We leave it to the reader to prove that for $n > d$, $v(n, d)$ is a polynomial in n of degree d.

14. Permutations with a given index and a given number of summits

Let i be the given index and s the given number of summits. Evidently then, the required number is the number of solutions of the Diophantine equation

$$sr_1 + (s-1)r_2 + \cdots + r_s = i, \quad n > r_1 + r_2 + \cdots + r_s, \tag{1}$$

in positive integers r. This is the same as the number of solutions of the Diophantine equation

$$sx_1 + (s-1)x_2 + \cdots + x_s = i - (s+1;2) \tag{2}$$

in nonnegative integers x. The left side of (2) provides a partition of $i - (s+1;2)$ into parts the largest of which does not exceed s or, what is the same thing, into at most s parts. Hence the required number is

$$p(i - (s+1;2), s).$$

Here we take $p(0, s) = 1$, and $p(i - (s+1;2), s) = 0$ if $i < (s+1;2)$.

15. Permutations with a given index set

Let $P_n(j_1, j_2, \ldots, j_s)$ denote the number of permutations of S_n which have the index set

$$J(s) = \{j_1, j_2, \ldots, j_s\},$$

where $1 \leqslant j_1 < j_2 < \cdots < j_s \leqslant n-1$.

When the index set is empty, we have

$$P_n(\emptyset) = 1.$$

THEOREM 8. *For any* $n > j_s$,

$$P_n(j_1, j_2, \ldots, j_s) + P_n(j_1, j_2, \ldots, j_{s-1})$$
$$= (n; j_s) \,.\, P_{j_s}(j_1, j_2, \ldots, j_{s-1}).$$

Proof. From the set of n symbols $1, 2, 3, \ldots, n$ select any $(n - j_s)$, arrange them in ascending order and take them to form the section

$$a_{j_s+1}\; a_{j_s+2} \ldots a_n$$

of some permutation (3.1) of S_n. Arrange the remaining j_s symbols so that they have summits at

$$a_{j_1}, a_{j_2}, \ldots, a_{j_{s-1}}.$$

If in the resulting permutation

$$a_{j_s} > a_{j_s+1},$$

then we get a permutation of S_n with the given index set. Otherwise, we get a permutation of S_n with index set

$$\{j_1, j_2, \ldots, j_{s-1}\}.$$

Hence

$$P_n(j_1, j_2, \ldots, j_s) + P_n(j_1, j_2, \ldots, j_{s-1})$$
$$= (n; n - j_s) \,.\, P_{j_s}(j_1, j_2, \ldots, j_{s-1}).$$

This is the theorem.

In particular, we have

$$P_n(j_1) = (n; j_1) - 1;$$

$$P_n(j_1, j_2) = (n; j_2) \,.\, P_{j_2}(j_1) - P_n(j_1)$$
$$= (n; j_2)(j_2; j_1) - (n; j_2) - (n; j_1) + 1.$$

Proceeding in this manner, one can easily obtain an explicit expression for $P_n(j_1, j_2, \ldots, j_s)$ for any s, in terms of combinatory functions. The general result can, however, be obtained directly as follows.

The s summits break up the permutation into $(s + 1)$ sections in each of which the elements are in an ascending order. The last element of each section is, however, strictly greater than the first element of the section on its right (if there is one).

From the set of n symbols $1, 2, 3, \ldots, n$ select $(n - j_s)$, arrange them in ascending order and let them form the last section. From the remaining j_s symbols, select $j_s - j_{s-1}$, arrange them in ascending order as before, and let them form the section preceding the last. Continue this process of selection and arranging in order to form a new section. Then the second section from the left is formed from $j_2 - j_1$ symbols selected from the j_2 available at that stage. The remaining j_1 symbols provide the first section from the left.

The procedure described above, will give us all those permutations of S_n, the index sets of which are subsets of $J(s)$, including of course the empty set and the set $J(s)$ itself, but no permutation the index set of which is not a subset of $J(s)$.

Hence, the number $N_n(J(s))$ of permutations of S_n, the index sets of which are subsets of $J(s)$, is given by

$$\begin{aligned} N_n(J(s)) &= (n; n - j_s)(j_s; j_s - j_{s-1}) \ldots (j_2; j_2 - j_1) \\ &= (n; j_s)(j_s; j_{s-1}) \ldots (j_2; j_1). \end{aligned} \tag{2}$$

Given any index set J^*, (2) enables us to give an expression for $N_n(J^*)$ – the number of those permutations of S_n the index sets of which belong to the class of subsets of J^*. If $|J^*| = k$, the number of such subsets is 2^k.

Letting J^* run through the subsets of $J(s)$, (2) gives

$$N_n(J(s)) = \sum_{J^*} P_n(J^*). \tag{3}$$

Using the inclusion and exclusion principle, we then have

$$P_n(J(s)) = \sum_{J^*} (-1)^{s-|J^*|} N_n(J^*). \tag{4}$$

This compares favourably with the corresponding formula in [2]. For $s = 3$, (4) gives

$$\begin{aligned} P_n(j_1, j_2, j_3) = {} & N_n(j_1, j_2, j_3) - N_n(j_1, j_2) - N_n(j_1, j_3) - N_n(j_2, j_3) \\ & + N_n(j_1) + N_n(j_2) + N_n(j_3) - 1 \\ = {} & (n; j_3)(j_3; j_2)(j_2; j_1) - (n; j_2)(j_2; j_1) \\ & - (n; j_3)(j_3; j_1) - (n; j_3)(j_3; j_2) \\ & + (n; j_1) + (n; j_2) + (n; j_3) - 1. \end{aligned}$$

15.1. Some important consequences of theorem 8

We have

$$P_n(j_1, j_2, \ldots, j_s) = (n; j_s) P_{j_s}(j_1, \ldots, j_{s-1}) - P_n(j_1, \ldots, j_{s-1}) \tag{1}$$

Evidently the expression on the right in (1) vanishes if we replace n by j_s in it.

Similarly, $P_n(j_1, \ldots, j_{s-1})$ vanishes for $n = j_{s-1}$. This in turn implies that the said expression vanishes also at $n = j_{s-1}$. The same reasoning will show that

$$P_n(j_1, j_2, \ldots, j_s) = 0 \quad (n = j_k, k = 1, 2, \ldots, s.) \tag{2}$$

Moreover, (1) implies that we can write

$$P_n(j_1, j_2, \ldots, j_s) = (n; j_s)q_s + \cdots + (-1)^k (n; j_{s-k})q_{s-k} + \cdots$$
$$\cdots + (-1)^{s-1}(n; j_1)q_1 + (-1)^s; \tag{3}$$

where the q's are independent of n. It is thus clear that $P_n(j_1, j_2, \ldots, j_s)$ is a polynomial of degree j_s in n which assumes the value zero for $n = j_1, j_2, \ldots, j_s$. This simple observation enables us to evaluate the q's in (3) when the j's are given. For example, we have

$$P_n(2, 4, 6, 8) = (n; 8)q_4 - (n; 6)q_3 + (n; 4)q_2 - (n; 2)q_1 + 1.$$

Replacing n by 2, 4, 6, 8 in turn, we get without difficulty

$$q_1 = 1; \quad q_2 = 5; \quad q_3 = 61; \quad q_4 = 1385.$$

Thus for each $n > 8$, we have

$$P_n(2, 4, 6, 8) = 1385(n; 8) - 61(n; 6) + 5(n; 4) - (n; 2) + 1.$$

For any prime $p > 8$, we thus have

$$P_p(2, 4, 6, 8) \equiv 1 \pmod{p}.$$

16. Permutations with a given index and a given degree

Let the given index of the required set of permutations of S_n be i and the given degree d. We can assume without any loss of generality that i and d are both positive. Let

$$a_1 a_2 \ldots a_n \tag{1}$$

represent a typical permutation of the set. Let

$$F = \{f_1, f_2, \ldots, f_n\}, \tag{2}$$

where the f's are non-negative integers, be its degree vector and

$$J(s) = \{j_1, j_2, \ldots, j_s\}, \tag{3}$$

where $0 < j_1 < j_2 < \cdots < j_s < n$, its index set. Then, we have

$$f_1 + f_2 + \cdots + f_n = d; \tag{4}$$

$$j_1 + j_2 + \cdots j_s = i. \tag{5}$$

If

$$R(s) = \{r_1, r_2, \ldots, r_s\} \tag{6}$$

is the interval set corresponding to (3), then we also have

$$sr_1 + (s-1)r_2 + (s-2)r_3 + \cdots + 2r_{s-1} + r_s = i. \tag{7}$$

It will be recalled that in (6) and (7), r's are positive integers.

We first assert that for $i + d = n$, there is at least one permutation (1) of S_n with index i, degree d and $a_n < n$.

Proof. The permutation is, in fact, given by

$$\begin{aligned} a_k &= k, \quad k \leqslant i-1;\\ &= n, \quad k = i;\\ &= k-1, \quad i < k \leqslant n. \end{aligned}$$

On the other hand for $n > (i + d)$, there is no permutation (1) of S_n with index i, degree d and $a_n < n$.

Proof. If in (1), $a_n < n$, then there exists a unique $k \leqslant j_s$ for which $a_k = n$. In view of (5), $k \leqslant i$. The f_k corresponding to this k is $n - k$. In view of (4), $f_k \leqslant d$. Hence

$$n \leqslant k + d \leqslant i + d.$$

This contradicts the assumption that $n > i + d$, and our assertion follows.

It is thus clear that the permutations with index i and degree d must be looked for among the permutations of S_n with $n \leqslant i + d$. Permitting a_n to have the value n, all the permutations of index i and degree d can be listed if we take $n = i + d$, and this we shall do in all that follows.

Let $A(i, d)$ denote the number of permutations of S_n, $n = i + d$, with index i and degree d. Then, we have merely to find a formula for $M_d(J(s))$ – the number of permutations of S_n with degree d and the index set $J(s)$, and then $A(i, d)$ will be determined by letting $J(s)$ run over all the partitions of i into one or more distinct summands in (5). The number of summands in any such partition cannot exceed h, where h is the largest integer for which

$$(h + 1; 2) \leqslant i. \tag{8}$$

In fact, we shall have

$$A(i, d) = \sum_{s \leqslant h} M_d(J(s))$$

with h as in (8).

Determination of $M_d(J(s))$

It is clear that $M_d(J(s))$ is the number of vectors F satisfying equation (4) and the conditions:

$$f_k > f_{k+1} \quad \text{if and only if} \quad k = j_1, j_2, \ldots, j_s. \tag{9}$$

Let

$$\begin{aligned} f_1 + \cdots + f_{j_1} &= d_1, \\ f_{j_1+1} + \cdots + f_{j_2} &= d_2, \\ &\vdots \\ f_{j_{s-1}+1} + \cdots + f_{j_s} &= d_s. \end{aligned} \tag{10}$$

Since

$$f_{j_s+1} = f_{j_s+2} = \cdots = f_n = 0$$

we must have

$$d_1 + d_2 + \cdots + d_s = d. \tag{11}$$

Here $d_1, d_2, \ldots, d_s$ are necessarily positive integers. Hence d must be greater than or equal to s. Noting that relations (10) provide partitions of numbers $d_1, d_2, \ldots, d_s$ into $r_1, r_2, \ldots, r_s$ non-negative parts respectively, one way of dealing with the problem in any given case will be the following.

List all the partitions of $d_1, d_2, \ldots, d_s$ into $r_1, r_2, \ldots, r_s$ parts respectively. Take any partition of d_1 from your list. To its right write down one of the partitions of d_2. To the right of this one put down a partition of d_3 and carry on till you have written down a partition of d_s. Now add $(n - j_s)$ 0's to the right. In this manner, you will obtain

$$p(d_1, r_1)\, p(d_2, r_2) \ldots p(d_s, r_s)$$

vectors F. Keep those that satisfy each and every one of the conditions in (9). The number of these acceptable vectors is the contribution of the solution $(d_1, d_2, \ldots, d_s)$ of (11) to $M_d(J(s))$. The sum of the contributions of all the $(s-1; d-1)$ solutions of (11) in positive integers is $M_d(J(s))$ for the given $J(s)$. A computer can easily handle the problem in this manner.

It is clear that

$$M_d(J(s)) \leqslant \sum_{D^*} p(d_1, r_1)\, p(d_2, r_2) \ldots p(d_s, r_s),$$

where D^* runs over all the solutions of (11) in positive integers. In particular when $J(s)$ is a one element set, we have

$$M_d(\{k\}) = p(d, k).$$

This shows that

$$M_k(\{k\}) = p(k),$$

so that

$$A(k, k) \geqslant p(k), \quad n \geqslant 2k.$$

This is an improvement on theorem 1 in [1].

Remark

The results presented in appendix 2 were obtained in the manner described above but without the aid of a computer.

A formula for $M_d(J(s))$

In (11), $d_1, d_2, \ldots, d_s$ had been allowed to run over all solutions of that equation in positive integers. We now relax this condition and let them run over all solutions of the equation

$$t_1 + t_2 + \cdots + t_s = d$$

in *non-negative* integers. Denote this set of solutions by T^*.

Write

$$W_d(J(s)) = \sum_{T^*} p(d_1, r_1)\, p(d_2, r_2) \ldots p(d_s, r_s),$$

where $p(0, r_k) = 1$ for each $k \leqslant s$. Then $W_d(J(s))$ counts all those permutations of S_n which have degree d and whose index sets are non-empty subsets of $J(s)$. Since for $d > 0$, we have $M_d(\emptyset) = 0$, it follows that

$$W_d(J(s)) = \sum_{J^*} M_d(J^*),$$

where J^* runs over all the subsets of $J(s)$. The inclusion and exclusion principle now gives

$$M_d(J(s)) = \sum_{J^*} (-1)^{s-|J^*|} W_d(J^*)$$

Here $W_d(\emptyset) = 1$, by definition. We are thus through.

Example

$$\begin{aligned}
M_5(\{1,3,6\}) &= W_5(\{1,3,6\}) - W_5(\{1,3\}) - W_5(\{1,6\}) - W_5(\{3,6\}) \\
&\quad + W_5(\{1\}) + W_5(\{3\}) + W_5(\{6\}) - 1 \\
&= \sum_{\substack{T^* \\ s=3 \\ d=5}} p(d_1, 1)\, p(d_2, 2)\, p(d_3, 3) \\
&\quad - \sum_{\substack{T^* \\ s=2 \\ d=5}} \{p(d_1, 1)\, p(d_2, 2) + p(d_1, 1)\, p(d_2, 5) + p(d_1, 3)\, p(d_2, 3)\} \\
&\quad + p(5, 1) + p(5, 3) + p(5, 6) - 1. \\
&= 58 - 60 + 1 + 5 + 7 - 1 = 10.
\end{aligned}$$

17. The generating function of $W_d(J(s))$

As a direct consequence of our definition of $W_d(J(s))$ in the preceding section, it will be clear that $W_d(J(s))$ is the coefficient of x^d in

$$X_{r_1} X_{r_2} \ldots X_{r_s},$$

where

$$X_k = \{(1-x)(1-x^2)\ldots(1-x^k)\}^{-1}.$$

It will be useful to remark here that

$$X_{r_1} X_{r_2} \ldots X_{r_s} = X_{j_s}\{X_{r_1} X_{r_2} \ldots X_{r_s}/X_{j_s}\} \tag{1}$$

The expression within the curly brackets on the right of (1) will be seen to be a polynomial in x, with integral coefficients, and of degree at most

$$(j_s + 1; 2) - \sum_{k=1}^{s} (r_k + 1; 2).$$

Observe further that for any $h < k$, we can write

$$X_h = (1 - x^{h+1})(1 - x^{h+2}) \ldots (1 - x^k) X_k. \tag{2}$$

The generating function of $A(u, d)$ – the number of permutations of S_n of any given index u and of degree d (any $d > 0$), can now be written out as in the following example with $u = 8$.

Index set	Generating function
$\{8\}$	$X_8 - 1$
$\{1, 7\}$	$X_1X_6 - X_1 - X_7 + 1$
$\{2, 6\}$	$X_2X_4 - X_2 - X_6 + 1$
$\{3, 5\}$	$X_3X_2 - X_3 - X_5 + 1$
$\{1, 2, 5\}$	$X_1X_1X_3 - X_1X_1 - X_1X_4 - X_2X_3 + X_1 + X_2 + X_5 - 1$
$\{1, 3, 4\}$	$X_1X_2X_1 - X_1X_2 - X_1X_3 - X_3X_1 + X_1 + X_3 + X_4 - 1$

Denoting by C_u the generating function of $A(u, d)$, we thus have

$$\begin{aligned} C_8 &= X_8 + (x + x^2 + x^3 + x^4 + x^5 + x^6)X_7 \\ &\quad + (2 + 2x + 3x^2 + 2x^3 + 3x^4 + 2x^5)X_6 \\ &\quad + (1 + 2x + 2x^2 + 3x^3 + 3x^4)X_5 \\ &\quad - (1 + x + 0x^2 - x^3)X_4 - (3 + 2x + x^2)X_3 \\ &\quad - (1 + x)X_2 + X_1 \\ &= X_8(x + 4x^2 + 7x^3 + 11x^4 + 11x^5 + 10x^6 + 5x^7 - x^8 \\ &\quad - 13x^9 - 21x^{10} - 27x^{11} - 27x^{12} - 21x^{13} - 11x^{14} \\ &\quad + 2x^{15} + 15x^{16} + 25x^{17} + 27x^{18} + 25x^{19} + 16x^{20} \\ &\quad + 7x^{21} - 4x^{22} - 11x^{23} - 15x^{24} - 13x^{25} - 9x^{26} \\ &\quad - 4x^{27} + 0x^{28} + 3x^{29} + 5x^{30} + 4x^{31} + 2x^{32} + 0x^{33} \\ &\quad - x^{34} - x^{35}). \end{aligned}$$

Example

$$A(8,8) = p(7) + 4p(6) + 7p(5) + 11p(4) + 11p(3) + 10p(2) + 5p(1) - p(0) = 15 + 44 + 49 + 55 + 33 + 20 + 5 - 1 = 220.$$

This is in agreement with the result in [1].

As an example of the use of (2), we give the following. We have (appendix 4)

$$C_3 = X_3(x + x^2 + x^3 - x^4 - x^5) \tag{3}$$

$$= X_6(x + x^2 + x^3 - x^4 - 2x^5 - 2x^6 - 3x^7 - x^8 + x^9 + 3x^{10} + 3x^{11} + 3x^{12} + x^{13} - x^{14} - 2x^{15} - 2x^{16} - x^{17} - x^{18} + x^{19} + x^{20}). \tag{4}$$

Using (3), we will get

$$A(3,6) = p(5,3) + p(4,3) + p(3,3) - p(2,3) - p(1,3) = 5 + 4 + 3 - 2 - 1 = 9.$$

From (4), we obtain

$$A(3,6) = p(5) + p(4) + p(3) - p(2) - 2p(1) - 2p(0) = 7 + 5 + 3 - 2 - 2 - 2 = 9.$$

The latter gives the result in terms of unrestricted partitions only. We have thus shown that C_u can be expressed in either of the forms:

$$C_u = X_u(c_{u,1}x + c_{u,2}x^2 + \cdots + c_{u,k}x^k), \quad k = (u+1;2),$$

or

$$C_u = X_v(c'_{v,1}x + c'_{v,2}x^2 + \cdots + c'_{v,h}x^h), \quad h = (v+1;2),$$

where the c and c' are integers, not necessarily positive or non-zero, and $v > u$.

18. The Alter–Curtz–Wang conjecture

In 1974, R. Alter, T. B. Curtz and C. C. Wang [1], conjectured that for each $n \leqslant (u+v)$,

$$A(u,v) = A(v,u). \tag{1}$$

For $n = u + v$, this means that the coefficient of x^v in C_u equals the coefficient of x^u in C_v. Assuming, without loss of generality, that $v > u$ and v is not too big, for any given u and v, (1) can be verified as follows. Let

$$C_u = X_v(c'_{v,1}x + c'_{v,2}x^2 + \cdots + c'_{v,v}x^v + \cdots),$$
$$C_v = X_v(c_{v,1}x + c_{v,2}x^2 + \cdots + c_{v,u}x^u + \cdots).$$

Then, all that we have to do is to check if the two expressions

$$c'_{v,1}\,p(v-1) + c'_{v,2}\,p(v-2) + \cdots + c'_{v,v}p(0),$$

and

$$c_{v,1}\,p(u-1) + c_{v,2}\,p(u-2) + \cdots + c_{v,u}p(0),$$

are or are not equal.

Example. For $u = 5$, $v = 8$, we have

$$C_5 = X_8(x + 2x^2 + 3x^3 + 2x^4 + x^5 - 3x^6 - 5x^7 - 7x^8 + \cdots)$$
$$C_8 = X_8(x + 4x^2 + 7x^3 + 11x^4 + 11x^5 + \cdots).$$

Now

$$p(7) + 2p(6) + 3p(5) + 2p(4) + p(3) - 3p(2) - 5p(1) - 7p(0)$$
$$= 15 + 22 + 21 + 10 + 3 - 6 - 5 - 7 = 53$$

and

$$p(4) + 4p(3) + 7p(2) + 11p(1) + 11p(0)$$
$$= 5 + 12 + 14 + 11 + 11 = 53.$$

Hence $A(5, 8) = A(8, 5)$.

We proceed to prove that the Alter–Curtz–Wang conjecture holds good for each $u \leqslant 4$ and all v. We make no use of the generating functions. All that we use is the method of direct computation suggested in section 16. Details are given for the case $u = 3$. For $u = 4$, only the contributions from different interval sets are noted. The cases $u = 1$ and $u = 2$ are simple and are left to the reader.

Proof for u = 3

Since the conjecture is easy to verify for small values of v, we take $v > 9$. We first show that

$$A(3, v) = [(v^2 + 3)/12] + v. \tag{2}$$

The contribution of the index set $\{3\}$ to $A(3, v)$ is $p(v, 3)$ and that of the index set $\{1, 2\}$ is the same as the number of partitions of v into two distinct summands. Since

$$p(v, 3) = [(v^2 + 3)/12] + [v/2] + 1,$$

and the number of partitions of v into two distinct summands is

$$[(v-1)/2],$$

(2) follows at once.

Now let $v = m + 9$. Then, the contribution of the index set $\{v\}$ to $A(v, 3)$ is $p(3)$.

The contribution of the two solutions (1, 2) and (2, 1) of

$$d_1 + d_2 = 3$$

for the index set $\{1, v - 1\}$ is 3. For the remaining index sets except the last one which is $\{(v - 1)/2, (v + 1)/2\}$ when v is odd, and $\{(v - 2)/2, (v + 2)/2\}$ when v is even, the contribution is 4. For the last index set it is 1 when v is odd and 3 when v is even. Hence the total contribution from the two element index sets is

$$2v - 6, \quad v > 9.$$

Finally, the solution (1, 1, 1) of the equation

$$d_1 + d_2 + d_3 = 3$$

makes a contribution of 1 or 0 to $A(v, 3)$, according as an interval set $\{r_1, r_2, r_3\}$ given by the equation

$$3r_1 + 2r_2 + r_3 = v \tag{3}$$

does or does not satisfy the three conditions

$$r_1 \geqslant 1, \quad r_2 \geqslant 2, \quad r_3 \geqslant 2.$$

The number of interval sets satisfying these conditions is the same as the number of partitions of $v - 9$ into at most three parts, or what is the same thing, the number of partitions of $(v - 6)$ into exactly three parts. This is

$$[((v - 6)^2 + 3)/12] = [((m + 3)^2 + 3)/12],$$

Thus, we have

$$A(v, 3) = p(3) + (2v - 6) + [(v^2 - 12v + 39)/12]$$
$$= v + [(v^2 + 3)/12],$$

and we are through.

Proof for $u = 4$

We take $v > 16$. For $A(4, v)$, the contributions for different interval sets are

Interval set	Contribution
$\{4\}$	$p(v, 4)$
$\{1, 2\}$	$[(v + 4)(v - 1)/6]$.

For $A(v, 4)$, these are

$\{v\}$		$p(4, v) = 5$
$\{r_1, r_2\}$	with $2r_1 + r_2 = v$	$2[(5v - 18)/2]$
$\{r_1, r_2, r_3\}$	with $3r_1 + 2r_2 + r_3 = v$	$(v^2 - 13v + 44)/2$
$\{r_1, r_2, r_3, r_4\}$	with $4r_1 + 3r_2 + 2r_3 + r_4 = v$	$p(v - 16, 4)$

To establish the equality of $A(4, v)$ and $A(v, 4)$, we use the identity

$$p(v, 4) = p(v - 16, 4) + p(v, 3) + p(v - 4, 3) + p(v - 8, 3) + p(v - 12, 3).$$

We might observe here that for any fixed u and v ($>u^2$) sufficiently large but in an arbitrarily fixed residue class modulo the l.c.m. of $1, 2, 3, \ldots, u$, the contribution to $A(u, v)$ from the k-element interval sets is a polynomial in v of degree $k-1$, not necessarily with integral coefficients. In the case of $A(v, u)$, it is the contribution from $(u+1-k)$-element interval sets which is a polynomial of degree $k-1$ in v. (The two polynomials are however not usually identical.) In fact, the major contribution to $A(u, v)$ comes from the one-element interval set $\{u\}$ and is given by $p(v, u)$, while the major contribution to $A(v, u)$ comes from the u-element interval sets $\{r_1, r_2, \ldots, r_u\}$ satisfying the relation

$$ur_1 + (u-1)r_2 + \cdots + r_u = v \tag{4}$$

from the solution $(1, 1, 1, \ldots, 1)$ of the equation

$$d_1 + d_2 + \cdots + d_u = u.$$

This contribution is the same as the number of solutions of (4) with

$$r_1 \geqslant 1; \quad r_2, r_3, \ldots, r_u \geqslant 2:$$

and is therefore given by $p(v-u^2, u)$. Since for v sufficiently large and u fixed,

$$p(v-u^2, u) \sim p(v, u),$$

it follows that

$$A(u, v) \sim A(v, u). \tag{5}$$

This strongly supports the Alter–Curtz–Wang conjecture.

19. A relation between A's and c's

Assume that $v \geqslant u$, and

$$C_v = X_v(c_{v,1}x + c_{v,2}x^2 + \cdots + c_{v,v}x^v + \cdots).$$

Then, as in section 17, we have

$$\begin{aligned}
A(v, 1) &= c_{v,1}\, p(0) \\
A(v, 2) &= c_{v,1}\, p(1) + c_{v,2}\, p(0) \\
A(v, 3) &= c_{v,1}\, p(2) + c_{v,2}\, p(1) + c_{v,3}\, p(0) \\
&\ \ \vdots \\
A(v, u) &= c_{v,1}\, p(u-1) + c_{v,2}\, p(u-2) + c_{v,3}\, p(u-3) + \cdots + c_{v,u}\, p(0)
\end{aligned}$$

Now, for each $t > 0$, we have

$$p(t) = \sum_{k \geqslant 1} (-1)^{k-1} \{p(t - \tfrac{1}{2}k(3k-1)) + p(t - \tfrac{1}{2}k(3k+1))\}$$

with $p(-q) = 0$ for each $q \geqslant 1$. This means that

$$p(t) - p(t-1) - p(t-2) + p(t-5) + p(t-7) - p(t-12) - \cdots = 0$$

where the numbers 1, 2, 5, 7, . . . are the pentagonal numbers

$$\tfrac{1}{2}k(3k-1) \quad \text{and} \quad \tfrac{1}{2}k(3k+1).$$

This immediately gives the nice relation

$$A(v,u) - A(v,u-1) - A(v,u-2) + A(v,u-5) + - - + + - - + \cdots = c_{v,u}$$

Given the c's, this enables us to compute $A(v, 1), A(v, 2), \ldots, A(v, v)$ in succession without the use of any partition table.

Example. For $v = 8$, we have

$$A(8,1) = 1;$$

$$A(8,2) - A(8,1) = 4, \quad A(8,2) = 5;$$

$$A(8,3) - A(8,2) - A(8,1) = 7, \quad A(8,3) = 13;$$

$$A(8,4) - A(8,3) - A(8,2) = 11, \quad A(8,4) = 29;$$

$$A(8,5) - A(8,4) - A(8,3) = 11, \quad A(8,5) = 53;$$

$$A(8,6) - A(8,5) - A(8,4) + A(8,1) = 10, \quad A(8,6) = 91;$$

$$A(8,7) - A(8,6) - A(8,5) + A(8,2) = 5, \quad A(8,7) = 144;$$

$$A(8,8) - A(8,7) - A(8,6) + A(8,3) + A(8,1) = -1, \quad A(8,8) = 220.$$

Note that we take $A(8, 0) = 0$.

One can compute the value of $A(u, v)$ for any given $v \geqslant u$ in the same manner. But the Alter–Curtz–Wang conjecture still evades proof except in the few cases considered in the preceding section.

Quite a lot of work has been done on permutations during recent years particularly by L. Carlitz. The study of this chapter will greatly facilitate the study of such work. Some of it is listed in the bibliography at the end of this book.

References

1. Ronald Alter, Thaddeus B. Curtz and Chung C. Wang: Permutations with fixed index and number of inversions. Proceedings of the S–E Conference on Combinatorics, Graph Theory and Computing, X (1974), 209–228.
2. H. O. Foulkes: Enumeration of permutations with prescribed up-down and inversion sequences. *Discrete Math.* **15** (1976), 235–252.

Appendix 1

A complete census of the F vectors of permutations with index 5 and degree 7, and those with index 7 and degree 5. The zero part at the end is dropped. The summits are indicated by crosses.

1	2	3	4	5	n
				×	
0	0	0	0	7	12
0	0	0	1	6	11
0	0	0	2	5	10
0	0	0	3	4	9
0	0	1	1	5	10
0	0	1	2	4	9
0	0	1	3	3	8
0	0	2	2	3	8
0	1	1	1	4	9
0	1	1	2	3	8
0	1	2	2	2	7
1	1	1	1	3	8
1	1	1	2	2	7
×			×		
1	0	0	6		10
1	0	1	5		9
1	0	2	4		8
1	0	3	3		7
2	0	0	5		9
2	0	1	4		8
2	0	2	3		7
2	1	1	3		7
2	1	2	2		6
3	0	0	4		8
3	0	1	3		7
3	0	2	2		6
3	1	1	2		6
4	0	0	3		7
4	0	1	2		6
4	1	1	1		5
5	0	0	2		6
5	0	1	1		6
6	0	0	1		7
	×	×			
0	6	1			8
0	5	2			7
0	4	3			6
1	5	1			7
1	4	2			6
2	4	1			6
2	3	2			5
3	3	1			5

1	2	3	4	5	6	7	n
						×	
0	0	0	0	0	0	5	12
0	0	0	0	0	1	4	11
0	0	0	0	0	2	3	10
0	0	0	0	1	1	3	10
0	0	0	0	1	2	2	9
0	0	0	1	1	1	2	9
0	0	1	1	1	1	1	8
×					×		
1	0	0	0	0	4		10
1	0	0	0	1	3		9
1	0	0	0	2	2		8
1	0	0	1	1	2		8
1	0	1	1	1	1		7
2	0	0	0	0	3		9
2	0	0	0	1	2		8
2	0	0	1	1	1		7
3	0	0	0	0	2		8
3	0	0	0	1	1		7
4	0	0	0	0	1		7
	×			×			
0	1	0	0	4			9
0	1	0	1	3			8
0	1	0	2	2			7
0	2	0	0	3			8
0	2	0	1	2			7
0	2	1	1	1			6
1	1	0	0	3			8
1	1	0	1	2			7
0	3	0	0	2			7
0	3	0	1	1			6
1	2	0	0	2			7
1	2	0	1	1			6
0	4	0	0	1			6
1	3	0	0	1			6
2	2	0	0	1			6
		×	×				
0	0	4	1				7
0	0	3	2				6
0	1	3	1				6
0	2	2	1				5
1	1	2	1				5
×	×		×				
3	1	0	1				5
2	1	0	2				6

$n = 5 \quad 6 \quad 7 \quad 8 \quad 9 \quad 10 \quad 11 \quad 12$

$A = 3 + 9 + 10 + 8 + 5 + 3 + 1 + 1 = 40$

Appendix 2

The following calculations support the Alter–Curtz–Wang conjecture

(i) $d = 6$, $i = 10$ $n \geqslant 16$ $d = 10$, $i = 6$

Index set	Contribution	Index set	Contribution
10	11	6	35
1, 9	18	1, 5	63
2, 8	32	2, 4	62
3, 7	36		
4, 6	24	1, 2, 3	4
1, 2, 7	7		
1, 3, 6	22		164
1, 4, 5	7		
2, 3, 5	7		
	164		

(ii) $d = 10$, $i = 7$ $n \geqslant 17$ $d = 7$, $i = 10$

Index set	Contribution	Index set	Contribution
7	38	10	15
1, 6	77	1, 9	29
2, 5	111	2, 8	53
3, 4	30	3, 7	61
		4, 6	40
1, 2, 4	22	1, 2, 7	14
		1, 3, 6	40
	278	1, 4, 5	13
		2, 3, 5	13
		1, 2, 3, 4	0
			278

(iii) $d = 6$, $i = 11$ $n \geqslant 17$ $d = 11$, $i = 6$

Index set	Contribution	Index set	Contribution
11	11	6	44
1, 10	18	1, 5	83
2, 9	32	2, 4	79
3, 8	38		
4, 7	34	1, 2, 3	5
5, 6	8		
1, 2, 8	7		211
1, 3, 7	24		
1, 4, 6	22		
2, 3, 6	10		
2, 4, 5	7		
	211		

Appendix 3

Generating functions C_u, for values of $u \leqslant 7$

$u = 1 \quad X_1 - 1$

$u = 2 \quad X_2 - 1$

$u = 3 \quad X_3 + xX_2 - X_1$

$u = 4 \quad X_4 + (x + x^2)X_3 - X_1$

$u = 5 \quad X_5 + (x + x^2 + x^3)X_4 + (x + x^2)X_3 - X_2 - X_1 + 1$

$u = 6 \quad X_6 + (x + x^2 + x^3 + x^4)X_5 + (1 + x + 2x^2 + x^3)X_4 - (2 + x)X_2$

$u = 7 \quad X_7 + (x + x^2 + x^3 + x^4 + x^5)X_6 + (1 + 2x + 2x^2 + 2x^3 + 2x^4)X_5$
$\quad + (x + x^2 + 2x^3)X_4 - (2 + x)X_3 - (1 + x)X_2 + 1.$

Appendix 4

Generating functions C_u in expanded form

$u = 1 \quad X_1 x$

$u = 2 \quad X_2(x + x^2 - x^3)$

$u = 3 \quad X_3(x + x^2 + x^3 - x^4 - x^5)$

$u = 4 \quad X_4(x + 2x^2 + x^3 + x^4 - 2x^5 - 2x^6 - x^7 + 0x^8 + x^9)$

$u = 5 \quad X_5(x + 2x^2 + 3x^3 + 2x^4 + x^5 - 3x^6 - 4x^7 - 4x^8 - 2x^9$
$\quad + x^{10} + 2x^{11} + 2x^{12} + x^{13} + 0x^{14} - x^{15})$

$u = 6 \quad X_6(x + 3x^2 + 4x^3 + 4x^4 + 2x^5 + x^6 - 5x^7 - 7x^8 - 7x^9$
$\quad - 5x^{10} - 2x^{11} + 2x^{12} + 5x^{13} + 4x^{14} + 3x^{15} + x^{16}$
$\quad + 0x^{17} - 2x^{18} - x^{19})$

$u = 7 \quad X_7(x + 3x^2 + 6x^3 + 7x^4 + 7x^5 + 3x^6 + 0x^7 - 8x^8 - 13x^9$
$\quad - 15x^{10} - 13x^{11} - 8x^{12} + 0x^{13} + 7x^{14} + 11x^{15}$
$\quad + 13x^{16} + 9x^{17} + 5x^{18} - x^{19} - 4x^{20} - 5x^{21} - 4x^{22}$
$\quad - 3x^{23} + 0x^{24} + x^{25} + 2x^{26} + x^{27} - x^{28}).$

Expressions for $u = 8$ have already been given in the text.

MISCELLANEOUS PROBLEMS

1. For given positive integers h, $k (h \neq k)$, there exist infinitely many n for which n, $n+h$, $n+k$ are all sums of two squares. (C. Hooley)
2. For any given positive integers a and k, does there always exist at least one n for which $\phi(n+k) = a\phi(n)$?
3. For all odd primes p and all natural numbers n, show that

$$\sum_{k=1}^{N} \binom{n}{(p-1)k} \equiv 0 \pmod{p}, \qquad N = [(n-1)/(p-1)]. \qquad \text{(Hans Riesel)}$$

4. Show that each of the numbers $m/(m, n)$ and $(m-n+1)/(m+1, n)$ divides $\binom{m}{n}$.
5. Solve the Diophantine equation

$$(x^2-1)^2 + (y^2-1)^2 = (z^2-1)^2.$$

6. Show that

$$\text{g.c.d. of } \binom{n-1}{k}, \binom{n}{k-1}, \binom{n+1}{k+1} = \text{g.c.d. of } \binom{n-1}{k-1}, \binom{n}{k+1}, \binom{n+1}{k}.$$

(H. W. Gould)

7. If $N = 2^t M$, where M has s distinct odd prime divisors, then show that the congruence

$$X^2 \equiv 1 \pmod{N}$$

has 2^s solutions if $t = 0$ or 1; 2^{s+1} solutions if $t = 2$; and 2^{s+2} solutions if $t \geqslant 3$.
8. If a prime $p \equiv 3 \pmod 4$ has a primitive root g such that $g^2 = g + 1$, then $(g+1)$ and $(g-1)$ are also primitive roots of p. (D. Shanks)
9. Prove the following recurrence relation for $q(n)$:

$$q(n) = \sum_{k \geqslant 0} (-1)^{k(k+1)/2} p\left(2n - \binom{k+1}{2}\right). \qquad \text{(John A. Ewell)}$$

(Here $p(n)$ and $q(n)$ denote respectively the numbers of unrestricted partitions of n and the partitions of n into distinct parts.)
10. If for each natural number n,

$$f(n) = \sum_{k \geqslant 0} p(k)p(n-2k-1)$$

then $f(n) \equiv 0 \pmod 3$ if $n \equiv 0 \pmod 3$. (John A. Ewell)
11. Let $p(n, k)$ denote the number of partitions of n into exactly k summands, then show that

$$p(n, k) = \sum_{j=0}^{k-1} a_j(k) n^j$$

where $a_j(k)$ depend on the residue of n modulo the l.c.m. of the integers: $1, 2, 3, \ldots, [k/(j+1)]$.

12. If p and $p+2$ are both primes, then show that

$$4 \,.\, (p-1)! + p + 4 \equiv 0 \pmod{p(p+2)}.$$

13. Find the largest integer h for which the Diophantine equation

$$3x_1 + 5x_2 + 7x_3 = h$$

has no solution in nonnegative integers x_1, x_2, x_3.

14. Show that

$$e^x = \prod_{n \geqslant 1} (1 - x^n)^{-\mu(n)/n}$$

where μ is the Möbius function. (Barry Brent)

15. Prove that

$$\sum_{a <' n} (a-1, n) = \phi(n)\, d(n)$$

where the symbols have their usual meaning and we take $(0, n) = n$.

(P. Kesava Menon)

16. Show that for $n > 1$,

$$\sum_{j=1}^{n-1} [[(n/j)]/(n/j)] = 1 \quad \text{or} \quad > 1$$

according as n is or is not a prime. (Stephen Regimbal)

17. Is it true that

$$x^2 - ay^2 = z^2 \quad \text{and} \quad x^2 + ay^2 = t^2$$

have integral solutions for x, y if $a \equiv 5, 6, 7 \pmod 8$? (N. M. Stephens)

18. Among any four consecutive integers > 11, there is always at least one which is divisible by a prime > 11. (E. F. Ecklund; R. B. Eggleton)

19. Are there any composite numbers n for which $\phi(n)$ divides $(n-1)$?

20. Solve the Diophantine equation

$$\sum_{j=1}^{n} (m+j)^2 = \sum_{j=1}^{n-1} (t+j)^2.$$

21. For what values of n does there exist a set of n integers such that all the $n(n-1)/2$ sums of the integers taken two at a time shall be squares? (P. Erdõs)

Example: $\{-4878, 4978, 6903, 12978, 31122\}$.

22. If p and q are odd primes with $p = 2q + 1$, then $(q+1)$ is a primitive root of p if $q \equiv 1 \pmod 4$; and q is a primitive root of p if $q \equiv 3 \pmod 4$. (A. Wilansky)

23. For which natural numbers n is the l.c.m. of the binomial coefficients $\binom{n}{k}$ equal to the product of all the primes $\leqslant n$? (Kurt Mahler)

24. Show that the only triangular numbers in the Fibonacci sequence are 0, 1, 3, 21, 55 and the only squares therein are 0, 1, 144.

25. Show that all the prime divisors of $n^2 - n - 1$ other than 5 are of the form $10m \pm 1$.

26. $(2k + 1)$ is a prime if and only if

$$\binom{2k-m}{n} \equiv (-1)^n \binom{m+n}{n} \pmod{2k+1}. \qquad \text{(S. Pigntaro)}$$

27. The Erdös–Straus conjecture is that for each $n > 1$, $4/n$ can be written as a sum of at most three fractions of the type $1/x$ with x an integer. Try if you can prove it.
28. Prove that $\sigma(n)\phi(n) \leqslant n^2$.
29. Show that $n(n + 1)/2$ can be represented as a sum of two positive triangular numbers if and only if $n^2 + (n + 1)^2$ is a composite number.
30. For any natural number n, the sequence

$$f_0(n) = n, \quad f_{2k+1}(n) = \sigma(f_{2k}(n)), \quad f_{2k}(n) = \phi(f_{2k-1}(n))$$

is periodic. Find if this conjecture is true. (Leon Alaoglu; P. Erdös)
31. An Erdös–Sierpinski problem: Is $\sigma(n) = \sigma(n + 1)$ infinitely often?
32. Show that the numbers

$$\frac{(4k+1)^{4k+1} - 1}{4k} \quad \text{and} \quad \frac{(4k+3)^{4k+3} + 1}{4k+4}$$

are composite for $k = 1, 2, 3, \ldots$.
33. Show that the number of directed partitions of n is 2^{n-1}.
34. Find if the conjecture in question 30 holds if ϕ is replaced by d (the number of divisors function) therein.

SOME INTERESTING FACTS OF NUMBER THEORY

1. Every integer >45 is the sum of distinct primes each of which is >11. (R. E. Dressler)
2. Every set of seven consecutive numbers >36 contains a multiple of some prime >41.
3. All permutations of the primes 113; 199; 337 are primes. List four digited primes with this property.
4. The largest twin-primes presently known are $76.3^{139} \pm 1$. (H. C. Williams; C. R. Zarnke)
5. The largest known prime is $2^{44497} - 1$ (a number with 13395 digits). (communicated by D. H. Lehmer)
6. For $n > 1$, there is always a prime between $3n$ and $4n$. (D. Hanson)
7. For $n > 6$, the $(n + 1)$th prime p_{n+1} is less than $2p_n - 10$. (R. E. Dressler)
8. The forms $x^2 + x + 41$ and $2x^2 - 199$ have an exceptionally high density of primes. (E. Karst)
9. $\phi(5186) = \phi(5187) = \phi(5188)$; $\phi(72) = \phi(78) = \phi(84) = \phi(90)$. (M. Lal; P. Gillard)
10. 17163 is the largest number which cannot be expressed as a sum of squares of distinct primes. 128 is the largest number which cannot be expressed as a sum of distinct squares. (R. E. Dressler; L. Pigno; R. Young)
11. Every sufficiently large even number can be expressed either as $p_1 + p_2$ or as $p_1 + p_2 p_3$ where p's are primes. (Ching Jun Ch'en)
12. 2 is not a primitive root of any prime of the form $3.2^n + 1$ with the exception of 13. (S. W. Golomb)
13. The smallest biperiodic square is

$$(36363636364)^2 = 13223140496\ 13223140496.$$

14. $30^4 + 120^4 + 272^4 + 315^4 = 353^4$.

K. Rose and S. Brudno (Math. Comp. 27 (1973), 491–494) give 82 sets of this type.
15. There are only four numbers: 1, 2, 145, 40585 which are equal to the sums of the factorials of their digits. Thus

$$145 = 1! + 4! + 5!.$$ (George D. Poole)

16. 1729 is the smallest number which can be expressed as a sum of two cubes in two different ways. (Ramanujan)

$$1729 = 10^3 + 9^3 = 12^3 + 1^3.$$

17. 1301 appears to be the only number which cannot be expressed as a sum of less than 13 cubes of primes (1 included among primes).

$$1301 = 3.7^3 + 2.5^3 + 2.2^3 + 6.1^3.$$ (H. Gupta)

18. The Hardy–Ramanujan formula gave

$p(14031) = 92$ 85303 04759 09931 69434 85156 67127
75089 29160 56358 46500 54568 28164
58081 50403 46756 75123 95895 59113
47418 88383 22063 43272 91599 91345
00745

(D. H. Lehmer)

BIBLIOGRAPHY

BOOKS

George E. Andrews:
Number Theory, W. B. Saunders Company, Philadelphia 1971.
Encyclopedia of Mathematics and its Applications – The Theory of Partitions, Addison-Wesley, Reading, Massachusetts 1976.
On the general Rogers–Ramanujan theorem, American Mathematical Society, Providence, R.I. 1974.

Albert H. Beiler:
Recreations in the Theory of Numbers, Dover Publications, New York 1964.

C. Berg:
Principles of Combinatorics, Academic Press, New York 1971.

Z. I. Borevich and I. R. Shafarevich:
Number Theory, Academic Press, New York 1966.

K. Chandrasekharan:
Introduction to Analytic Number Theory, Springer-Verlag, New York 1968.

H. H. Crapo and G.-C. Rota:
On the foundations of Combinatorial Theory – Combinatorial Geometries, M.I.T. Press, Cambridge, Mass. 1970.

H. Davenport:
The Higher Arithmetic, Hutchinson's University Library, London W.1. 1952.
Multiplicative Number Theory, Markham, Chicago, Ill. 1967.

L. E. Dickson:
History of the Theory of Numbers, Chelsea, New York 1966.

Paul Erdös:
Problems and results on Combinatorial Number Theory, North Holland, Amsterdam, 1973.

Albert Gloden:
Mehrgradige Gleichungen, P. Noordhoff N. V., Verlagsbuchhandlung, Groningen 1944.

Emil Grosswald:
Topics from the Theory of Numbers, The Macmillan Company, New York 1966.

H. Halberstam and H.-E. Richert:
Sieve Methods, Academic Press, New York 1974.

Marshall Hall, Jr.:
Combinatorial Theory, Blaisdell, Waltham, Mass. 1967.

G. H. Hardy:
Ramanujan, Chelsea, New York 1940.

G. H. Hardy; P. V. Seshu Aiyar and B. M. Wilson:
Collected Papers of Srinivasa Ramanujan, Chelsea, New York 1962.

G. H. Hardy and E. M. Wright:
An Introduction to the Theory of Numbers, The Clarendon Press, Oxford, Fourth Edition 1960.

C. Jordan:
Calculus of Finite Differences, Chelsea, New York 1965.

D. H. Lehmer:
Guide to Tables in the Theory of Numbers, The National Research Council, National Academy of Sciences, Washington, D.C. 1941.

William Judson LeVeque:
Topics in Number Theory Vols. I & II, Addison Wesley, Reading, Mass. 1956.

C. L. Liu:
Introduction to Combinatorial Mathematics, McGraw-Hill, New York 1968.

Percy A. MacMahon:
Combinatorial Analysis, Chelsea, New York 1960.

L. J. Mordell:
Diophantine Equations, Academic Press, London 1969.

Tryve Nagell:
Introduction to Number Theory, Wiley, New York 1951.

Guiseppe Palama':
I problemi di Escott-Tarry e di Prouhet-Tarry, Editori · Pellerano-Del Gaudio · Napoli 1956.

Hans Rademacher:
Lectures on Elementary Number Theory, Blaisdell, New York 1964.
Topics in Analytic Number Theory, Springer-Verlag, New York 1973.

Hans Rademacher and Emil Grosswald:
Dedekind Sums, The Mathematical Association of America, Washington, D.C. 1972.

John Riordan:
An Introduction to Combinatorial Analysis, Wiley, New York 1958.
Combinatorial Identities, Wiley, New York 1968.

Ivan Niven and Herbert S. Zuckerman:
An Introduction to the Theory of Numbers, Wiley, New York 1960.

H. J. Ryser:
Combinatorial Mathematics, Wiley, New York 1963.

I. M. Vinogradov:
Elements of Number Theory, Dover Publications, New York 1954.

W. A. Whitworth:
Choice and Chance, Hafner, New York 1958.

PAPERS – I

Abramson, Morton.

1) Some generalized Eulerian numbers. Proc. Fifth South-eastern Conf. on Combinatorics, Graph Theory and Computing. Congressus Numerantium, No. X, *Utilitas Math.*, Winnipeg, Man., 1974, pp. 159–172.
2) Enumeration of sequences by levels and rises. *Discrete Math.* **12** (1975), 101–112.
3) A simple solution of Simon Newcomb's problem. *J. Combinatorial Theory Ser* **A 18** (1975), 223–225.
4) Sequences by number of *w*-rises. *Canad. Math. Bull.* **18** (1975), 317–319.

5) A note on permutations with fixed pattern. *J. Combinatorial Theory Ser.* **A 19** (1975), 237–239.

Alderson, Helen Popova.
1) On the quintic character of 2. *Mathematika* **11** (1964), 125–130.

Alladi, Krishnaswamy.
1) A Farey sequence of Fibonacci numbers. *Fibonacci Quart* **13** (1975), 1–10.
2) A rapid method to form Farey–Fibonacci fractions. *Fibonacci Quart.* **13** (1975), 31–32.

Alter, Ronald; Kubota, K. K.
1) Prime and prime power divisibility of Catalan numbers. *J. Combinatorial Theory Ser.* **A 15** (1973), 243–256.

Alter, Ronald; Curtz, Thaddeus B.
1) On binary non-associative products and their relation to a classical problem of Euler. *Comment. Math. Prace Mat.* **17** (1973), 1–8.

Bambah, R. P.
1) On complete primitive residue sets. *Bull. Calcutta Math. Soc.* **38** (1946), 113–116.

——; Chowla, S.
1) On numbers which can be expressed as a sum of two squares. *Proc. Nat. Inst. Sci. India* **13** (1947), 101–103.

Barnes, C. W.
1) Primes of the form $4n + 1$ which can be written as the sum of two squares of consecutive integers. *J. Elisha Mitchell Sci. Soc.* **89** (1973), 226–227.

Barnett, I. A.; Mendel, C. W.
1) On equal sums of squares. *Amer. Math. Monthly* **49** (1942), 157–170.

Bateman, Paul T.
1) On the representations of a number as sum of three squares. *Trans. Amer. Math. Soc.* **71** (1951), 70–101.

Bauer, M.
1) Zur Theorie der identischen Kongruenzen. *J. London Math. Soc.* **15** (1940), 82–84.

Bell, E. T.
1) Note on a certain type of Diophantine system. *Bull. Amer. Math. Soc.* **47** (1941), 155–159.

Beiler, A. H.
1) A peculiar property of the primitive roots of 13. *Amer. Math. Monthly* **48** (1941), 185–187.

Bhaskaran, M.
1) On some theorems on congruence. *Math. Student* **25** (1957), 40–41.

Brandt, Heinrich.
1) Über das quadratische Reziprozitätsgesetz. *Ber. Verh. Sächs. Akad. Wiss. Leipzig Math. Nat. Kl.* **99** (1951), no. 1, 17pp.

Browkin, J.
1) Certain property of triangular numbers. *Wiadom. Mat.* (2) **2** (1959) 253–255.

de Bruijn, N. G.; Zaring, W. M.
1) On invariants of G.C.D. algorithms. *Nieuw Arch. Wiskunde* (3) **1** (1953), 105–112.

Burde, Klaus.
1) Ein rationales biquadratisches Reziprozitätsgesetz. *J. Reine Angew. Math.* **235** (1969), 175–184.

Carlitz, L.
1) Note on a theorem of Glaisher. *J. London Math. Soc.* **28** (1953), 245–246.
2) A theorem on congruences. *J. Indian Math. Soc.* (N.S.) **17** (1953), 43–45.
3) On the representation of an integer as the sum of twenty-four squares. *Nederl. Akad. Wetensch. Proc. Ser.* **A 58** = *Indag. Math.* **17** (1955), 504–506.
4) A sum connected with quadratic residues. *Nagoya Math. J.* **10** (1956), 1–7.
5) Sets of primitive roots. *Compositio Math.* **13** (1956), 65–70.
6) Note on sums of four and six squares. *Proc. Amer. Math. Soc.* **8** (1957), 120–124.
7) A note on Gauss' first proof of the quadratic reciprocity theorem. *Proc. Amer. Math. Soc.* **11** (1960), 563–565.
8) A note on the generalized Wilson's theorem. *Amer. Math. Monthly* **68** (1961), 251–253.
9) An extension of the congruences of Bauer and Lubelski. *Scripta Math.* **27** (1964), 213–222 (1965).
10) A short proof of Jacobi's four square theorem. *Proc. Amer. Math. Soc.* **17** (1966), 768–769.
11) A theorem of Schemmel. *Math. Mag.* **39** (1966), 86–87.
12) Note on numbers of Jordan and Ward. *Duke Math. J.* **38** (1971), 783–790.
13) A reciprocity and four-term relation for generalized Dedekind sums. *Nederl. Akad. Wetensch. Proc. Ser.* **A 77** = *Indag. Math.* **36** (1974), 413–422.
14) Enumeration of permutations by rises and cycle structure. *J. Reine Angew. Math.* **262/263** (1973), 220–233.
15) Enumeration of a special class of permutations by rises. *Univ. Beograd. Publ. Elektrotehn. Fak. Ser. Mat. Fiz. no.* **412–460** (1973), 189–196.
16) Generating functions for a special class of permutations. *Proc. Amer. Math. Soc.* **47** (1975), 251–256.
17) Some sums involving fractional parts. *Amer. Math. Monthly* **82** (1975), 264–269.
18) Permutations and sequences. *Advances in Math.* **14** (1974), 92–120.
19) Permutations, sequences and special functions. *SIAM Rev.* **17** (1975), 298–322.
20) The reciprocity theorem for Dedekind–Rademacher sums. *Acta Arith.* **29** (1976), 309–313.
21) Some sums containing the greatest integer function. *Fibonacci Quart.* **15** (1977), 78–84.

——; Scoville, R.
1) Enumeration of permutations by rises, falls, rising maxima and falling maxima. *Acta Math. Acad. Sci. Hungar.* **25** (1974), 269–277.

——; Scoville, Richard; Vaugham, Teresa.
1) Enumeration of pairs of permutations and sequences. *Bull. Amer. Math. Soc.* **80** (1974), 881–884.

Chang, Teh-Hsien.
1) Lösung der Kongruenz $x^2 \equiv a \pmod p$ nach einen Primzahl modul $p = 4n + 1$. *Math. Nachr.* **22** (1960), 136–142.

Chidambaraswamy, J.
1) Totients with respect to a polynomial. *Indian J. Pure Appl. Math.* **5** (1974), 345–346.

2) On a conjecture of Makowski. *Math. Student* **31** (1963), 5–6 (1964).

Chowla, S.
1) Proof of a theorem of Lerch and P. Kesava Menon. *Math. Student* **15** (1947), 4 (1948).

——; Nathanson, M. B.
1) Mellin's formula and some combinatorial identities. *Monatsh. Math.* **81** (1976), 261–265.

——; Chowla, P.
1) On the integer points on some special hyper-elliptic curves over a finite field. *J. Number Theory* **8** (1976), 280–281.

Comtet, Louis; Fiolet, Michel.
1) Sur le mode des nombres de Stirling de seconde espèce. *C.R. Acad. Sci. Paris Ser.* **A-B 280** (1975) Ai, A1263–A1265.

Corrádi, Keresztély; Kátai, Imre.
1) On the theory of quadratic residues (Hungarian). *Mat. Lapok* **18** (1967), 75–81.

Cosgrave, J. B.
1) A new proof of Wilson's theorem. *Math. Gaz.* **51** (1967), 47–49.

Daykin, D. E.
1) An addition algorithm for greatest common divisor. *Fibonacci Quart.* **8** (1970), 347–349.

Dumont, Dominique.
1) Interprétations combinatoires des nombres de Genocchi. *Duke Math. J.* **41** (1974), 305–318.

Erdös, Paul.
1) On the least primitive root of a prime. *Bull. Amer. Math. Soc.* **51** (1945), 131–132.
2) On the converse of Fermat's theorem. *Amer. Math. Monthly* **56** (1949), 623–624.
3) Some remarks on Euler's ϕ function. *Acta Arith.* **4** (1958), 10–19.
4) On some divisibility properties of $\binom{2n}{n}$. *Canad. Math. Bull.* **7** (1964), 513–518.
5) Problems and results on number-theoretic properties of consecutive integers and related questions. Proc. Fifth Manitoba Conf. on numerical Math., *Utilitas Math.* Publ. Winnipeg, Man. 1976.

——; Graham, R. L.
1) On a linear Diophantine problem of Frobenius. *Acta Arith.* **21** (1972), 399–408.

——; Selfridge, J. L.
1) The product of consecutive integers is never a power. *Illinois J. Math.* **19** (1975), 292–301.

Ernvall, Reijo.
1) A formula for the least prime greater than a given integer. *Elem. Math.* **30** (1975), 13–14.

Elliot, P. D. T. A.
1) Some notes on k-th power residues. *Acta Arith.* **14** (1967/68), 153–162.

Estermann, T.
1) On the representations of a number as a sum of three squares. *Proc. London Math. Soc.* (3) **9** (1959), 575–594.

Fielder, Daniel C.
1) Generation of Stirling numbers by means of special partitions of numbers. *Fibonacci Quart.* **6** (1968), 1–10.

Foulkes, H. O.
1) Enumeration of permutations with prescribed up-down and inversion sequences. *Discrete Math.* **15** (1976), 235–252.
2) Tangent and secant numbers and representations of symmetric groups. *Discrete Math.* **15** (1976), 311–324.

Fraenkel, Aviezri S.
1) New proof of the Chinese Remainder Theorem. *Proc. Amer. Math. Soc.* **14** (1963), 790–791.

Fröberg, Carl-Erik.
1) On sums and products of quadratic residues. *Nordisk Tidskr Informations-behandling* (BIT) **11** (1971), 389–398.

Furquim de Almeida, F.
1) The law of quadratic reciprocity. *Bol. Soc. Mat. São Paulo* **3** (1948) 3–8 (1951).

Gandhi, J. M.
1) Congruences for $\sigma(n)$. *Proc. Nat. Acad. Sci. India Sect.* **A 37** (1967), 193–194.

Gangadharan, K. S.
1) On some identities and a generalization of Wilson's theorem. *Math. Student* **12** (1945), 75–77.

Gerst, I.
1) On the theory of nth power residues and a conjecture of Kronecker. *Acta Arith.* **17** (1970), 121–139.

Gloden, A.
1) Table des solutions de la congruence $X^4 + 1 \equiv 0 \pmod p$ pour $800000 < p < 1000000$. Chez l'auteur, Luxembourg 1959. (He has numerous other tables of the same type.)

Goldman, Jay R.
1) An identity for fixed points of permutations. *Aequationes Math.* **13** (1975), 155–156.

Golomb, Solomon W.
1) Formulas for the next prime. *Pacific J. Math.* **63** (1976), 401–404.

Goodman, A. W.; Zaring, W. M.
1) Euclid's algorithm and the least remainder algorithm. *Amer. Math. Monthly* **59** (1952), 156–159.

Grimson, R. C.
1) Formulas involving the greatest integer function and some applications. *Delta* (*Waukesha*) **3** (1972–73), 18–32.
2) Reciprocity theorem for Dedekind sums. *Amer. Math. Monthly* **81** (1974), 747–749.

Grossman, Edward H.
1) On the prime ideal divisors of $(a^n - b^n)$. *Pacific J. Math.* **54** (1974), 73–83.

Grosswald, Emil.
1) Contribution to the theory of Euler's function $\phi(x)$. *Bull. Amer. Math. Soc.* **79** (1973), 337–341.

Gupta, Hansraj.
1) A problem in Diophantine Analysis. *Amer. J. Math.* **56** (1934), 269–274.
2) Decompositions into squares of primes. *Proc. Indian Acad. Sci.* **1** (1935), 789–794.
3) Two new perfect numbers. *Amer. Math. Monthly* **42** (1935), 163–164.
4) On linear quotient sequences. *Math. Student* **3** (1935), 132–137.
5) On sums of powers. *Proc. Indian Acad. Sci.* **4** (1936), 571–574.
6) On the Diophantine equation $m^2 = n! + 1$. *Amer. Math. Monthly* **43** (1936), 32–34.
7) Minimum partitions into specified parts. *Amer. J. Math.* **58** (1936), 573–576.
8) Decompositions into cubes of primes. *Proc. Indian Acad. Sci.* **4** (1936), 216–221.
9) On a conjecture of Chowla. *Proc. Indian Acad. Sci.* **5** (1937), 381–384.
10) Decompositions into cubes of primes. *Tôhoku Math. J.* **43** (1937), 11–16.
11) Waring's theorem for powers of primes. *J. Indian Math. Soc.* **3** (1938), 136–145.
12) On a problem of arrangements. *Proc. Indian Acad. Sci.* **9** (1938), 396–398.
13) A note on some Diophantine equations. *Math. Student* **7** (1938), 29–30.
14) Waring's problem for powers of primes. *J. Indian Math. Soc.* **4** (1940), 71–79.
15) On the absolute weight of an integer. *Proc. Indian Acad. Sci.* **12** (1940), 60–62.
16) Some idiosyncratic numbers of Ramanujan. *Proc. Indian Acad. Sci.* **13** (1941), 519–520.
17) On numbers of the form $4^a(8b + 7)$. *J. Indian Math. Soc.* **5** (1941), 192–202.
18) On the class-numbers of binary quadratic forms. *Univ. Nac. Tucuman, Revista* **A 3** (1942), 283–299.
19) Cylindrical projection and rolling. *Math. Student* **10** (1942), 175–176.
20) On residue chains. *Math. Student* **11** (1943), 54–55.
21) A formula for $L(n)$. *J. Indian Math. Soc.* **7** (1943), 68–71.
22) Congruence properties of $\sigma(n)$. *Math. Student* **13** (1945), 25–29.
23) On nth power residues. *Quart. J. Math. Oxford* **18** (1947), 253–256.
24) A table of values of $N_3(t)$. *Proc. Nat. Inst. Sci. India* **13** (1947), 35–63.
25) On Kummer's identity in combinatory functions. *Math. Student* **15** (1947), 93–95.
26) On $N_q(r)$ in the Tarry–Escott problem. *Proc. Nat. Inst. Sci. India* **14** (1948), 335–336.
27) On a conjecture of Miller. *J. Indian Math. Soc.* **13** (1949), 85–90.
28) A solution of the Tarry–Escott problem of degree r. *Proc. Nat. Inst. Sci. India* **15** (1949), 37–39.
29) Tables of distributions. *Res. Bull. Panjab Univ. No.* **2** (1950), 13–44. (Values of Stirling's numbers of the second kind.)
30) On a problem of Erdös. *Amer. Math. Monthly* **57** (1950), 326–329.
31) A note on sums of powers. *Math. Student* **19** (1951), 117.
32) A table of values of $N_2(t)$. *Res. Bull. Panjab Univ.* **No. 20** (1951), 13–93.
33) On triangular numbers in arithmetical progression. *Math. Student* **22** (1954), 141–143.
34) Some properties of quadratic residues. *Math. Student* **23** (1955), 105–107.
35) A congruence property of Euler's ϕ-function. *J. London Math. Soc.* **39** (1964), 303–306.
36) Classes of divisors modulo 24. *Proc. Nat. Inst. Sci. India* **30** (1964), 758–766.

37) An identity. *Res. Bull. Panjab Univ.* **15** (1964), 347–349.
38) A system of equations having no non-trivial solutions. *J. Res. Nat. Bur. Standards*, **71B** (1966), 181–182.
39) A sum involving the Möbius function. *Proc. Amer. Math. Soc.* **19** (1968), 445–447.
40) Number of topologies on a finite set. *Res. Bull. Panjab Univ.* **19** (1968), 231–241.
41) An arithmetical sum. *Indian J. Math.* **10** (1968), 83–86.
42) Enumeration of symmetric matrices. *Duke Math. J.* **35** (1968), 653–660.
43) On a problem in parity. *Indian J. Math.* **11** (1969), 157–163.
44) Three combinatorial problems. *Res. Bull. Panjab Univ.* **20** (1969), 443–448.
45) Meanings of non-associative expressions. *Res. Bull. Panjab Univ.* **21** (1970), 255–256.
46) Chains of quadratic residues. *Math. Comp.* **25** (1971), 379–382.
47) On the enumeration of symmetric matrices. *Duke Math. J.* **38** (1971), 709–710.
48) A combinatorial identity. *Indian J. Math.* **13** (1971), 139–140.
49) The Legendre and Jacobi symbols for k-ic residues. *J. Number Theory* **4** (1972), 219–222.
50) A problem in permutations and Stirling's numbers. *Math. Student* **39** (1971), 341–345.
51) Frequency distribution of digits over the first N natural numbers. *Math. Student* **41** (1973), 447–448.
52) The combinatorial recurrence. *Indian J. Pure Appl. Math.* **4** (1973), 529–532.
53) A restricted Diophantine equation. *J. Indian Math. Soc.* **37** (1973), 71–77.
54) Product decomposition of complete residue sets. *Indian J. Math.* **15** (1973), 191–195.
55) An algorithm for finding the prime divisors of $\binom{n}{k}$. *Univ. Beograd. Publ. Elektrotehn. Fak. Ser. Mat. Fiz. no.* **512** (1975), 77–83.
56) A direct method of obtaining Farey–Fibonacci sequences. *Fibonacci Quart.* **14** (1976), 389–391.
57) Ramanujan's ternary quadratic form $x^2 + y^2 + 10z^2$. *Res. Bull. Panjab Univ.* **24** (1973), 57 (1977).
58) On permutation cubes and Latin cubes. *Indian J. Pure Appl. Math.* **5** (1974), 1003–1021 (1978).
59) Primary factorizations of complete residue sets. *Indian J. Pure Appl. Math.* **5** (1974), 1085–1092 (1978).
60) Some sequences with distinct sums. *Indian J. Pure Appl. Math.* **5** (1974), 1093–1109 (1978).

——; Khare, S. P.
1) On $\binom{k^2}{k}$ and the product of the first k primes. *Univ. Beograd. Publ. Elektrotehn. Fak. Ser. Mat. Fiz. no.* 577–598 (1977), 25–29.

——; Cheema, Mohinder Singh; Gupta, Om Prakash.
1) On Möbius means. *Res. Bull. Panjab Univ.* **No. 42** (1954), 1–16.

——; Vaidya, A. M.
1) The number of representations of a number as a sum of two squares. *Amer. Math. Monthly* **70** (1963), 1081–1082.

——; Anand, Harsh; Dumir, Vishwa Chander.
1) A combinatorial distribution problem. *Duke Math. J.* **33** (1966), 757–770.

——; Srinivasan, S.

1) Cycles of quadratic congruences. *Res. Bull. Panjab Univ.* **22** (1971), 401–404.

2) The number of 3 by 3 magic matrices. *Res. Bull. Panjab Univ.* **22** (1971), 525–526.

——; Nath, G. Baikunth.

1) Enumeration of stochastic cubes. *Indian J. Pure Appl. Math.* **4** (1973), 545–567.

——; Erdös, P.; Khare, S. P.

1) On the number of distinct prime divisors of $\binom{n}{k}$. *Utilitas Math.* **10** (1976), 229–240.

——; Bhattacharjya, G.

1) A sum involving the greatest integer function. *Res. Bull. Panjab Univ.* **24** (1973), 51–53. (1977).

Guy, Richard K.; Hayashi, Elmer K.

1) Factoring integers whose digits are all ones. *Math. Mag.* **49** (1976), 19–22.

Habicht, Walter.

1) Ein elementares Beweis des kubischen Reziprozitätsgesetzes. *Math. Ann.* **139** (1960), 343–365.

Hagis, Peter; Lord, Graham.

1) Unitary harmonic numbers. *Proc. Amer. Math. Soc.* **51** (1975), 1–7.

Halter-Koch, Franz.

1) Kriterien zum 8. Potenzcharakter der Reste 3, 5 und 7. *Math. Nachr.* **44** (1970), 129–144.

Hansen, Rodney T.

1) Greatest integer function identities. *New Zealand Math. Mag.* **12** (1975), 23–26.

Hanzlik, Stefan.

1) On the values of Euler's function. *Acta Fac. Rerum Natur. Univ./ Comenian Math. Publ.* **25** (1971), 85–92.

Harris, V. C.; Subbarao, M. V.

1) Congruence properties of $\sigma_r(N)$. *Pacific J. Math.* **12** (1962), 925–928.

Hartig, K.; Surányi, J.

1) Combinatorial and geometrical investigations in elementary number theory. *Period. Math. Hungar.* **6** (1975), 235–240.

Haselgrove, C. B.

1) A disproof of a conjecture of Pólya. *Mathematika* **5** (1958), 141–145.

Hausman, Miriam.

1) Primitive roots satisfying a co-prime condition. *Amer. Math. Monthly* **83** (1976), 720–723.

Hayashi, Heima.

1) A simple proof of Eisenstein's (cubic) reciprocity law. *Mem. Fac. Sci. Kyushu Univ. Ser.* **A 28** (1974), 93–99.

Hoggatt, Verner E., Jr.; Alladi, Krishnaswami.

1) Generalized Fibonacci tiling. *Fibonacci Quart.* **13** (1975), 137–145.

Hua, Loo-keng.
1) On the least primitive root of a prime. *Bull. Amer. Math. Soc.* **48** (1942), 726–730.

Hudson, Richard H.
1) On the distribution of the kth power nonresidues. *Duke Math. J.* **39** (1972), 85–88.

Jackson, D. M.; Reilly, J. W.
1) Permutations with a prescribed number of p-runs. *Ars Combinatoria* **1** (1976), 297–305.

Kelisky, R. P.
1) Concerning the Euclidean algorithm. *Fibonacci Quart.* **3** (1965), 219–223.

Kelly, John B.
1) A characteristic property of quadratic residues. *Proc. Amer. Math. Soc.* **5** (1954), 38–46.

Kesava, Menon P.
1) An enumeration theorem for symmetric groups. *Math. Student* **39** (1971), 421–423.

Kleiman, H.
1) A note on the Tarry–Escott problem. *J. Reine Angew. Math.* **278/279** (1975), 48–51.

Kloosterman, H. D.
1) The law of quadratic reciprocity. *Nederl. Akad. Wetensch. Proc. Ser.* **A 68** = *Indag. Math.* **27** (1965), 163–164.

Knopfmacher, J.
1) A note on perfect numbers. *Math. Gaz.* **44**(1960), 45.

Koschmieder, L.
1) Zu Eisensteins transzendentem Beweis des quadratischen Reziprozitätsgesetzes. *Comment. Math. Helv.* **37** (1962/63), 235–239.

Kraitchik, Maurice.
1) On certain rational cuboids. *Scripta Math.* **11** (1945), 317–326.

Krajňáková, Dorota.
1) Remark on the theory of power residues (mod p^a). *Mat.-Fyz. Časopis Slovensk. Akad. Vied* **4** (1954), 212–217.

Kudaev, M. B.
1) From the problems of prime number theory. *Colloq. Math.* **24** (1971/72), 121–123.

Lal, M.; Blundon, W. J.
1) Solutions of the Diophantine equations $x^2 + y^2 = l^2, y^2 + z^2 = m^2, z^2 + x^2 = n^2$. *Math. Comp.* **20** (1966), 144–147.

Lehmer, D. H.
1) Multisectioned moments of Stirling numbers of the second kind. *J. Combinatorial Theory Ser.* **A 15** (1973), 210–224.
2) A low energy proof of the reciprocity law. *Amer. Math. Monthly* **64** (1957), 103–106.

Lehmer, Emma.
1) The quintic character of 2 and 3. *Duke Math. J.* **18** (1951), 11–18.
2) Criteria for cubic and quartic residuacity. *Mathematika* **5** (1958), 20–29.

Leonard, Philip A.
1) On factoring quartics (mod p), *J. Number Theory* **1** (1969), 113–115.

——; Mortimer, Brian C.; Williams, Kenneth S.
1) The eleventh power character of 2. *J. Reine. Angew. Math.* **286/287** (1976), 213–222.

——; Williams, Kenneth S.
1) The septic character of 2, 3, 5 and 7. *Pacific J. Math.* **52** (1974), 143–147.

Lewin, Mordechai.
1) An algorithm for a solution of a problem of Frobenius. *J. Reine Angew. Math.* **276** (1975), 68–82.

Longo, Carmelo.
1) Sui residui n^{ici} d'un modulo primo p. *Atti Accad. Ligure* **2** (1942), 83–86.

Makowski, A.
1) On the equation $13^x - 3^y = 10$. *Math. Student* **28** (1960), 87 (1962).
2) Remark on a conjecture of Erdös on binomial coefficients. *Math. Comp.* **24** (1970), 705.
3) Remark on multiplicative functions. *Elem. Math.* **27** (1972), 132–133.

Maxfield, John; Maxfield, Margaret.
1) The existence of integers less than p belonging to ep^{r-1} (mod p^r). *Math. Mag.* **33** (1959/60), 219–220.

Menon, V. V.
1) On the maximum of Stirling's numbers of the second kind. *J. Combinatorial Theory Ser.* **A 15** (1973), 11–24.

Miller, J. C. P.
1) On factorization, with a suggested new approach. *Math. Comp.* **29** (1975), 155–172.

Mordell, L. J.
1) Rational functions representing all residues modulo p. *J. London Math. Soc.* (2) **5** (1972), 166–168; *Proc. Amer. Math. Soc.* **35** (1972), 411–412.
2) The sign of Bernoulli numbers. *Amer. Math. Monthly* **80** (1973), 547–549.

Moser, L.
1) Notes on number theory. V. Insolvability of $\binom{2n}{n} = \binom{2a}{a}\binom{2b}{b}$. *Canad. Math. Bull.* **6** (1963), 167–169.

Muskat, J. B.
1) On the solvability of $x^e \equiv e$ (mod p). *Pacific J. Math.* **14** (1964), 257–260.

Niven, Ivan; Powell, Barry.
1) Primes in certain arithmetical progressions. *Amer. Math. Monthly* **83** (1976), 467–469.

Olds, C. D.
1) On the number of representations of the square of an integer as the sum of three squares. *Amer. J. Math.* **63** (1941), 763–767.

Olson, F. R.
1) An extension of a theorem of Nielsen. *Portugal Math.* **25** (1966), 63–66.

Paasche, Ivan.
1) Drei Noten über Stirlingzahlen. *Univ. Beograd. Publ. Elektrotehn. Fak. Ser. Mat. Fiz.* **331** (1970), 17–21.

Parker, R. V.
1) The complete polynomial grid. *Mat. Vesnik* **10** (25) (1973), 181–203.

Pólya, G.; Alexanderson, G. L.
1) Gaussian binomial coefficients. *Elem. Math.* **26** (1971), 102–109.

Pomerance, Carl.
1) On multiply perfect numbers with a special property. *Pacific J. Math.* **57** (1975), 511–517.

Price, H. E.
1) Solution of the Diophantine equation $ax + by = c$. *Math. Gaz.* **50** (1966), 144–147.

Ramanathan, K. G.
1) Multiplicative arithmetic functions. *J. Indian Math. Soc.* (*N.S.*) **7** (1943), 111–116.
2) Congruence properties of $\sigma_a(n)$. *Math. Student* **13** (1945), 30; *Proc. Indian Acad. Sci.* **A 25** (1947), 314–321.

Rédei, L.
1) Einfacher Beweis des quadratischen Reziprozitätssatzes. *Math. Z.* **54** (1951), 25–26.

Reitan, L.
1) Some considerations on a problem in number theory. *Norsk. Mat. Tidsskr.* **22** (1940), 110–115. (Norwegian).

Riesel, Hans.
1) Note on the congruence $a^{p-1} \equiv 1 \pmod{p^2}$. *Math. Comp.* **18** (1964), 149–150.

Rotkiewicz, André.
1) On the numbers $\phi(a^n \pm b^n)$. *Proc. Amer. Math. Soc.* **12** (1961), 419–421.
2) On pseudo prime numbers. *Publ. Math. Debrecen* **14** (1967), 69–74.
3) On the prime factors of the number $2^{p-1} - 1$. *Glasgow Math J.* **9** (1968), 83–86.
4) On a problem of W. Sièrpiński. *Elem. Math.* **27** (1972), 83–85; *Acta Arith.* **21** (1972), 251–259.

Santos, Bernardo Racamán.
1) Twelve and its totitives. *Math. Mag.* **49** (1976), 239–240.

Satyanarayana, U. V.
1) On the inversion property of the Möbius μ-function. I, II. *Math. Gaz.* **47** (1963), 38–42; **49** (1965), 171–178.

Shah, S. M.
1) On triangular numbers which are sums of two triangular numbers. *Vidya* **9** (1966), 161–163.

Shanks, Daniel; Wrench, John W., Jr.
1) Brun's constant. *Math. Comp.* **28** (1974), 293–299.

Sièrpiński, W.
1) Sur les nombres triangulaires qui sont sommes de deux nombres triangulaires. *Elem. Math.* **17** (1962), 63–65.
2) Remarques sur la décomposition des nombres en sommes des carrés de nombres impairs. *Colloquim Math.* **2** (1949), 52–55.

Singh, Sahib.
1) Bounds of quadratic residues in arithmetical progression. *J. Number Theory* **2** (1970), 162–167.

Singmaster, David.

1) Repeated binomial coefficients and Fibonacci numbers. *Fibonacci Quart.* **13** (1975), 295–298.

Sinha, T. N.

1) A note on a theorem of Lehmer. *J. London Math. Soc.* (2) **4** (1971/72), 541–544. (Tarry–Escott problem).

Skolem, Th.

1) Remarks on proofs by cyclotomic formulas of reciprocity laws for proper residues. *Math. Scand.* **9** (1961), 229–242.
2) A proof of the quadratic law of reciprocity with proofs of two so-called "Ergänzungssätze". *Norske Vid. Selsk. Forh. Trondheim* **34** (1961), 18–24.

Steiger, Franz.

1) Über die Grundlösungen der Gleichung $a^2 + b^2 + c^2 = d^2$. *Elem. Math.* **11** (1956), 105–108.

Steuerwald, Rudolf.

1) Über die Kongruenz $2^{n-1} \equiv 1 \pmod{n}$. *Bayer. Akad. Wiss. Math.-Natur. Kl. S.-B.* **1947**, 177 (1949).

Stewart, C. L.

1) The greatest prime factor of $a^n - b^n$. *Acta Arith.* **26** (1974/75), 427–433.

Subbarao, M. V.

1) The algebra of biquadratic residues. *J. Madras Univ. Sect. B* **30** (1960), 123–131.

Surányi, János.

1) Une remarque sur le théorème fondamental de la théorie des nombres. *Mat. Lapok* **11** (1960), 40–45. (Hungarian).

Tanny, Stephen M.

1) On some numbers related to Bell numbers. *Canad. Bull. Math.* **17** (1974/75), 733–738.
2) Permutations and successions. *J. Combinatorial Theory Ser.* **A 21** (1976), 196–202.

Uchiyama, Saburô.

1) Three primes in arithmetical progression. *Proc. Japan Acad.* **37** (1961), 329–330.

Vaidya, A. M.

1) Some congruence properties of indices. *Vidya* **10** (1967), 179–181.
2) On factorization mod n of the set $\{1, 2, 3, \ldots, n-1\}$. *Indian J. Math.* **11** (1969), 67–74.
3) On representing an integer as a sum of two triangular numbers. *Vidya* **B 15** (1972), 104–105.

Vaidyanathaswamy, R.

1) The Algebra of cubic residues. *J. Indian Math. Soc. (N.S.)* **21** (1957), 57–66. (1958).

Vegh, Emanuel.

1) Arithmetic progressions of primitive roots of a prime. I–III. *J. Reine Angew. Math.* **235** (1969), 185–188; **244** (1970), 108–111; **256** (1972), 130–137.

Vijayaraghavan, T.; Chowla, S.

1) On complete residue sets. *Quart. J. Math. Oxford Ser.* **19** (1948), 193–199.

Vitek, Yehoshua.
1) Bounds for a linear Diophantine problem of Frobenius. I, II. *J. London Math. Soc.* (2) **10** (1975), 79–85; *Canad. J. Math.* **28** (1976), 1280–1288.

Weaver, Milo.
1) Zero integral homogeneous symmetric functions in certain rings. *Amer. Math. Monthly* **63** (1956), 387–391.

Weintraub, Sol.
1) A compact prime listing. *Math. Comp.* **28** (1974), 323–326.

Whiteman, A. L.
1) The sixteenth power residue character of 2. *Canad. J. Math.* **6** (1954), 364–373,

Williams, Ian S.
1) On a problem of Kurt Mahler concerning binomial coefficients. *Bull. Austral. Math. Soc.* **14** (1976), 299–302.

Williams, Kenneth S.
1) Quadratic polynomials with the same residues. *Amer. Math. Monthly* **75** (1968), 969–973.
2) An elementary number theoretic formula. *Math. Student* **35** (1967), 47–50 (1969).
3) A quadratic partition of primes $\equiv 1 (\bmod 7)$. *Math. Comp.* **28** (1974), 1133–1136.
4) On Euler's criterion for cubic non-residues. *Proc. Amer. Math. Soc.* **49** (1975), 277–283.
5) 3 as a ninth power (mod p). *Math. Scand.* **35** (1974), 309–317.
6) Note on the supplement to the law of cubic reciprocity. *Proc. Amer. Math. Soc.* **47** (1975), 333–334.
7) 2 as a ninth power (mod p). *J. Indian Math. Soc.* **39** (1975), 167–172.
8) On Euler's criterion for quintic non-residues. *Pacific J. Math.* **61** (1975), 543–550.
9) Explicit criteria for quintic residuacity. *Math. Comp.* **30** (1976), 847–853.
10) A rational octic reciprocity law. *Pacific J. Math.* **63** (1976), 563–570.
11) On the supplement to the law of biquadratic reciprocity. *Proc. Amer. Math. Soc.* **59** (1976), 147–153.
12) Cubic nonresidues (mod p). *Delta (Waukesha)* **6** (1976), 23–28.
13) The quadratic character of 2 (mod p). *Math. Mag.* **49** (1976), 89–90.

Wilson, Neil Y.
1) Conjectures as to a factor of $2^p \pm 1$. *Edinburgh Math. Notes, no.* **39** (1954), 6–9.

Zachariou, Andreas; Zachariou, Eleni.
1) Perfect, semiperfect and Ore numbers. *Bull. Soc. Math. Greece (N.S.)* **13** (1972), 1–2, 12–22.

Zee, Yun Cheng.
1) Some sufficient conditions for quintic residuacity. *Proc. Amer. Math. Soc.* **54** (1976), 8–10.

PAPERS–II

PARTITIONS

Alder, Henry L.
1) Generalizations of Rogers–Ramanujan identities. *Pacific J. Math.* **4** (1954), 161–168.

2) Partition identities – from Euler to the present. *Amer. Math. Monthly* **76** (1969), 733–746.

——; Muwafi, Amin A.

1) Identities relating the number of partitions into an even and odd number of parts. *Fibonacci Quart.* **13** (1975), 147–149.
2) Proof of Andrews' conjecture on partition identities. *Proc. Amer. Math. Soc.* **22** (1969), 688–689.

Andrews, George E.

1) A simple proof of Jacobi's triple product identity. *Proc. Amer. Math. Soc.* **16** (1965), 333–334.
2) On basic hypergeometric series, mock theta functions, and partitions. *Quart. J. Math. Oxford Ser.* (2) **17** (1966), 64–80.
3) An analytic proof of the Rogers–Ramanujan–Gordon identities. *Amer. J. Math.* **88** (1966), 844–846.
4) On generalizations of Euler's partition theorem. *Michigan Math. J.* **13** (1966), 491–498.
5) A generalization of a partition theorem of MacMahon. *J. Combinatorial Theory* **3** (1967), 100–101.
6) Some new partition theorems. *J. Combinatorial Theory* **2** (1967), 431–436.
7) Partition theorem related to the Rogers–Ramanujan identities. *J. Combinatorial Theory* **2** (1967), 422–430.
8) A generalization of the Göllnitz–Gordon partition theorems. *Proc. Amer. Math. Soc.* **18** (1967), 945–952.
9) On Schur's second partition theorem. *Glasgow Math. J.* **8** (1967), 127–132.
10) On partition functions related to Schur's second partition theorem. *Proc. Amer. Math. Soc.* **19** (1968), 441–444.
11) A new generalization of Schur's second partition theorem. *Acta Arith.* **14** (1967/68), 429–434.
12) On a calculus of partition functions. *Pacific J. Math.* **31** (1969), 555–562.
13) Two theorems of Euler and a general partition theorem. *Proc. Amer. Math. Soc.* **20** (1969), 499–502.
14) A general theorem on partitions with difference conditions. *Amer. J. Math.* **91** (1969), 18–24.
15) Some new partition theorems. *J. Combinatorial Theory* **7** (1969), 262–263.
16) On a partition theorem of Göllnitz and related formulae. *J. Reine Angew. Math.* **236** (1969), 37–42.
17) A generalization of the classical partition theorems. *Trans. Amer. Math. Soc.* **145** (1969), 205–221.
18) On a partition problem of J. J. Sylvester. *J. London Math. Soc.* (2) **2** (1970), 365–376.
19) Lambert series, false theta functions, and partitions. *Proc. Second Chapel Hill Conf.* on Combinatorial Mathematics and its applications (Univ. North Carolina, Chapel Hill, N.C. 1970) pp. 1–11.
20) On a partition problem of H. L. Alder. *Pacific J. Math.* **36** (1971), 279–284.
21) Congruence properties of the m-ary partition function. *J. Number Theory* **3** (1971), 104–110.
22) Sieves for theorems of Euler, Rogers and Ramanujan. *Proc. Conf.* Western Michigan Univ., Kalamazoo, Mich. 1971, 1–20.
23) Two theorems of Gauss and allied identities proved arithmetically. *Pacific J. Math.* **41** (1972), 563–578.
24) Sieves in the theory of partitions. *Amer. J. Math.* **94** (1972), 1214–1230. (An important paper.)

25) A general theory of identities of the Rogers–Ramanujan type. *Bull. Amer. Math. Soc.* **80** (1974), 1033–1052.
26) An analytical generalization of the Rogers–Ramanujan identities for odd moduli. *Proc. Nat. Acad. Sci. U.S.A.* **71** (1974), 4082–4085.
27) On Rogers–Ramanujan type identities related to the modulus 11. *Proc. London Math. Soc.* (3) **30** (1975), 330–346.
28) On identities implying the Rogers–Ramanujan identities. *Houston J. Math.* **2** (1976), 289–298.
29) On the Alder polynomials and a new generalization of the Rogers–Ramanujan identities. *Trans. Amer. Math. Soc.* **204** (1975), 42–64.

——; Subbarao, M. V.; Vidyasagar, M.
1) A family of combinatorial identities. *Canad. Math. Bull.* **15** (1972), 11–18.

Arkin, Joseph.
1) Congruences for the coefficients of the kth powers of a power series. I, II. *Amer. Math. Monthly* **71** (1964), 899–900; **72** (1965), 390–393.
2) An extension of a theorem of Ramanujan. *Amer. Math. Monthly* **73** (1966), 1087–1090.
3) A note on a theorem of Jacobi. *Fibonacci Quart.* **4** (1966), 359–362.
4) Researches on partitions. *Duke Math. J.* **37** (1970), 403–409.
5) Researches on some classical problems. *Proc. Washington State Univ. Conf.* on number theory. Pullman, Wash. 1971.

Atkin, A. O. L.
1) A note on ranks and conjugacy of partitions. *Quart. J. Math. Oxford Ser.* (2) **17** (1966), 335–338.
2) Proof of a conjecture of Ramanujan. *Glasgow Math. J.* **8** (1967), 14–32.

——; Swinnerton-Dyer, P.
1) Some properties of partitions. *Proc. London Math. Soc.* (3) **4** (1954), 84–106.

——; Hussain, S. M.
1) Some properties of partitions. *Trans. Amer. Math. Soc.* **89** (1958), 184–200.

——; O'Brien, J. N.
1) Some properties of $p(n)$ and $c(n)$ modulo powers of 13. *Trans. Amer. Math. Soc.* **126** (1967), 442–459.

Auluck, F. C.
1) On some new types of partitions associated with generalized Ferrers graphs. *Proc. Cambridge Philos. Soc.* **47** (1951), 679–686.
2) On partitions of bipartite numbers. *Proc. Nat. Inst. Sci.* **19** (1953), 313–314.
3) An asymptotic formula for $p_k(n)$. *J. Indian Math. Soc. (N.S.)* **6** (1942), 113–114.

——; Chowla, S.; Gupta, H.
1) On the maximum value of the number of partitions of n into k parts. *J. Indian Math. Soc. (N.S.)* **6** (1942), 105–112.

Bailey, Wilfred Norman.
1) On the simplification of some identities of the Rogers–Ramanujan type. *Proc. London Math. Soc.* (3) **1** (1951), 217–221.
2) A note on two of Ramanujan's formulae. *Quart. J. Math. Oxford Ser.* (2) **3** (1952), 29–31.

Bambah, R. P.
Properties of Ramanujan's function $\tau(n)$ are considered in the following papers:
1) *J. London Math. Soc.* **21** (1946), 91–93;
2) *Bull. Amer. Math. Soc.* **53** (1947), 764–765;
3) *Proc. Nat. Inst. Sci. India* **12** (1946), 431–432;

——; Chowla, S.
4) *Proc. Nat. Inst. Sci. India* **12** (1946), no. 8, 1 p.;
5) *Proc. Nat. Inst. Sci. India* **12** (1946), 433;
6) *Math. Student* **14** (1946), 24–26;
7) *Quart. J. Math. Oxford Ser.* **18** (1947), 122–123;
8) *Bull. Amer. Math. Soc.* **53** (1947), 768–769;
9) *Bull. Amer. Math. Soc.* **53** (1947), 950–955;
10) *J. London Math. Soc.* **22** (1947), 140–147;

——; Chowla, S.; Gupta, H.
11) *Bull. Amer. Math. Soc.* **53** (1947), 766–767;

——; Chowla, S.; Gupta, H.; Lahiri, D. B.
12) *Quart. J. Math. Oxford Ser.* **18** (1947), 101–103.

Bateman, Paul T.; Erdös, P.
1) Monotonicity of partition function. *Mathematika* **3** (1956), 1–14.
2) Partitions into primes. *Publ. Math. Debrecen* **4** (1956), 198–200.

Bender, Edward A.; Knuth, Donald E.
1) Enumeration of plane partitions. *J. Combinatorial Theory Ser.* **A 13** (1972), 40–54.

Berg, Lother.
1) Über eine Differenzengleichung aus der Theorie der partitionen. *Wiss. Z. Univ. Rostock. Math. Nat. Reihe* **5** (1955/56), 269–278.

Blakley, G. R.
1) Combinatorial remarks on partitions of a multipartite number. *Duke Math. J.* **31** (1964), 335–340.

Carlitz, Leonard.
1) Note on some partition identities. *Proc. Amer. Math. Soc.* **4** (1953), 530–534.
2) A note on the Jacobi theta formula. *Bull. Amer. Math. Soc.* **68** (1962), 591–592.
3) A problem in partitions. *Duke Math. J.* **30** (1963), 203–213.
4) Note on some continued fractions of the Rogers–Ramanujan type. *Duke Math. J.* **32** (1965), 713–720.
5) Some identities in combinatorial analysis. *Duke Math. J.* **39** (1972), 31–37.
6) Enumeration of two-line arrays. *Fibonacci Quart.* **11** (1973), 113–130.
7) Restricted compositions. *Fibonacci Quart.* **14** (1976), 254–264.

——; Scoville, Richard.
1) A generating function for triangular partitions. *Math. Comp.* **29** (1975), 67–77.

Cheema, Mohindar Singh.
1) Vector partitions and combinatorial identities. *Math. Comp.* **18** (1964), 414–420.
2) Duality in the theory of partitions. *Res. Bull. Panjab Univ.* **20** (1969), 201–206 (1970).

——; Gordon, Basil.
1) Some remarks on two and three-line partitions. *Duke Math. J.* **31** (1964), 267–273.

——; Gupta, Hansraj.
1) The maxima of $P_r(n_1, n_2)$. *Math. Comp.* **22** (1968), 199–200.

——; Conway, W. E.
1) Numerical investigation of certain asymptotic results in the theory of partitions. *Math. Comp.* **26** (1972), 999–1005.

Churchhouse, R. F.
1) Congruence properties of the binary partition function. *Proc. Cambridge Philos. Soc.* **66** (1969), 371–376.

Connar, Willard G.
1) Partition theorems related to some identities of Rogers and Watson. *Trans. Amer. Math. Soc.* **214** (1975), 95–111.

Dirdal, Gunnar.
1) On restricted *m*-ary partitions. *Math. Scand.* **37** (1975), 51–60.
2) Congruences for *m*-ary partitions. *Math. Scand.* **37** (1975), 76–82.
3) Congruence properties for a class of arithmetical functions. *Math. Scand.* **38** (1976), 247–261.

Dutta, Mahadev.
1) On new partitions of numbers. *Rend. Sem. Mat. Univ. Padova* **25** (1956), 138–143.
2) On new partitions of numbers. *Bull. Calcutta Math. Soc.* **49** (1957) 221–224.

——; Debnath, Lokenath.
1) On new partitions of numbers. II. *Bull. Calcutta Math. Soc.* **51** (1959), 77–78.

Erdös, Paul.
1) On an elementary proof of some asymptotic formulas in the theory of partitions. *Ann. of Math.* (2) **43** (1942), 437–450.
2) On some asymptotic formulas in the theory of partitions. *Bull. Amer. Math. Soc.* **52** (1946), 185–188.

——; Lehner, Joseph.
1) The distribution of the number of summands in the partitions of a positive integer. *Duke Math. J.* **8** (1941), 335–345.

——; Turán, P.
1) On some general problems in the theory of partitions. I. *Acta Arith.* **18** (1971), 53–62.

Ewell, John A.
1) Partition recurrences. *J. Combinatorial Theory Ser.* **A 14** (1973), 125–127.
2) Consequences of a Gaussian identity. *Indian J. Pure Appl. Math.* **5** (1974). 509–512.

Fielder, Daniel C.
1) Enumeration of partitions subject to limitations on size of members. *Fibonacci Quart.* **4** (1966), 209–216.

Fine, Nathan J.
1) Some new results on partitions. *Proc. Nat. Acad. Sci. U.S.A.* **34** (1948), 616–618.

Gandhi, J. M.
1) Congruences for $p_r(n)$ and Ramanujan's τ function. *Amer. Math. Monthly* **70** (1963), 265–274.
2) Generalization of Ramanujan's congruences $p(5m + 4) \equiv 0 \pmod 5$ and $p(7m + 5) \equiv 0 \pmod 7$. *Monatsh. Math.* **69** (1965), 389–392.
3) Congruences for restricted partitions of a number. *Math. Student* **35** (1967), 69–71 (1969).
4) Congruences for new types of partitions. *Proc. Nat. Inst. Sci. India* Part **A 32** (1966), 141–145.
5) Some congruences for k-line partitions of a number. *Amer. Math. Monthly* **74** (1967), 179–181.

Göllnitz, Heinz.
1) Partitionen mit differenzenbedingungen. *J. Reine Angew. Math.* **225** (1967), 154–190.

Gordon, Basil.
1) A combinatorial generalization of the Rogers–Ramanujan identities. *Amer. J. Math.* **83** (1961), 393–399.
2) Some identities in combinatorial analysis. *Quart. J. Math. Oxford Ser.* (2) **12** (1961), 285–290.
3) Two new representations of the partition function. *Proc. Amer. Math. Soc.* **13** (1962), 869–873.
4) Two theorems on multipartite partitions. *J. London Math. Soc.* **38** (1963), 459–464.
5) Some continued fractions of the Rogers–Ramanujan type. *Duke Math. J.* **32** (1965), 741–748.
6) Notes on plane partitions:
I, II. *J. Combinatorial Theory* **4** (1968), 72–80; 81–89.
III. *Duke Math. J.* **36** (1969), 801–824.
IV. Combinatorics, *Amer. Math. Soc.*, Providence, R.I. 1971.
V. *J. Combinatorial Theory Ser.* **B 11** (1971), 157–168.

Grosswald, E.
1) Some theorems concerning partitions. *Trans. Amer. Math. Soc.* **89** (1958), 113–128. Corrections **95** (1960), 190.
2) Results new and old, in the theory of partitions. (Spanish). *Rev. Un. Mat. Argentina* **20** (1962), 40–57.
3) Elementary proofs in the theory of partitions. *Math. Z.* **81** (1963), 52–61.

Gupta, Hansraj.
1) A table of partitions. *Proc. London Math. Soc.* **39** (1935), 142–149.
2) On partitions of n. *J. London Math. Soc.* **11** (1936), 278–280.
3) A table of partitions (II). *Proc. London Math. Soc.* **42** (1937), 546–549.
4) On a conjecture of Ramanujan. *Proc. Indian Acad. Sci.* **4** (1936), 625–629.
5) A formula in partitions. *J. Indian Math. Soc.* **6** (1942), 115–117.
6) On an asymptotic formula in partitions. *Proc. Indian Acad. Sci.* **16** (1942), 101–102.
7) An inequality in partitions. *J. Bombay Univ.* **11** (1942), 16–18.
8) On the maximum values of $p_k(n)$ and $\pi_k(n)$. *J. Indian Math. Soc.* **7** (1943), 72–75.
9) Congruence properties of $\tau(n)$. *Proc. Benares Math. Soc.* **5** (1943), 17–22.
10) A note on the parity of $p(n)$. *J. Indian Math. Soc.* **10** (1946), 25–29.
11) An asymptotic formula in partitions. *J. Indian Math. Soc.* **10** (1946), 73–76.
12) A congruence property of $\tau(n)$. *Proc. Indian Acad. Sci.* **24** (1946) 441–442.

13) On the vanishing of Ramanujan's function $\tau(n)$. *Current Science* **17** (1948), 180.
14) A generalization of the partition function. *Proc. Nat. Inst. Sci. India* **17** (1951), 231–238.
15) On a generating function in partition theory. *Proc. Nat. Inst. Sci. India* **20** (1954), 582–586.
16) Partitions into distinct primes. *Proc. Nat. Inst. Sci. India* **21** (1955), 185–187.
17) Partitions in general. *Res. Bull. Panjab Univ. no.* **67** (1955), 31–38.
18) Partitions in terms of combinatory functions. *Res. Bull. Panjab Univ. no.* **94** (1956), 153–159.
19) Certain averages connected with partitions. *Res. Bull. Panjab Univ. no.* **124** (1957), 427–430.
20) Partitions of j-partite numbers into k summands. *J. London Math. Soc.* **33** (1958), 403–405.
21) Partitions of j-partite numbers. *Res. Bull. Panjab Univ. no.* **146** (1958), 119–121.
22) Graphic representation of a partition of a j-partite number. *Res. Bull. Panjab Univ.* **10** (1959), 189–196.
23) On the partition of j-partite numbers. *Proc. Nat. Inst. Sci. India* **27** (1961), 579–587.
24) An inequality for $P(N_j, k)$. *Res. Bull. Panjab Univ.* **13** (1962), 173–178.
25) On the coefficients of the powers of Dedekind's modular form. *J. London Math. Soc.* **39** (1964), 433–440.
26) Highly restricted partitions. *J. Res. Nat. Bur. Standards* **73B** (1969), 329–350.
27) Partitions – A survey. *J. Res. Nat. Bur. Standards* **74B** (1970), 1–29.
28) Products of parts in partitions into primes. *Res. Bull. Panjab Univ.* **21** (1970), 251–253.
29) Proof of the Churchhouse conjecture concerning binary partitions. *Proc. Cambridge Philos. Soc.* **70** (1971), 53–56.
30) On Sylvester's theorem in partitions. *Indian J. Pure Appl. Math.* **2** (1971), 740–748.
31) On partitions of n into k summands. *Proc. Edinburgh Math. Soc.* **17** (1971), 337–339.
32) On m-ary partitions. *Proc. Cambridge Philos. Soc.* **71** (1972), 343–345.
33) Two theorems in partitions. *Indian J. Math.* **14** (1972), 7–8.
34) Restricted solid partitions. *J. Combinatorial Theory* **13** (1972), 140–144.
35) A simple proof of the Churchhouse conjecture concerning binary partitions. *Indian J. Pure Appl. Math.* **3** (1972), 791–794.
36) A partition theorem of Subbarao. *Canad. Math. Bull.* **17** (1974), 121–123.
37) A technique in partitions. *Univ. Beograd Publ. Eletrotehn. Fak. Ser. Mat. Fiz. no.* **511** (1975), 73–76.
38) Combinatorial proof of a theorem on partitions into an Even or Odd number of parts. *J. Combinatorial Theory* **21** (1976), 100–103.
39) A direct proof of the Churchhouse conjecture concerning binary partitions. *Indian J. Math.* **18** (1976), 1–6.
40) Partitions embedded in a rectangle. *Utilitas Math.* **10** (1976), 229–240.
41) Partitions of n into divisors of m. *Indian J. Pure Appl. Math.* **6** (1975), 1276–1286.
42) Three-dimensional models of partitions of bipartite numbers. *Indian J. Pure Appl. Math.* **6** (1975), 1287–1308.
43) Magic partitions. *Indian J. Pure Appl. Math.* **6** (1975), 1159–1166.

Gupta, Om Prakash.
1) Partitions into exactly k distinct primes. *Res. Bull. Panjab Univ. no.* **107** (1957), 283–290.
2) Partitions of j-partite numbers. *Proc. Nat. Inst. Sci. India Part* **A 26** (1960), 656–660.

——; Luthra, S.
1) Partitions into primes. *Proc. Nat. Inst. Sci. India Part* **A 21** (1955), 181–184.

Haberzetle, Mary.
1) On some partition functions. *Amer. J. Math.* **63** (1941), 589–599.

Hagis, Peter, Jr.
1) Partitions into odd summands. *Amer. J. Math.* **85** (1963), 213–222.
2) Partitions into odd and unequal parts. *Amer. J. Math.* **86** (1964), 317–329.
3) Some theorems concerning partitions into odd summands. *Amer. J. Math.* **88** (1966), 664–681.

Hall, R. R.; Wild, K.
1) On a problem of Turán concerning partitions. *J. London Math. Soc.* (2) **13** (1976), 472–474.

Haselgrove, C. B.; Temperley, H. N. V.
1) Asymptotic formulae in the theory of partitions. *Proc. Cambridge Philos. Soc.* **50** (1954), 225–241.

Herrman, Oskar.
1) Kongruenzeigenschaften der Partitionenfunktion. *J. Number Theory* **1** (1969), 431–458.

Hickerson, Dean R.
1) Identities relating the number of partitions into an even and odd number of parts. *J. Combinatorial Theory Ser.* **A 15** (1973), 351–353.
2) Recursion-type formulae for partitions into distinct parts. *Fibonacci Quart.* **11** (1973), 307–311.
3) A note on the congruence properties of certain restricted partitions. *Math. Mag.* **48** (1975), 102.

Hirschhorn, M. D.
1) Sylvester's partition theorem and a related result. *Mich. Math. J.* **21**(1974), 133–136.
2) Simple proofs of identities of MacMahon and Jacobi. *Discrete Math.* **16** (1976), 161–162.

——; Loxton, J. H.
1) Congruence properties of the binary partition function. *Math. Proc. Cambridge Philos. Soc.* **78** (1975), 437–442.

Hjelle, T.; Klove, T.
1) Congruence properties and density problems for the Fourier coefficients of modular forms. *Math. Scand.* **23** (1968), 160–166 (1969).

Houten, Lorne.
1) A note on solid partitions. *Acta Arith.* **15** (1968), 71–76.

Hua, Loo-Keng.
1) On the number of partitions of a number into unequal parts. *Trans. Amer. Math. Soc.* **51** (1942), 194–201.

Ingham, A. E.
1) A Tauberian theorem for partitions. *Ann. of Math.* (2) **42** (1941), 1075–1090.

Intrator, Jakub.
1) Partitions. I. *Czechoslovak Math. J.* **18** (93) (1968), 16–24.

Iseki, Kaneshiro.
1) Ein Theorem der Zahlentheorie. *Tôhoku Math. J.* **48** (1941), 6063.

Karpe, Robert.
1) Some relations between combinatorial numbers $p(n)$. (Czech). *Casopis Pest. Mat.* **94** (1969), 108–114.

Kelly, John B.
1) Partitions with equal products. *Proc. Amer. Math. Soc.* **15** (1964), 987–990.

Kempner, A. J.
1) The development of "partitio numerorum" with particular reference to the work of Messrs. Hardy, Littlewood and Ramanujan. I, II. *Amer. Math. Monthly* **30** (1923), 354–369; 416–425.

Klove, Torleiv.
1) Recurrence formulae for the coefficients of modular forms and congruences for the partition function and for the coefficients of $j(\tau)$, $(j(\tau) - 1728)^{1/2}$ and $(j(\tau))^{1/3}$. *Math. Scand.* **23** (1968), 133–159 (1969).
2) Density problems for $p(n)$. *J. London Math. Soc.* (2) **2** (1970), 504–508.

Kolberg, O.
1) Identities involving the partition functions $q(n)$ and $q_0(n)$. *Math. Scand.* **6** (1958), 80–86.
2) Note on the parity of the partition function. *Math. Scand.* **7** (1959), 377–378.
3) Congruences involving the partition function for the moduli 17, 19 and 23. *Univ. Bergen Arbok Naturvit Rekke* 1959, #15, 10 pp. (1960).

Koshti, M. E.
1) Partitions and the method of differences. *Math. Student* **32** (1964), 83–85.

Kruyswijk, D.
1) On some well-known properties of the partition function $p(n)$ and Euler's infinite product. *Nieuw Arch. Wisk.* (2) **23** (1950), 97–107.

Lahiri, D. B.
1) On a type of series involving the partition function with applications to certain congruence relations. *Bull. Calcutta Math. Soc.* **38** (1946), 125–132.
2) Some non-Ramanujan congruence properties of the partition function. *Proc. Nat. Inst. Sci. India* **14** (1948), 337–338.
3) Some restricted partition functions. Congruences modulo 11. *Pacific J. Math.* **38** (1971), 103–116.

Lehmer, D. H.
1) On the remainders and convergence of the series for the partition function. *Trans. Amer. Math. Soc.* **46** (1939), 362–373.
2) A triangular number formula for the partition function. *Scripta Math.* **17** (1951), 17–19.
3) On reciprocally weighted partitions. *Acta Arith.* **21** (1972), 379–388.

Lehner, Joseph.

1) Proof of Ramanujan's partition congruences for the modulus 11^a. *Proc. Amer. Math. Soc.* **1** (1950), 172–181.

Luthra, S. M.

1) Partitions of bipartite numbers when the summands are unequal. *Proc. Nat. Inst. Sci. India Part* **A 23** (1957), 370–376.

MacMahon, P. A.

1) The theory of modular partitions. *Proc. Cambridge Philos. Soc.* **21** (1922–23), 197–204.
2) The parity of $p(n)$, the number of partitions of n, when $n \leqslant 1000$. *J. London Math. Soc.* **1** (1926), 225–226.

Manzur, Hussain S.

1) On the function $c(n) = q_k^e(n) - q_k^o(n)$ and some inequalities in partition theory. *Panjab Univ. J. Math. (Lahore),* **3** (1970), 49–57.

Meinardus, G.

1) Asymptotische aussagen über Partitionen. *Math. Z.* **59** (1954), 388–398.

Metropolis, N. N.; Stein, P. R.

1) An elementary solution to a problem in restricted partitions. *J. Combinatorial Theory* **9** (1970), 365–376.

Mian, A. M.; Chowla, S.

1) The differential equations satisfied by certain functions. *J. Indian Math. Soc. (N.S.)* **8** (1944), 27–28.

Motzkin, Theodor.

1) Ordered and cyclic partitions. *Riveon Lematematika* **1** (1947), 61–67.

Nanda, V. S.

1) Tables of solid partitions. *Proc. Nat. Inst. Sci. India* **19** (1953), 313–314.
2) Bipartite partitions. *Proc. Cambridge Philos. Soc.* **53** (1957), 273–277.

Newman, D. J.

1) The evaluation of the constant in the formula for the number of partitions of *n*. *Amer. J. Math.* **73** (1951), 599–601.
2) A simplified proof of the partition formula. *Michigan Math. J.* **9** (1962), 283–287.

Newman, Morris.

1) The coefficients of certain infinite products. *Proc. Amer. Math. Soc.* **4** (1953), 435–439.
2) Congruences for the coefficients of modular forms and some new congruences for the partition function. *Canad. J. Math.* **9** (1957), 549–552.
3) Further identities and congruences for the coefficients of modular forms. *Canad. J. Math.* **10** (1958), 577–586.
4) Periodicity modulo *m* and divisibility properties of partition function. *Trans. Amer. Math. Soc.* **97** (1960), 225–236.
5) Congruences for the partition function to composite moduli. *Illinois J. Math.* **6** (1962), 59–63.
6) Note on the partition function. *Amer. Math. Monthly* **71** (1964), 1022.

Niebur, Douglas.

1) A formula for Ramanujan's function. *Illinois J. Math.* **19** (1975), 448–449.

Niven, Ivan.

1) On a certain partition function. *Amer. J. Math.* **62** (1940), 353–364.

Oderfeld, J.; Pleszczyńska, E.
1) On some applications of partitions. (Polish; Russian and English summaries). *Zastos. Math.* **6** (1962), 189–198.

Pagni, Plinio.
1) Studio sulle partizioni numeriche. I, II. III. *Period. Mat.* (4) **32** (1954), 172–183; 199–211; 294–301.

Rademacher, Hans
1) On the expansion of the partition function in a series. *Ann. of Math.* (2) **44** (1943), 416–422.
2) On the Selberg formula for $A_k(n)$. *J. Indian Math. Soc. (N.S.)* **21** (1957), 41–55 (1958).

Ramamani, V.; Venkatachaliengar, K.
1) On a partition theorem of Sylvester. *Michigan Math. J.* **19** (1972), 137–140.

Rieger, G. J.
1) Über Partitionen. *Math. Ann.* **138** (1959), 356–362.

Richmond, L. B.
1) Asymptotic relations for partitions. *J. Number Theory* **7** (1975), 389–405.

Robertson, M. M.
1) Partitions of large multipartites. *Amer. J. Math.* **84** (1962), 16–34.
2) Asymptotic formulae for the number of partitions of a multipartite number. *Proc. Edinburgh Math. Soc.* (2) **12** (1960–61); 31–40.
3) Partitions with congruence conditions. *Proc. Amer. Math. Soc.* **57** (1976), 45–49.

Rödseth, Öystein.
1) Some arithmetical properties of m-ary partitions. *Proc. Cambridge Philos. Soc.* **68** (1970), 447–453.

Roth, K. F.; Szekeres, G.
1) Some asymptotic formulae in the theory of partitions. *Quart. J. Math. Oxford Ser.* (2) **5** (1954), 241–249.

Rosselle, D. P.
1) Coefficients associated with the expansion of certain products. *Proc. Amer. Math. Soc.* **45** (1974), 144–150.
2) Restricted k-partite partitions. *Math. Nachr.* **32** (1966), 139–148.

Rushforth, J. M.
1) Congruence properties of the partition function and associated functions. *Proc. Cambridge Philos. Soc.* **48** (1952), 402–413.

Shanks, Daniel.
1) A short proof of an identity of Euler. *Proc. Amer. Math. Soc.* **2** (1951), 747–749.
2) Two theorems of Gauss. *Pacific J. Math.* **8** (1958), 609–612.

Simons, William H.
1) Congruences involving the partition function $p(n)$. *Bull. Amer. Math. Soc.* **37** (1945), 113–114.

Sprague, Roland.

1) Über additive Zerlegungen in lauter verschiedene Glieder einer Teilfolge der natürlichen Zahlenreihe. *Math. Z.* **56** (1952), 258–260.

Stanley, Richard P.

1) Theory and application of plane partitions. I, II. *Studies in Applied Math.* **50** (1971), 167–188; 259–279.
2) The conjugate trace and trace of a plane partition. *J. Combinatorial Theory Ser.* **A 14** (1973), 53–65.

Straus, E. G.

1) On a problem in the theory of partitions. *Proc. Amer. Math. Soc.* **13** (1962), 192–196.

Subbarao, M. V.

1) Some remarks on the partition function. *Amer. Math. Monthly* **73** (1966), 851–854.
2) Two new combinatorial identities. *Norskevid. Selsk. Skr.* (Trondheim) 1972; 4 pp.
3) Combinatorial proofs of some identities. *Proc. Washington State University Conf. on number theory.* Pullman, Wash. 1971, pp. 80–91.

——; Vidyasagar, M.

1) On Watson's quintuple product identity. *Proc. Amer. Math. Soc.* **26** (1970), 23–27.

——; Subrahmanyasastri, V. V.

1) A transformation formula for products arising in partition theory. *Rocky Mountain J. Math.* **6** (1976), 345–356.

Subrahmanyasastri, V. V.

1) Some results involving partition functions. *J. Indian Math. Soc. (N.S.)* **26** (1962), 97–113.
2) On some properties of partitions. *Math. Student* **35** (1967), 177–192.
3) Partitions with congruence conditions. *J. Indian Math. Soc. (N.S.)* **36** (1972), 177–194.

Sudler, Culbreth, Jr.

1) A direct proof of two theorems on two-line partitions. *Proc. Amer. Math. Soc.* **16** (1965), 161–168.

Szekeres, G.

1) An asymptotic formula in the theory of partitions. *Quart. J. Math. Oxford Ser.* (2) **2** (1951), 85–108.
2) Some asymptotic formulae in the theory of partitions. II. *Quart. J. Math. Oxford Ser.* (2) **4** (1953), 96–111.

Uppuluri, V. R. R.; Carpenter, J. A.

1) A problem in restricted partitions. *Naval Res. Logist. Quart.* **21** (1974), 201–205.

Vaidya, A. M.

1) A congruence property of $c_k(n)$. *Amer. Math. Monthly* **72** (1965), 38–40.

Wall, H. S.

1) A continued fraction related to some partition formulas of Euler. *Amer. Math. Monthly* **48** (1941), 102–108.

Watson, G. N.

1) A new proof of the Rogers–Ramanujan identities. *J. London Math. Soc.* **4** (1929), 4–9.

2) Two tables of partitions. *Proc. London Math. Soc.* (2) **42** (1937), 550–556.
3) Ramanujan's Vermutung über Zerfallungszahlen. *J. Reine Angew. Math.* **179** (1938), 97–128.

Whiteman, A. L.
1) A sum connected with the partition function. *Bull. Amer. Math. Soc.* **53** (1947), 598–603.
2) A sum connected with the series for the partition function. *Pacific J. Math.* **6** (1956), 159–176.

Winquist, Lasse.
1) An elementary proof of $p(11m + 6) \equiv 0 \pmod{11}$. *J. Combinatorial Theory* **6** (1969), 56–59.

Wright, E. M.
1) Partitions of multipartite numbers. *Proc. Amer. Math. Soc.* **7** (1956), 880–890.
2) The number of partitions of a large bipartite number. *Proc. London Math. Soc.* (3) **7** (1957), 150–160.
3) Partitions of large bipartites. *Amer. J. Math.* **80** (1958), 643–658.
4) The asymptotic behaviour of the generating functions of partitions of multipartites. *Quart. J. Math. Oxford Ser.* (2) **10** (1959), 60–69.
5) Partitions into k parts. *Math. Ann.* **142** (1960/61), 311–316.
6) Partitions of multipartite numbers into a fixed number of parts. *Proc. London Math. Soc.* (3) **11** (1961), 499–510.
7) A simple proof of a known result in partitions. *Amer. Math. Monthly* **68** (1961), 144–145.
8) Direct proof of the basic theorem on multipartite partitions. *Proc. Amer. Math. Soc.* **15** (1964), 469–472.
9) Partition of multipartite numbers into k parts. *J. Reine Angew. Math.* **216** (1964), 101–112.
10) Proof of a conjecture of Sudler's. *Quart. J. Math. Oxford Ser.* (2) **15** (1964), 11–15.
11) An extension of a theorem of Gordon. *Proc. Glasgow Math. Assocn.* **7** (1965), 39–41.
12) The generating function of solid partitions. *Proc. Royal Soc. Edinburgh Sec.* **A 67** (1965/67), 185–195.
13) An enumerative proof of an identity of Jacobi. *J. London Math. Soc.* **40** (1965), 55–57.
14) Rotatable partitions. *J. London Math. Soc.* **43** (1968), 501–505.

Zolnowsky, John.
1) A direct combinatorial proof of the Jacobi identity. *Discrete Math.* **9** (1974), 293–298.

Zubrzycki, S.
1) A recursive formula for the number of restricted partitions. (Polish: Russian and English summaries). *Zastos. Mat.* **6** (1962), 231–234.

Zuckerman, H. S.
1) Identities analogous to Ramanujan's identities involving the partition function. *Duke Math. J.* **5** (1939), 88–110.
2) The computation of the smaller coefficients of $J(\tau)$. *Bull. Amer. Math. Soc.* **45** (1939), 917–919.

INDEX

(excluding Bibliography)

Addition formula for Stirling's numbers 152, 155
Alaoglu, Leon 363
Alter, R. 353, 357
Alter–Curtz–Wang conjecture 353, 357
Andrews, George E. 222, 223, 224, 231, 253
Arithmetic functions: $d(n)$, $\sigma(n)$, $\phi(n)$, $\mu(n)$ etc. 16, 19, 20, 21
Asymptotic formula for $P(N_j, k)$ 301
Atkin, A. O. L. 242, 253
Auluck, F. C. 256, 282, 303
Auluck–Chowla–Gupta conjecture 256

Balanced n-sets 167
Bauer's Theorems 179
Bell, E. T. 331
Bernoulli numbers 138, 149, 174
Bhaskara 315
Binary partitions 247
Biquadratic residues 134
Birkhoff, G. D. 68
Block addition 198
Bounds for $P(N_j, k)$ 298
Bounds for $M(r)$ and $N(r)$ 328, 330
Brahmagupta's identity 96, 97, 100
Brenner, J. L. 91
Brent, Barry 362

Canonical form 11
Carlitz, L. 84, 211, 212, 215, 229, 231, 357
Carlitz's equivalent of Jacobi's identity 211, 212
Catalan numbers 321
Cheema, M. S. 198, 213, 214, 231
Cheema's graphical proof of Jacobi's identity 213, 214
Chinese remainder theorem 25
Chowla, S. 71, 101, 241, 253, 256, 282, 303, 331
Churchhouse, R. F. 247, 253
Churchhouse conjecture 247
Classes of k-ic residues and non-residues 119, 120
Class matrix 124
Coefficients in the polynomial for $G(n, r)$ 162
Combinatory function $\binom{n}{r}$ and its generalization $\begin{bmatrix} n \\ r \end{bmatrix}_a$ 180
Congruences for self-contained balanced sets 168
Congruence properties of $q_r(n)$ 237
Conjecture regarding $p(N_j, k)$ 303
Conjugate classes 122
Cubic residues 136
Curtz, Thaddeus B. 353, 357

Davenport, H. 10
Decompositions 188
Dedekind, R. 36
Degree of an operator 286
Degree of a permutation 335
Degree-vectors of a permutation 334
Determination of a permutation from its degree-vectors 336
Difference operator Δ 138
Diophantine equations:
$x_1 x_2 \ldots x_n = y_1 y_2 \ldots y_m$ 305
$x^n = y_1 y_2 \ldots y_m$ 315
$x_1 + x_2 + \cdots + x_m = n$ 314
$a_1 x_1 + a_2 x_2 + \cdots + a_m x_m = n$ 321
$x_1 + 2x_2 + 3x_3 + \cdots + m x_m = n,$ $x_m \geqslant 1$ 323

Diophantine equations in partitions 191
Dirichlet, P. L. 81
Divisors 9, 16
Durfee's identity 209

Ecklund, E. F. 362
Eggleton, R. B. 362
Eisenstein, F. G. M. 84
Eisenstein technique in summation 33
Equinumerous partitions and Euler-pairs 222
Erdös, P. 233, 235, 256, 281, 282, 323, 331, 362, 363
Erdös–Sierpinski problem 363
Erdös–Straus conjecture 363
Euler's criterion 78
Euler–Fermat Theorem 24, 62
Euler's identity 206
Euler, L. 16, 84, 206, 222, 224, 260
Euler–Maclaurin summation formula 260
Euler's totient function $\phi(n)$ 16
Ewell, John A. 361
Explicit expressions for $P(Nj, k)$ and $Q(Nj, k)$ 290
Exponent 37
Extended frequency matrix 130

Ferrars graphs 201
Foulkes, H. O. 357
Franklin, F. 206
Franklin's proof of Euler's identity 206
Frequency 120, 128
Frequency matrix 124
Frobenius, G. 84
Functions:

$G(n, r), F(r, j), H(r, j)$ 141, 142, 151

$G(-n, r), G(-n, -r)$ 144, 145

$S(n, r)$ 147

$G'(n, r), S'(n, r)$ 172

$q_r(n)$ 236

$r(n), r_2(n), r_4(n)$ 107, 111, 113

Fundamental theorem of arithmetic 9

Gandhi, J. M. 237, 254
Gauss, C. F. 56, 81, 84, 118, 160, 215
Gauss' identity 215
Gauss' lemma 81
Gauss' Theorem 160
g.c.d. of a and b: (a, b) 16
Generalizations of well-known theorems 155
Generating functions in partitions 191
Generating functions in permutations 359
Generating polynomials for equal sums of powers 325
Glaisher, J. W. L. 113, 186, 253
$G(n, r)$ in terms of combinatory functions 141, 150
Gordon, Basil 210, 220, 221, 224, 231
Gordon's identity 210, 221
Gordon's Theorem 220

Gould, H. W. 361
Graph of a partition of Nj 302
Graph of a permutation 339
Greatest integer function $[x]$ 12
Greatest primitive divisor of $a^n - b^n$, $b <' a$ 63
Greatest primitive factor of $a^n - b^n$ 64
Gupta, H. 112, 237, 247, 255, 256

Hardy–Ramanujan (divergent) expansion for $p(n)$ 255, 281
Hardy–Ramanujan–Rademacher (convergent) expansion for $p(n)$ 256
Hardy, G. H. 117, 255, 256, 281, 282
Hat function Λ 23
Highly restricted partitions 224
Hjelle, T. 254
Hooley, C. 361

Identical congruences 176, 182
Inclusion and exclusion principle 29
Indices 38, 39, 43, 44, 49, 50, 71
Index of a summit 333
Index of a permutation 333
Index-set of a permutation 333
Inequality for $p(Nj, k)$ and $q(Nj, k)$ 303
Infinitude of primes 11, 67
Interval-set of a permutation 334
Inverse function 32

Jacobi, K. C. J. 113, 210, 211, 214, 215
Jacobi's identity 210, 215
Jacobi's symbol 87, 88, 91
Jacobsthal's formula 101
j-partite numbers 283

k-ic residues and non-residues 70, 74
k-ic residue character of small primes 131
Kemmer's identity 249
Klove, T. 246, 254
Kolberg, O. 237, 246

Lagrange, J. L. 15
Lattice points 33
Legendre's symbol 78, 88, 91, 92, 109
Legendre, A. M. 78, 88, 91, 92, 109
Lehmer, D. H. 13, 84, 91, 241, 252, 254, 256, 282
Lehmer's property of Jacobi's symbol 91
Lehmer's theorem in partitions 252
Lehner, J. 233, 235, 256, 281, 282, 323, 331
Lerch 186
Leudesdorf's Theorem 173, 174
Linear congruence 24

Liouville's function 20, 113
Listing of primes 35
Listing of partitions 189
Lucas, E. 16

MacMahon, P. A. 245
Mahler, Kurt 362
Maximum number of frequencies 128
Menon, P. Kesava 362
Method of descent 97
Method of steepest descent 257
Methods of finding primitive roots 52
Möbius function 19, 56
Möbius inversion formula 21
Moller, Raymond 56
Multiplicative functions 17, 20
Multiset 119
Murad Ismail 98

Natural numbers and their classification 9
Newman, M. 237, 246, 254
Newman's conjecture 246
Non-associative products 320
Non-existence of primitive roots 45
Number of permutations:

(i) with s summits 342
(ii) with a given index 342
(iii) with a given degree 343
(iv) with a given degree and a given number of summits 345
(v) with a given index set 346
(vi) with a given degree and a given index 348

Number of primitive roots 41, 43
Number of k-ic residues 70
Numbers N_j, M_j 165, 167
Number of parameters 305, 315
Number of solutions of certain Diophantine equations 314, 321
Number of roots of a polynomial congruence 28

Operators σ_r 284
Ordering of permutations 332

Parameters 305, 315
Parity of $p(n)$ 245
Partition functions: $p(n), q(n), p(n, m), P(n, m), q(n, m), p(n, A), q(n, A)$ 190
Partitions with and without restrictions 188
Partitions of j-partite numbers 283
Permutations of the first n natural numbers 332
Pigntaro, S. 363
$P(N_j, k)$ as a determinant 291

Polya, G. 21
Polya's conjecture 21
Position of a permutation 337
p-potency of n ($\text{pot}_p(n)$) 12
p-potency of $n!$ 12
p-potencies of $\binom{p^c}{r}$ and $\binom{n}{r}$ 13, 14
p-piquancy of n ($\text{piq}_p\, n$) 47
Primes 9, 11, 67, 364
Primitive roots and indices 39, 40, 41, 45, 47, 55, 58
Primitive roots of unity 58
Primitive divisors and factors of $a^n - b^n$, $b <' a$ 59, 61, 63
Proper k-ic residues 72
Pseudo primitive roots 50, 59

Quadratic reciprocity law 84
Quadratic residues 78, 132

Rademacher, H. 256, 282
Ramanujan, S. 59, 113, 175, 237, 255
Ramanujan's identity 115
Ramanujan's function $\tau(n)$ 236, 244
Ramanujan's conjecture 241
Ramanujan's trigonometric sum 59
Recurrence relations for $g(n, m, h, k)$ 225
Recurrence relations for partition functions 196
Recurrence relation for Stirling's numbers 138
Recurrence formulae for $p(N_j, k)$ and $q(N_j, k)$ 284, 285
Reduction formula for $P(n, k)$ 198
Regimbal, Stephen 362
Relation between $P(n, k)$ and $p(n, k)$ 199
Relationships between frequencies in class algebra 126
Residue systems 24
Riesel, Hans 361
Rödseth, O. 247, 254
Rotkiewicz, A. 68

Saddle point 258
Sastry, S. 331
Scales of notation 14
Sets self-contained modulo n 167
Shanks, D. 361
Sierpinski, W. 363
Stephens, N. M. 362
Stirling's formula 280
Stirling's numbers of three kinds 137
Straus, E. G. 363
Subbarao, M. V. 215, 223
Sum of squares 95
Summits of a permutation 333

Sylvester's Theorem 229
Szekeres, George 256, 282
Szekeres' formula for $p(n, k)$ and $q(n, k)$ 276, 279

Table of values of χ, χ^* 282
Tables of values of $p(n)$, $q(n)$, $q_0(n)$ 209, 230, 231
Tarry–Escott problem 323
t^*-partitions 284
Two-line partitions 218
Two theorems concerning $g(n, m, h, k)$ 227
Type of a permutation 335
Type of a derived permutation 340

Vaidya, A. M. 112
Vandiver, H. S. 68
Von Staudt's Theorem 175

Wang, C. C. 353, 357
Ward, Morgan 143, 151
Watson, G. N. 241, 254
Weight of an operator 286
Wilansky, A. 362
Williams, Ian S. 35
Wilson's quotient 185
Wilson's Theorem 79, 342
Winquist, L. 215, 231
Winquist's identity 215
Wolstenholme's Theorem 157, 172
Wright, E. M. 117, 304, 331